A·N·N·U·A·L E·D·I·T·I·O·N·S

Global Issues

Twentieth Edition

04/05

D0218370

EDITOR

Robert M. Jackson

California State University, Chico

Robert M. Jackson is a professor of political science and dean of the School of Graduate, International, and Sponsored Programs at California State University, Chico. In addition to teaching, he has published articles on the international political economy, international relations simulations, and political behavior. His special research interest is in the way northern California is becoming increasingly linked to the Pacific Basin. His travels include China, Japan, Hong Kong, Taiwan, Singapore, Malaysia, Portugal, Spain, Morocco, Costa Rica, El Salvador, Honduras, Guatemala, Mexico, Germany, Belgium, the Netherlands, Russia, and Czechoslovakia.

McGraw-Hill/Dushkin

2460 Kerper Blvd., Dubuque, IA 52001

Visit us on the Internet
http://www.dushkin.com

Credits

1. **Global Issues in the Twenty-First Century: An Overview**
   Unit photo—© 2004 by Photo Disc, Inc.
2. **Population and Food Production**
   Unit photo—United Nations photo.
3. **The Global Environment and Natural Resources Utilization**
   Unit photo—United Nations photo.
4. **Political Economy**
   Unit photo—Courtesy of Novosti Photo.
5. **Conflict**
   Unit photo—AP/Wide World photo.
6. **Cooperation**
   Unit photo—United Nations photo.
7. **Values and Visions**
   Unit photo—United Nations photo.

Copyright

Cataloging in Publication Data
Main entry under title: Annual Editions: Global Issues. 2004/2005.
1. Global Issues—Periodicals. I. Jackson, Robert M., comp. II. Title: Global Issues.
ISBN 0–07–286151–7   658'.05   ISSN 1093–278X

Twentieth Edition

Cover image © Stock Trek/Getty Images
Printed in the United States of America   1234567890QPDQPD987654   Printed on Recycled Paper

Editors/Advisory Board

Members of the Advisory Board are instrumental in the final selection of articles for each edition of ANNUAL EDITIONS. Their review of articles for content, level, currentness, and appropriateness provides critical direction to the editor and staff. We think that you will find their careful consideration well reflected in this volume.

EDITOR

Robert M. Jackson
California State University, Chico

ADVISORY BOARD

Peter K. Angstadt
Wesley College

Thomas E. Arcaro
Elon University

Diane N. Barnes
University of Southern Maine

Eric N. Budd
Fitchburg State College

H. Thomas Collins
George Washington University

E. Gene DeFelice
Purdue University - Calumet

Dennis R. Gordon
Santa Clara University

Dorith Grant-Wisdom
University of Maryland, College Park

Elizabeth Crump Hanson
University of Connecticut, Storrs

James E. Harf
University of Tampa

Fayyaz Hussain
Michigan State University

Sadat Kazi
Vanier College

Steven L. Lamy
University of Southern California

T. David Mason
University of Memphis

Louis L. Ortmayer
Davidson College

Guy Poitras
Trinity University

Helen E. Purkitt
U.S. Naval Academy

Christian Søe
California State University, Long Beach

John A. Tures
LaGrange College

John H. P. Williams
East Carolina University

Kenneth L. Wise
Creighton University

Rodger Yeager
West Virginia University

Staff

To the Reader

In publishing ANNUAL EDITIONS we recognize the enormous role played by the magazines, newspapers, and journals of the public press in providing current, first-rate educational information in a broad spectrum of interest areas. Many of these articles are appropriate for students, researchers, and professionals seeking accurate, current material to help bridge the gap between principles and theories and the real world. These articles, however, become more useful for study when those of lasting value are carefully collected, organized, indexed, and reproduced in a low-cost format, which provides easy and permanent access when the material is needed. That is the role played by ANNUAL EDITIONS.

The beginning of the new millennium was celebrated with considerable fanfare. The prevailing mood in much of the world was that there was a great deal for which we could congratulate ourselves. The very act of sequentially watching on television live celebrations from one time zone to the next was proclaimed as a testimonial to globalization and the benefits of modern technology. The tragic events of September 11, 2001, however, were a stark reminder of the intense emotions and methods of destruction available to those determined to challenge the status quo. The subsequent wars in Afghanistan and Iraq along with continuing acts of terror have dampened the optimism that was expressed at the outset of the twenty-first century.

While the mass media may focus on the latest crisis for a few weeks or months, the broad forces that are shaping the world are seldom given the in-depth analysis that they warrant. Scholarly research about these historic forces of change can be found in a wide variety of publications, but these are not readily accessible. In addition, students just beginning to study global issues can be discouraged by the terminology and abstract concepts that characterize much of the scholarly literature. In selecting and organizing the materials for this book, we have been mindful of the needs of beginning students and have, thus, selected articles that invite the student into the subject matter.

Each unit begins with an introductory article providing a broad overview of the area to be explored. The remaining articles examine in more detail some of the issues presented. The unit then concludes with an article (or two) that not only identifies a problem but also suggests positive steps that are being taken to improve the situation. Recent events are a continual reminder that the world faces many serious challenges, the magnitude of which would discourage even the most stouthearted individual. While identifying problems is easier than solving them, it is encouraging to know that many are being addressed.

Perhaps the most striking feature of the study of contemporary global issues is the absence of any single, widely held theory that explains what is taking place. As a result, we have made a conscious effort to present a wide variety of points of view. The most important consideration has been to present global issues from an international perspective, rather than from a purely American or Western point of view. By encompassing materials originally published in different countries and written by authors of various nationalities, the anthology represents the great diversity of opinions that people hold. Two writers examining the same phenomenon may reach very different conclusions. It is not just a question of who is right or wrong, but rather understanding that people from different vantage points can have differing perspectives on an issue.

Another major consideration when organizing these materials was to explore the complex interrelationship of factors that produce social problems such as poverty. Too often, discussions of this problem (and others like it) are reduced to arguments about the fallacies of not following the correct economic policy or not having the correct form of government. As a result, many people overlook the interplay of historic, cultural, environmental, economic, and political factors that form complex webs that bring about many different problems. Every effort has been made to select materials that illustrate this complex interaction of factors, stimulating the beginning student to consider realistic rather than overly simplistic approaches to the pressing problems that threaten the existence of civilization.

In addition to an annotated *table of contents* and a *topic guide,* included in this edition of *Annual Editions: Global Issues* are *World Wide Web* sites that can be used to further explore topics addressed in the articles.

Finally, materials in this book were selected for both their intellectual insights and readability. Timely and well-written materials should stimulate good classroom lectures and discussions. I hope that students and teachers will enjoy using this book. Readers can have input into the next edition by completing and returning the postage-paid *article rating form* in the back of the book.

This is the twentieth edition of *Annual Editions: Global Issues.* When looking back over two decades of work, a great deal has taken place in world affairs, and the contents and organization of the book reflect these changes. Nonetheless there is one underlying constant. It is my continuing goal to work with the editors and staff at McGraw-Hill/Dushkin to provide materials that encourage the readers of this book to develop a life-long appreciation of the complex and rapidly changing world in which we live. This collection of articles is an invitation to further explore the global issues of the twenty-first century and become personally involved in the great issues of our time.

Robert M. Jackson
Editor

Contents

UNIT 1
Global Issues in the Twenty-First Century: An Overview

Four articles in this section present distinct views on the present and future state of life on Earth.

UNIT 2
Population and Food Production

Three selections in this section discuss the contributing factors in the world's population growth and the challenge of providing food for this added strain on the world's capacity.

The concepts in bold italics are developed in the article. For further expansion, please refer to the Topic Guide and the Index.

UNIT 3
The Global Environment and Natural Resources Utilization

Four articles in this section discuss natural resources and their effects on the world's environment.

The concepts in bold italics are developed in the article. For further expansion, please refer to the Topic Guide and the Index.

UNIT 4
Political Economy

Eleven articles present various views on economic and social development in the non-industrial and industrial nations.

The concepts in bold italics are developed in the article. For further expansion, please refer to the Topic Guide and the Index.

UNIT 5
Conflict

Six articles in this section discuss the basis for world conflict and the current state of order and disorder in the international community.

Unit Overview   108

The concepts in bold italics are developed in the article. For further expansion, please refer to the Topic Guide and the Index.

UNIT 6
Cooperation

Five selections in this section examine patterns of international cooperation and the social structures that support this cooperation.

The concepts in bold italics are developed in the article. For further expansion, please refer to the Topic Guide and the Index.

UNIT 7
Values and Visions

Seven articles discuss human rights, ethics, values, and new ideas.

The concepts in bold italics are developed in the article. For further expansion, please refer to the Topic Guide and the Index.

Topic Guide

This topic guide suggests how the selections in this book relate to the subjects covered in your course. You may want to use the topics listed on these pages to search the Web more easily.

On the following pages a number of Web sites have been gathered specifically for this book. They are arranged to reflect the units of this *Annual Edition*. You can link to these sites by going to the DUSHKIN ONLINE support site at *http://www.dushkin.com/online/*.

ALL THE ARTICLES THAT RELATE TO EACH TOPIC ARE LISTED BELOW THE BOLD-FACED TERM.

Agriculture
1. A Special Moment in History
6. SARS Has Lots of Company as Emerging Disease
7. Bittersweet Harvest: The Debate Over Genetically Modified Crops
11. We *Can* Build a Sustainable Economy

Communication
2. Clash of Globalizations
12. The Complexities and Contradictions of Globalization
14. Will Globalization Go Bankrupt?
21. What's Wrong With This Picture?
33. Aerial War Against Disease: Satellite Tracking of Epidemics Is Soaring

Conservation
1. A Special Moment in History
4. Mr. Order Meets Mr. Chaos
7. Bittersweet Harvest: The Debate Over Genetically Modified Crops
8. The Challenges We Face
9. Shifting the Pain: World's Resources Feed California's Growing Appetite
10. Water Scarcity Could Overwhelm the Next Generation
11. We *Can* Build a Sustainable Economy
18. Thirty Years of Petro-Politics
22. Leasing the Rain
33. Aerial War Against Disease: Satellite Tracking of Epidemics Is Soaring
36. Why Environmental Ethics Matters to International Relations

Cultural customs and values
2. Clash of Globalizations
11. We *Can* Build a Sustainable Economy
12. The Complexities and Contradictions of Globalization
19. India and China: Two Systems, One Grand Rivalry
21. What's Wrong With This Picture?
22. Leasing the Rain
25. "Why Do They Hate Us?"
34. Are Human Rights Universal?
35. The Grameen Bank
36. Why Environmental Ethics Matters to International Relations
38. Women Waging Peace
39. The Next Christianity
40. Modernization's Challenge to Traditional Values: Who's Afraid of Ronald McDonald?

Demographics
1. A Special Moment in History
5. The Big Crunch
6. SARS Has Lots of Company as Emerging Disease

Dependencies, international
2. Clash of Globalizations
7. Bittersweet Harvest: The Debate Over Genetically Modified Crops
20. Overcoming Japan's China Syndrome
22. Leasing the Rain

Development, economic
2. Clash of Globalizations
4. Mr. Order Meets Mr. Chaos
5. The Big Crunch
7. Bittersweet Harvest: The Debate Over Genetically Modified Crops
9. Shifting the Pain: World's Resources Feed California's Growing Appetite
10. Water Scarcity Could Overwhelm the Next Generation
11. We *Can* Build a Sustainable Economy
12. The Complexities and Contradictions of Globalization
13. The Five Wars of Globalization
14. Will Globalization Go Bankrupt?
16. Render Unto Caesar: Putin and the Oligarchs
19. India and China: Two Systems, One Grand Rivalry
22. Leasing the Rain
32. Countdown to Eradication
34. Are Human Rights Universal?
35. The Grameen Bank
40. Modernization's Challenge to Traditional Values: Who's Afraid of Ronald McDonald?

Development, social
2. Clash of Globalizations
4. Mr. Order Meets Mr. Chaos
5. The Big Crunch
7. Bittersweet Harvest: The Debate Over Genetically Modified Crops
9. Shifting the Pain: World's Resources Feed California's Growing Appetite
10. Water Scarcity Could Overwhelm the Next Generation
11. We *Can* Build a Sustainable Economy
12. The Complexities and Contradictions of Globalization
14. Will Globalization Go Bankrupt?
19. India and China: Two Systems, One Grand Rivalry
22. Leasing the Rain
32. Countdown to Eradication
34. Are Human Rights Universal?
35. The Grameen Bank
40. Modernization's Challenge to Traditional Values: Who's Afraid of Ronald McDonald?

Ecology
1. A Special Moment in History
4. Mr. Order Meets Mr. Chaos
7. Bittersweet Harvest: The Debate Over Genetically Modified Crops
8. The Challenges We Face
9. Shifting the Pain: World's Resources Feed California's Growing Appetite
10. Water Scarcity Could Overwhelm the Next Generation
11. We *Can* Build a Sustainable Economy
18. Thirty Years of Petro-Politics
22. Leasing the Rain
33. Aerial War Against Disease: Satellite Tracking of Epidemics Is Soaring
36. Why Environmental Ethics Matters to International Relations

Economics
1. A Special Moment in History
2. Clash of Globalizations

World Wide Web Sites

The following World Wide Web sites have been carefully researched and selected to support the articles found in this reader. The easiest way to access these selected sites is to go to our DUSHKIN ONLINE support site at *http://www.dushkin.com/online/*.

AE: Global Issues 04/05

The following sites were available at the time of publication. Visit our Web site—we update DUSHKIN ONLINE regularly to reflect any changes.

General Sources

U.S. Information Agency (USIA)
http://usinfo.state.gov

USIA's home page provides definitions, related documentation, and discussions of topics of concern to students of global issues. The site addresses today's Hot Topics as well as ongoing issues that form the foundation of the field.

World Wide Web Virtual Library: International Affairs Resources
http://www.etown.edu/vl/

Surf this site and its extensive links to learn about specific countries and regions, to research various think tanks and international organizations, and to study such vital topics as international law, development, the international economy, human rights, and peacekeeping.

UNIT 1: Global Issues int he Twenty-First Century: An Overview

The Henry L. Stimson Center
http://www.stimson.org

The Stimson Center, a nonpartisan organization, focuses on issues where policy, technology, and politics intersect. Use this site to find varying assessments of U.S. foreign policy in the post–cold war world and to research other topics.

The Heritage Foundation
http://www.heritage.org

This page offers discussion about and links to many sites having to do with foreign policy and foreign affairs, including news and commentary, policy review, events, and a resource bank.

IISDnet
http://www.iisd.org/default.asp

The International Institute for Sustainable Development presents information through links to business, sustainable development, and developing ideas. "Linkages" is its multimedia resource for policymakers.

The North-South Institute
http://www.nsi-ins.ca/ensi/index.html

Searching this site of the North-South Institute, which works to strengthen international development cooperation and enhance gender and social equity, will help you find information and debates on a variety of global issues.

UNIT 2: Population and Food Production

The Hunger Project
http://www.thp.org

Browse through this nonprofit organization's site, whose goal is the sustainable end to global hunger through leadership at all levels of society. The Hunger Project contends that the persistence of hunger is at the heart of the major security issues threatening our planet.

Penn Library: Resources by Subject
http://www.library.upenn.edu/cgi-bin/res/sr.cgi

This vast site is rich in links to information about subjects of interest to students of global issues. Its extensive population and demography resources address such concerns as migration, family planning, and health and nutrition in various world regions.

World Health Organization
http://www.who.int

This home page of the World Health Organization will provide you with links to a wealth of statistical and analytical information about health and the environment in the developing world.

WWW Virtual Library: Demography & Population Studies
http://demography.anu.edu.au/VirtualLibrary/

A definitive guide to demography and population studies can be found at this site. It contains a multitude of important links to information about global poverty and hunger.

UNIT 3: The Global Environment and Natural Resources Utilization

National Geographic Society
http://www.nationalgeographic.com

This site provides links to material related to the atmosphere, the oceans, and other environmental topics.

National Oceanic and Atmospheric Administration (NOAA)
http://www.noaa.gov

Through this home page of NOAA, part of the U.S. Department of Commerce, you can find information about coastal issues, fisheries, climate, and more. The site provides many links to research materials and to other Web resources.

Public Utilities Commission of Ohio (PUCO)
http://www.puc.state.oh.us/consumer/gcc/index.html

PUCO's site serves as a clearinghouse of information about global climate change. Its links explain the science and chronology of global climate change.

SocioSite: Sociological Subject Areas
http://www.pscw.uva.nl/sociosite/TOPICS/

This huge site provides many references of interest to those interested in global issues, such as links to information on ecology and the impact of consumerism.

United Nations Environment Programme (UNEP)
http://www.unep.ch

Consult this home page of UNEP for links to critical topics of concern to students of global issues, including desertification, migratory species, and the impact of trade on the environment.

UNIT 4: Political Economy

Belfer Center for Science and International Affairs (BCSIA)

http://ksgwww.harvard.edu/csia/

BCSIA is the hub of Harvard University's John F. Kennedy School of Government's research, teaching, and training in international affairs related to security, environment, and technology.

U.S. Agency for International Development

http://www.info.usaid.gov

Broad and overlapping issues such as democracy, population and health, economic growth, and development are covered on this Web site. It provides specific information about different regions and countries.

Virtual Seminar in Global Political Economy/Global Cities & Social Movements

http://csf.colorado.edu/gpe/gpe95b/resources.html

This site of Internet resources is rich in links to subjects of interest in regional environmental studies, covering topics such as sustainable cities, megacities, and urban planning. Links to many international nongovernmental organizations are included.

The World Bank Group

http://www.worldbank.org

News, press releases, summaries of new projects, speeches, publications, and coverage of numerous topics regarding development, countries, and regions are provided at this World Bank site. It also contains links to other important global financial organizations.

UNIT 5: Conflict

DefenseLINK

http://www.defenselink.mil

Learn about security news and research-related publications at this U.S. Department of Defense site. Links to related sites of interest are provided. The information systems BosniaLINK and GulfLINK can also be found here. Use the search function to investigate such issues as land mines.

Federation of American Scientists (FAS)

http://www.fas.org

FAS, a nonprofit policy organization, maintains this site to provide coverage of and links to such topics as global security, peace, and governance in the post–cold war world. It notes a variety of resources of value to students of global issues.

ISN International Relations and Security Network

http://www.isn.ethz.ch

This site, maintained by the Center for Security Studies and Conflict Research, is a clearinghouse for information on international relations and security policy. Topics are listed by category (Traditional Dimensions of Security, New Dimensions of Security, and Related Fields) and by major world region.

The NATO Integrated Data Service (NIDS)

http://www.nato.int/structur/nids/nids.htm

NIDS was created to bring information on security-related matters to within easy reach of the widest possible audience. Check out this Web site to review North Atlantic Treaty Organization documentation of all kinds, to read *NATO Review,* and to explore key issues in the field of European security and transatlantic cooperation.

UNIT 6: Cooperation

Carnegie Endowment for International Peace

http://www.ceip.org

An important goal of this organization is to stimulate discussion and learning among both experts and the public at large on a wide range of international issues. The site provides links to *Foreign Policy,* to the Moscow Center, to descriptions of various programs, and much more.

Commission on Global Governance

http://www.sovereignty.net/p/gov/gganalysis.htm

This site provides access to *The Report of the Commission on Global Governance,* produced by an international group of leaders who want to find ways in which the global community can better manage its affairs.

OECD/FDI Statistics

http://www.oecd.org/statistics/

Explore world trade and investment trends and statistics on this site from the Organization for Economic Cooperation and Development. It provides links to many related topics and addresses the issues on a country-by-country basis.

U.S. Institute of Peace

http://www.usip.org

USIP, which was created by the U.S. Congress to promote peaceful resolution of international conflicts, seeks to educate people and to disseminate information on how to achieve peace. Click on Highlights, Publications, Events, Research Areas, and Library and Links.

UNIT 7: Values and Visions

Human Rights Web

http://www.hrweb.org

The history of the human rights movement, text on seminal figures, landmark legal and political documents, and ideas on how individuals can get involved in helping to protect human rights around the world can be found in this valuable site.

InterAction

http://www.interaction.org

InterAction encourages grassroots action and engages government policymakers on advocacy issues. The organization's Advocacy Committee provides this site to inform people on its initiatives to expand international humanitarian relief, refugee, and development-assistance programs.

We highly recommend that you review our Web site for expanded information and our other product lines. We are continually updating and adding links to our Web site in order to offer you the most usable and useful information that will support and expand the value of your Annual Editions. You can reach us at: *http://www.dushkin.com/annualeditions/.*

World Map

Scale: 1 to 125,000,000

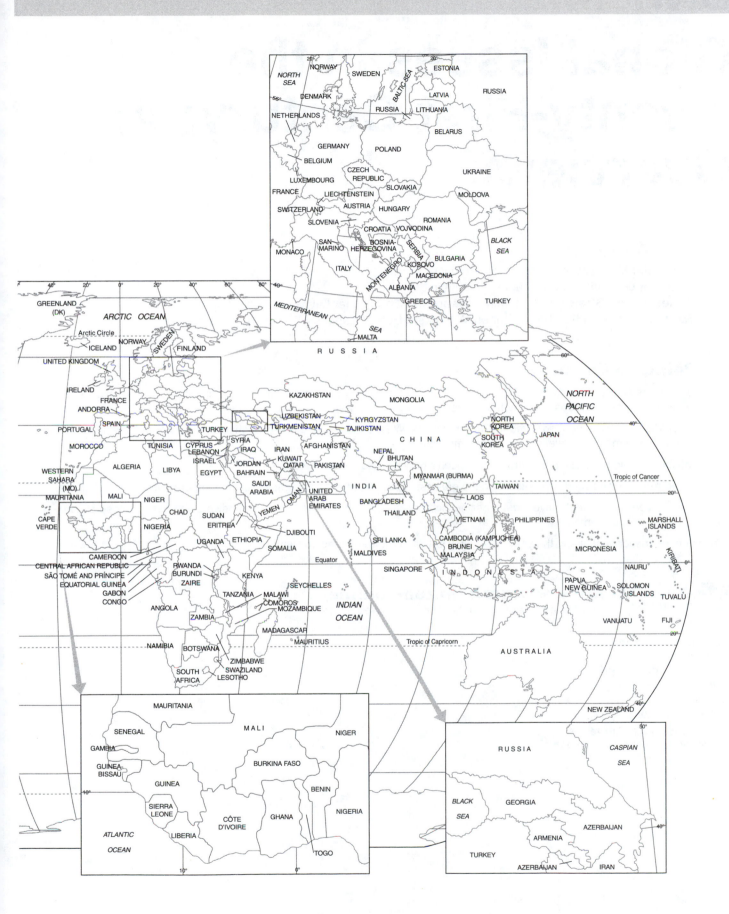

UNIT 1

Global Issues in the Twenty-First Century: An Overview

Unit Selections

1. **A Special Moment in History**, Bill McKibben
2. **Clash of Globalizations**, Stanley Hoffmann
3. **A New Grand Strategy**, Benjamin Schwarz and Christopher Layne
4. **Mr. Order Meets Mr. Chaos**, Robert Wright and Robert Kaplan

Key Points to Consider

- Do the analyses of any of the authors in this section employ the assumptions implicit in the allegory of the balloon? If so, how? If not, how are the assumptions of the authors different?

- All the authors point to interactions among different factors. What are some of the relationships that they cite? How do the authors differ in terms of the relationships they emphasize?

- What assets that did not exist 100 years ago do people now have to solve problems?

- What events during the twentieth century had the greatest impact on shaping the realities of contemporary international affairs?

- What do you consider to be the five most pressing global problems of today? How do your answers compare to those of your family, friends, and classmates?

- Describe international affairs in the year 2050. How is it different from today and why?

 Links: www.dushkin.com/online/
These sites are annotated in the World Wide Web pages.

The Henry L. Stimson Center
http://www.stimson.org

The Heritage Foundation
http://www.heritage.org

IISDnet
http://www.iisd.org/default.asp

The North-South Institute
http://www.nsi-ins.ca/ensi/index.html

Imagine yellow paint being brushed onto an inflated, clear balloon. The yellow color, for purposes of this allegory, represents *people.* In many ways the study of global issues is first and foremost the study of people. Today, there are more human beings occupying Earth than ever before. In addition, we are in the midst of a period of unprecedented population growth. Not only are there many countries where the majority of people are under age 16, but also due to improved health care, there are more older people alive than ever before. The effect of a growing global population, however, goes beyond sheer numbers, for this trend has unprecedented impacts on natural resources and social services. An examination of population trends and the related topic of food production is a good place to begin an in-depth study of global issues.

Imagine that our fictional artist next dips the brush into a container of blue paint to represent *nature.* The natural world plays an important role in setting the international agenda. Shortages of raw materials, drought, and pollution of waterways are just a few examples of how natural resources can have global implications.

Adding blue paint to the balloon reveals one of the most important underlying concepts found in this book. Although the balloon originally was covered by both yellow and blue paint (people and nature as separate conceptual entities), the two combined produce an entirely different color: green. Talking about nature as a separate entity or people as though they were somehow removed from the forces of the natural world is a serious intellectual error. The people-nature relationship is one of the keys to understanding many of today's most important global issues.

The third color to be added to the balloon is red. This color represents *social structures.* Factors falling into this category include whether a society is urban or rural, industrial or agrarian, and consumer-oriented or dedicated to the needs of the state. The relationship between this component and the others is extremely important. The impact of political decisions on the environment, for example, is one of the most significant features of the contemporary world. Will the whales or bald eagles survive? Historically, the forces of nature determined which species survived or perished. Today, survival depends on political decisions—or indecision. Understanding the complex relationship between social structure and nature (known as "ecopolitics") is central to the study of global issues.

Added to the three primary colors is the fourth and final color of white. It represents the *meta* component (i.e., those qualities that make human beings different from other life forms). These include new ideas and inventions, culture and values, religion and spirituality, and art and literature. The addition of the white paint immediately changes the intensity and shade of the mixture of colors, again emphasizing the relationship among all four factors.

If the painter continues to ply the paintbrush over the miniature globe, a marbling effect becomes evident. From one area to the next, the shading varies because one element is more dominant than another. Further, the miniature system appears dynamic. Nothing is static; relationships are continually changing. This leads to a number of important insights: (1) there are no such things as separate elements, only connections or relationships; (2) changes in one area (such as the weather) will result in changes in all other areas; and (3) complex and dynamic relationships make it difficult to predict events accurately, so observers and policy makers are often surprised by unexpected events.

This book is organized along the basic lines of the balloon allegory. The first unit provides a broad overview of a variety of perspectives on the major forces that are shaping the world of the twenty-first century. From this "big picture" perspective more in-depth analyses follow. Unit 2, for example, focuses on population and food production. Unit 3 examines the environment and related natural resource issues. The next three units look at different aspects of the world's social structures. They explore issues of economics, national security, conflict, and international cooperation. In the final unit, a number of "meta" factors are presented.

The reader should keep in mind that, just as it was impossible to keep the individual colors from blending into new colors on the balloon, it is also impossible to separate global issues into discrete chapters in a book. Any discussion of agriculture, for example, must take into account the impact of a growing population on soil and water resources, as well as new scientific breakthroughs in food production. Therefore, the organization of this book focuses attention on issue areas; it does not mean to imply that these factors are somehow separate.

With the collapse of the Soviet empire and the end of the cold war, the outlines of a new global agenda have begun to emerge. Rather than being based on the ideology and interests of the two superpowers, new political, economic, environmental, cultural and security issues are interacting in an unprecedented fashion. Rapid population growth, environmental decline, uneven economic progress, and global terrorist networks are all parts of a complex state of affairs for which there is no historic parallel. As we begin the twenty-first century, signs abound that we are entering a new era. In the words of Abraham Lincoln, "As our case is new, so we must think anew." Compounding this situation, however, is a whole series of old problems such as ethnic and religious rivalries.

The authors in this first unit provide a variety of perspectives on the trends that they believe are the most important to understanding the historic changes at work at the global level. This discussion is then pursued in greater detail in the following units.

It is important for the reader to note that although the authors look at the same world, they often come to different conclusions. This raises an important issue of values and beliefs, for it can be argued that there really is no objective reality, only differing perspectives. In short, the study of global issues will challenge each thoughtful reader to examine her or his own values and beliefs.

A Special Moment in History

Bill McKibben

We may live in the strangest, most thoroughly different moment since human beings took up farming, 10,000 years ago, and time more or less commenced. Since then time has flowed in one direction—toward *more*, which we have taken to be progress. At first the momentum was gradual, almost imperceptible, checked by wars and the Dark Ages and plagues and taboos; but in recent centuries it has accelerated, the curve of every graph steepening like the Himalayas rising from the Asian steppe....

But now—now may be the special time. So special that in the Western world we might each of us consider, among many other things, having only one child—that is, reproducing at a rate as low as that at which human beings have ever voluntarily reproduced. Is this really necessary? Are we finally running up against some limits?

To try to answer this question, we need to ask another: *How many of us will there be in the near future?* Here is a piece of news that may alter the way we see the planet—an indication that we live at a special moment. At least at first blush the news is hopeful. *New demographic evidence shows that it is at least possible that a child born today will live long enough to see the peak of human population.*

Around the world people are choosing to have fewer and fewer children—not just in China, where the government forces it on them, but in almost every nation outside the poorest parts of Africa.... If this keeps up, the population of the world will not quite double again; United Nations analysts offer as their mid-range projection that it will top out at 10 to 11 billion, up from just under six billion at the moment....

The good news is that we won't grow forever. The bad news is that there are six billion of us already, a number the world strains to support. One more near-doubling—four or five billion more people—will nearly double that strain. Will these be the five billion straws that break the camel's back?...

LOOKING AT LIMITS

The case that the next doubling, the one we're now experiencing, might be the difficult one can begin as readily with the Stanford biologist Peter Vitousek as with anyone else. In 1986 Vitousek decided to calculate how much of the earth's "primary productivity" went to support human beings. He added together the grain we ate, the corn we fed our cows, and the forests we cut for timber and paper; he added the losses in food as we overgrazed grassland and turned it into desert. And when he was finished adding, the number he came up with was 38.8 percent. We use 38.8 percent of everything the world's plants don't need to keep themselves alive; directly or indirectly, we consume 38.8 percent of what it is possible to eat. "That's a relatively large number," Vitousek says. "It should give pause to people who think we are far from any limits." Though he never drops the measured tone of an academic, Vitousek speaks with considerable emphasis: "There's a sense among some economists that we're *so* far from any biophysical limits. I think that's not supported by the evidence."

For another antidote to the good cheer of someone like Julian Simon, sit down with the Cornell biologist David Pimentel. He believes that we're in big trouble. Odd facts stud his conversation—for example, a nice head of iceberg lettuce is 95 percent water and contains just fifty calories of energy, but it takes 400 calories of energy to grow that head of lettuce in California's Central Valley, and another 1,800 to ship it east. ("There's practically no nutrition in the damn stuff anyway," Pimentel says. "Cabbage is a lot better, and we can grow it in upstate New York.") Pimentel has devoted the past three decades to tracking the planet's capacity, and he believes that we're already too crowded—that the earth can support only two billion people over the long run at a middle-class standard of living, and that trying to support more is doing damage. He has spent considerable time studying soil erosion, for instance. Every raindrop that hits exposed ground is like a small explosion, launching soil particles into the air. On a slope, more than half of the soil contained in those splashes is carried downhill. If crop residue—cornstalks, say—is left in the field after harvest, it helps to shield the soil: the raindrop doesn't hit hard. But in the developing world, where firewood is scarce, peasants burn those cornstalks for cooking fuel. About 60 percent of crop residues in China and 90 percent in Bangladesh are removed and burned, Pimentel says. When planting season comes, dry soils simply blow away. "Our measuring stations pick up African soils in the wind when they start to plough."

The very things that made the Green Revolution so stunning—that made the last doubling possible—now cause trouble. Irrigation ditches, for instance, water 27 percent of all arable land and help to produce a third of all crops. But when flooded soils are baked by the sun, the water evaporates and the minerals in the irrigation water are deposited on the land. A hectare (2.47 acres) can accumulate two to five tons of salt annually, and eventually plants won't grow there. Maybe 10 percent of all irrigated land is affected.

… [F]ood production grew even faster than population after the Second World War. Year after year the yield of wheat and corn and rice rocketed up about three percent annually. It's a favorite statistic of the eternal optimists. In Julian Simon's book *The Ultimate Resource* (1981) charts show just how fast the growth was, and how it continually cut the cost of food. Simon wrote, "The obvious implication of this historical trend toward cheaper food—a trend that probably extends back to the beginning of agriculture—is that real prices for food will continue to drop…. It is a fact that portends more drops in price and even less scarcity in the future."

A few years after Simon's book was published, however, the data curve began to change. That rocketing growth in grain production ceased; now the gains were coming in tiny increments, too small to keep pace with population growth. The world reaped its largest harvest of grain per capita in 1984; since then the amount of corn and wheat and rice per person has fallen by six percent. Grain stockpiles have shrunk to less than two months' supply.

No one knows quite why. The collapse of the Soviet Union contributed to the trend—cooperative farms suddenly found the fertilizer supply shut off and spare parts for the tractor hard to come by. But there were other causes, too, all around the world—the salinization of irrigated fields, the erosion of topsoil, and all the other things that environmentalists had been warning about for years. It's possible that we'll still turn production around and start it rocketing again. Charles C. Mann, writing in *Science*, quotes experts who believe that in the future a "gigantic, multi-year, multi-billion-dollar scientific effort, a

kind of agricultural 'person-on the-moon project,'" might do the trick. The next great hope of the optimists is genetic engineering, and scientists have indeed managed to induce resistance to pests and disease in some plants. To get more yield, though, a cornstalk must be made to put out another ear, and conventional breeding may have exhausted the possibilities. There's a sense that we're running into walls.

… What we are running out of is what the scientists call "sinks"—places to put the by-products of our large appetites. Not garbage dumps (we could go on using Pampers till the end of time and still have empty space left to toss them away) but the atmospheric equivalent of garbage dumps.

It wasn't hard to figure out that there were limits on how much coal smoke we could pour into the air of a single city. It took a while longer to figure out that building ever higher smokestacks merely lofted the haze farther afield, raining down acid on whatever mountain range lay to the east. Even that, however, we are slowly fixing, with scrubbers and different mixtures of fuel. We can't so easily repair the new kinds of pollution. These do not come from something going wrong—some engine without a catalytic converter, some waste-water pipe without a filter, some smokestack without a scrubber. New kinds of pollution come instead from things going as they're supposed to go—but at such a high volume that they overwhelm the planet. They come from normal human life—but there are so many of us living those normal lives that something abnormal is happening. And that something is different from the old forms of pollution that it confuses the issue even to use the word.

Consider nitrogen, for instance. But before plants can absorb it, it must become "fixed"—bonded with carbon, hydrogen, or oxygen. Nature does this trick with certain kinds of algae and soil bacteria, and with lightning. Before human beings began to alter the nitrogen cycle, these mechanisms provided 90–150 million metric tons of nitrogen a year. Now human activity adds 130–150 million more tons. Nitrogen isn't pollution—it's essential. And we are using more of it all the time. Half the industrial nitrogen fertilizer used in human history has been applied since 1984. As a result, coastal waters and estuaries bloom with toxic algae while oxygen concentrations dwindle, killing fish; as a result, nitrous oxide traps solar heat. And once the gas is in the air, it stays there for a century or more.

Or consider methane, which comes out of the back of a cow or the top of a termite mound or the bottom of a rice paddy. As a result of our determination to raise more cattle, cut down more tropical forest (thereby causing termite populations to explode), and grow more rice, methane concentrations in the atmosphere are more than twice as high as they have been for most of the past 160,000 years. And methane traps heat—very efficiently.

Or consider carbon dioxide. In fact, concentrate on carbon dioxide. If we had to pick one problem to obsess about over the next fifty years, we'd do well to make it CO_2—which is not pollution either. Carbon *mon*oxide is pollution: it kills you if you breathe enough of it. But carbon *di*oxide, carbon with two oxygen atoms, can't do a blessed thing to you. If you're reading this indoors, you're breathing more CO_2 than you'll ever get outside. For generations, in fact, engineers said that an engine burned clean if it produced only water vapor and carbon dioxide.

Here's the catch: that engine produces a *lot* of CO_2. A gallon of gas weighs about eight pounds. When it's burned in a car, about five and a half pounds of carbon, in the form of carbon dioxide, come spewing out the back. It doesn't matter if the car is a 1958 Chevy or a 1998 Saab. And no filter can reduce that flow—it's an inevitable by-product of fossil-fuel combustion, which is why CO_2 has been piling up in the atmosphere ever since the Industrial Revolution. Before we started burning oil and coal and gas, the atmosphere contained about 280 parts CO_2 per million. Now the figure is about 360. Unless we do everything we can think of to eliminate fossil fuels from our diet, the air will test out at more than 500 parts per million fifty or sixty years from now, whether it's sampled in the South Bronx or at the South Pole.

This matters because, as we all know by now, the molecular structure of this clean, natural, common element that we are adding to every cubic foot of the atmosphere surrounding us traps heat that would otherwise radiate back out to space. Far more than even methane and nitrous oxide, CO_2 causes global warming—the greenhouse effect—and climate change. Far more than any other single factor, it is turning the earth we were born on into a new planet.

… For ten years, with heavy funding from governments around the world, scientists launched satellites, monitored weather balloons, studied clouds. Their work culminated in a long-awaited report from the UN's Intergovernmental Panel on Climate Change, released in the fall of 1995. The panel's 2,000 scientists, from every corner of the globe, summed up their findings in this dry but historic bit of understatement: "The balance of evidence suggests that there is a discernible human influence on global climate." That is to say, we are heating up the planet—substantially. If we don't reduce emissions of carbon dioxide and other gases, the panel warned, temperatures will probably rise 3.6° Fahrenheit by 2100, and perhaps as much as 6.3°.

You may think you've already heard a lot about global warming. But most of our sense of the problem is behind the curve. Here's the current news: the changes are already well under way. When politicians and businessmen talk about "future risks," their rhetoric is outdated. This is not a problem for the distant future, or even for the near future. The planet has already heated up by a degree or more. We are perhaps a quarter of the way into the greenhouse era, and the effects are already being felt. From a new heaven, filled with nitrogen, methane, and carbon, a new earth is being born. If some alien astronomer is watching us, she's doubtless puzzled. This is the most obvious effect of our numbers and our appetites, and the key to understanding why the size of our population suddenly poses such a risk.

STORMY AND WARM

What does this new world feel like? For one thing, it's stormier than the old one. Data analyzed last year by Thomas Karl, of the National Oceanic and Atmospheric Administration, showed that total winter precipitation in the United States has increased by 10 percent since 1900 and that "extreme precipitation events"—rainstorms that dumped more than two inches of water in twenty-four hours and blizzards—had increased by 20 percent. That's because warmer air holds more water vapor than the colder atmosphere of the old earth; more water evaporates from the ocean, meaning more clouds, more rain, more snow. Engineers designing storm sewers, bridges, and culverts used to plan for what they called the "hundred-year storm." That is, they built to withstand the worst flooding or wind that history led them to expect in the course of a century. Since that history no longer applies, Karl says, "there isn't really a hundred-year event anymore… we seem to be getting these storms of the century every couple of years." When Grand Forks, North Dakota, disappeared beneath the Red River in the spring of last year, some meteorologists referred to it as "a 500-year flood"—meaning, essentially, that all bets are off. Meaning that these aren't acts of God. "If you look out your window, part of what you see in terms of weather is produced by ourselves," Karl says. "If you look out the window fifty years from now, we're going to be responsible for more of it."

Twenty percent more bad storms, 10 percent more winter precipitation—these are enormous numbers. It's like opening the newspaper to read that the average American is smarter by 30 IQ points. And the same data showed increases in drought, too. With more water in the atmosphere, there's less in the soil, according to Kevin Trenberth, of the National Center for Atmospheric Research. Those parts of the continent that are normally dry—the eastern sides of mountains, the plains and deserts—are even drier, as the higher average temperatures evaporate more of what rain does fall. "You get wilting plants and eventually drought faster than you would otherwise," Trenberth says. And when the rain does come, it's often so intense that much of it runs off before it can soak into the soil.

So—wetter and drier. *Different*….

The effects of… warming can be found in the largest phenomena. The oceans that cover most of the planet's surface are clearly rising, both because of melting glaciers and because water expands as it warms. As a result, low-lying Pacific islands already report surges of water washing across the atolls. "It's nice weather and all of a sudden water is pouring into your living room," one Marshall Islands resident told a newspaper reporter. "It's very clear that something is happening in the Pacific, and these islands are feeling it." Global warming will be like a much more powerful version of El Niño that covers the entire globe and lasts forever, or at least until the next big asteroid strikes.

If you want to scare yourself with guesses about what might happen in the near future, there's no shortage of possibilities. Scientists have already observed large-scale shifts in the duration of the El Niño ocean warming, for instance. The Arctic tundra has warmed so much that in some places it now gives off more carbon dioxide than it absorbs—a switch that could trigger a potent feedback loop, making warming ever worse. And researchers studying glacial cores from the Greenland Ice Sheet recently concluded that local climate shifts have occurred with incredible rapidity in the past—18° in one three-year stretch. Other scientists worry that such a shift might be enough to flood the oceans with fresh water and reroute or shut off currents like the Gulf Stream and the North Atlantic, which keep Europe far warmer than it would otherwise be. (See "The Great Climate Flip-flop," by William H. Calvin, January *Atlantic*.) In the words of Wallace Broecker, of Columbia University, a pioneer in the field, "Climate is an angry beast, and we are poking it with sticks."

But we don't need worst-case scenarios: best-case scenarios make the point. The population of the earth is going to nearly double one more time. That will bring it to a level that even the reliable old earth we were born on would be hard-pressed to support. Just at the moment when we need everything to be working as smoothly as possible, we find ourselves inhabiting a new planet, whose carrying capacity we cannot conceivably estimate. We have no idea how much wheat this planet can grow. We don't know what its politics will be like: not if there are going to be heat waves like the one that killed more than 700 Chicagoans in 1995; not if rising sea levels and other effects of climate change create tens of millions of environmental refugees; not if a 1.5° jump in India's temperature could reduce the country's wheat crop by 10 percent or divert its monsoons....

We have gotten very large and very powerful, and for the foreseeable future we're stuck with the results. The glaciers won't grow back again anytime soon; the oceans won't drop. We've already done deep and systemic damage. To use a human analogy, we've already said the angry and unforgivable words that will haunt our marriage till its end. And yet we can't simply walk out the door. There's no place to go. We have to salvage what we can of our relationship with the earth, to keep things from getting any worse than they have to be.

If we can bring our various emissions quickly and sharply under control, we *can* limit the damage, reduce dramatically the chance of horrible surprises, preserve more of the biology we were born into. But do not under-estimate the task. The UN's Intergovernmental Panel on Climate Change projects that an immediate 60 percent reduction in fossil-fuel use is necessary just to stabilize climate at the current level of disruption. Nature may still meet us halfway, but halfway is a long way from where we are now. What's more, we can't delay. If we wait a few decades to get started, we may as well not even begin. It's not like poverty, a concern that's always there for civilizations to address. This is a timed test, like the SAT: two or three decades, and we lay our pencils down. It's *the* test for our generations, and population is a part of the answer....

The numbers are so daunting that they're almost unimaginable. Say, just for argument's sake, that we decided to cut world fossil-fuel use by 60 percent—the amount that the UN panel says would stabilize world climate. And then say that we shared the remaining fossil fuel equally. Each human being would get to produce 1.69 metric tons of carbon dioxide annually—which would allow you to drive an average American car nine miles a day. By the time the population increased to 8.5 billion, in about 2025, you'd be down to six miles a day. If you carpooled, you'd have about three pounds of CO_2 left in your daily ration—enough to run a highly efficient refrigerator. Forget your computer, your TV, your stereo, your stove, your dishwasher, your water heater, your microwave, your water pump, your clock. Forget your light bulbs, compact fluorescent or not.

I'm not trying to say that conservation, efficiency, and new technology won't help. They will—but the help will be slow and expensive. The tremendous momentum of growth will work against it. Say that someone invented a new furnace tomorrow that used half as much oil as old furnaces. How many years would it be before a substantial number of American homes had the new device? And what if it cost more? And if oil stays cheaper per gallon than bottled water? Changing basic fuels—to hydrogen, say—would be even more expensive. It's not like running out of white wine and switching to red. Yes, we'll get new technologies. One day last fall *The New York Times* ran a special section on energy, featuring many up-and-coming improvements: solar shingles, basement fuel cells. But the same day, on the front page, William K. Stevens reported that international negotiators had all but given up on preventing a doubling of the atmospheric concentration of CO_2. The momentum of growth was so great, the negotiators said, that making the changes required to slow global warming significantly would be like "trying to turn a supertanker in a sea of syrup."

There are no silver bullets to take care of a problem like this. Electric cars won't by themselves save us, though they would help. We simply won't live efficiently enough soon enough to solve the problem. Vegetarianism won't cure our ills, though it would help. We simply won't live simply enough soon enough to solve the problem.

Reducing the birth rate won't end all our troubles either. That, too, is no silver bullet. But it would help.

There's no more practical decision than how many children to have. (And no more mystical decision, either.)

The bottom-line argument goes like this: The next fifty years are a special time. They will decide how strong and healthy the planet will be for centuries to come. Between now and 2050 we'll see the zenith, or very nearly, of human population. With luck we'll never see any greater production of carbon dioxide or toxic chemicals. We'll never see more species extinction or soil erosion. Greenpeace recently announced a campaign to phase out fossil fuels entirely by mid-century, which sounds utterly quixotic but could—if everything went just right—happen.

So it's the task of those of us alive right now to deal with this special phase, to squeeze us through these next fifty years. That's not fair—any more than it was fair that earlier generations had to deal with the Second World War or the Civil War or the Revolution or the Depression or slavery. It's just reality. We need in these fifty years to be working simultaneously on all parts of the equation—on our ways of life, on our technologies, and on our population.

As Gregg Easterbrook pointed out in his book *A Moment on the Earth* (1995), if the planet does manage to reduce its fertility, "the period in which human numbers threaten the biosphere on a general scale will turn out to have been much, much more brief" than periods of natural threats like the Ice Ages. True enough. But the period in question happens to be our time. That's what makes this moment special, and what makes this moment hard.

Bill McKibben is the author of several books about the environment, including *The End of Nature* (1989) and *Hope, Human and Wild* (1995). His article in this issue will appear in somewhat different form in his book *Maybe One: A Personal and Environmental Argument for Single-Child Families*, published in 1998 by Simon & Schuster.

Clash of Globalizations

Stanley Hoffmann

A NEW PARADIGM?

WHAT IS THE STATE OF international relations today? In the 1990s, specialists concentrated on the partial disintegration of the global order's traditional foundations: states. During that decade, many countries, often those born of decolonization, revealed themselves to be no more than pseudostates, without solid institutions, internal cohesion, or national consciousness. The end of communist coercion in the former Soviet Union and in the former Yugoslavia also revealed long-hidden ethnic tensions. Minorities that were or considered themselves oppressed demanded independence. In Iraq, Sudan, Afghanistan, and Haiti, rulers waged open warfare against their subjects. These wars increased the importance of humanitarian interventions, which came at the expense of the hallowed principles of national sovereignty and nonintervention. Thus the dominant tension of the decade was the clash between the fragmentation of states (and the state system) and the progress of economic, cultural, and political integration—in other words, globalization.

Everybody has understood the events of September 11 as the beginning of a new era. But what does this break mean? In the conventional approach to international relations, war took place among states. But in September, poorly armed individuals suddenly challenged, surprised, and wounded the world's dominant superpower. The attacks also showed that, for all its accomplishments, globalization makes an awful form of violence easily accessible to hopeless fanatics. Terrorism is the bloody link between interstate relations and global society. As countless individuals and groups are becoming global actors along with states, insecurity and vulnerability are rising. To assess today's bleak state of affairs, therefore, several questions are necessary. What concepts help explain the new global order? What is the condition of the interstate part of international relations? And what does the emerging global civil society contribute to world order?

SOUND AND FURY

TWO MODELS made a great deal of noise in the 1990s. The first one—Francis Fukuyama's "End of History" thesis—was not vindicated by events. To be sure, his argument predicted the end of ideological conflicts, not history itself, and the triumph of political and economic liberalism. That point is correct in a narrow sense: the "secular religions" that fought each other so bloodily in the last century are now dead. But Fukuyama failed to note that nationalism remains very much alive. Moreover, he ignored the explosive potential of religious wars that has extended to a large part of the Islamic world.

Fukuyama's academic mentor, the political scientist Samuel Huntington, provided a few years later a gloomier account that saw a very different world. Huntington predicted that violence resulting from international anarchy and the absence of common values and institutions would erupt among civilizations rather than among states or ideologies. But Huntington's conception of what constitutes a civilization was hazy. He failed to take into account sufficiently conflicts within each so-called civilization, and he overestimated the importance of religion in the behavior of non-Western elites, who are often secularized and Westernized. Hence he could not clearly define the link between a civilization and the foreign policies of its member states.

Other, less sensational models still have adherents. The "realist" orthodoxy insists that nothing has changed in international relations since Thucydides and Machiavelli: a state's military and economic power determines its fate; interdependence and international institutions are secondary and fragile phenomena; and states' objectives are imposed by the threats to their survival or security. Such is the world described by Henry Kissinger. Unfortunately, this venerable model has trouble integrating change, especially globalization and the rise of nonstate actors. Moreover, it overlooks the need for international cooperation that results from such new threats as the proliferation of weapons of mass destruction (WMD). And it ignores what the scholar Raymond Aron called the "germ of a universal consciousness": the liberal, promarket norms that developed states have come to hold in common.

Taking Aron's point, many scholars today interpret the world in terms of a triumphant globalization that submerges borders through new means of information and communication. In this universe, a state choosing to stay closed invariably faces decline and growing discontent among its subjects, who are eager for material progress. But if it opens up, it must accept a reduced role that is mainly limited to social protection, physical protection against aggression or civil war, and maintaining national identity. The champion of this epic without heroes is *The New York Times* columnist Thomas Friedman. He contrasts barriers with open vistas, obsolescence with modernity, state control with free markets. He sees in globalization the light of dawn, the "golden straitjacket" that will force contentious publics to understand that the logic of globalization is that of peace (since war would interrupt globalization and therefore progress) and democracy (because new technologies increase individual autonomy and encourage initiative).

BACK TO REALITY

THESE MODELS come up hard against three realities. First, rivalries among great powers (and the capacity of smaller states to exploit such tensions) have most certainly not disappeared. For a while now, however, the existence of nuclear weapons has produced a certain degree of prudence among the powers that have them. The risk of destruction that these weapons hold has moderated the game and turned nuclear arms into instruments of last resort. But the game could heat up as more states seek other WMD as a way of narrowing the gap between the nuclear club and the other powers. The sale of such weapons thus becomes a hugely contentious issue, and efforts to slow down the spread of all WMD, especially to dangerous "rogue" states, can paradoxically become new causes of violence.

Second, if wars between states are becoming less common, wars within them are on the rise—as seen in the former Yugoslavia, Iraq, much of Africa, and Sri Lanka. Uninvolved states first tend to hesitate to get engaged in these complex conflicts, but they then (sometimes) intervene to prevent these conflicts from turning into regional catastrophes. The interveners, in turn, seek the help of the United Nations or regional organizations to rebuild these states, promote stability, and prevent future fragmentation and misery.

Third, states' foreign policies are shaped not only by realist geopolitical factors such as economics and military power but by domestic politics. Even in undemocratic regimes, forces such as xenophobic passions, economic grievances, and transnational ethnic solidarity can make policymaking far more complex and less predictable. Many states—especially the United States—have to grapple with the frequent interplay of competing government branches. And the importance of individual leaders and their personalities is often underestimated in the study of international affairs.

For realists, then, transnational terrorism creates a formidable dilemma. If a state is the victim of private actors such as terrorists, it will try to eliminate these groups by depriving them of sanctuaries and punishing the states that harbor them. The national interest of the attacked state will therefore require either armed interventions against governments supporting terrorists or a course of prudence and discreet pressure on other governments to bring these terrorists to justice. Either option requires a questioning of sovereignty—the holy concept of realist theo-

ries. The classical realist universe of Hans Morgenthau and Aron may therefore still be very much alive in a world of states, but it has increasingly hazy contours and offers only difficult choices when it faces the threat of terrorism.

At the same time, the real universe of globalization does not resemble the one that Friedman celebrates. In fact, globalization has three forms, each with its own problems. First is economic globalization, which results from recent revolutions in technology, information, trade, foreign investment, and international business. The main actors are companies, investors, banks, and private services industries, as well as states and international organizations. This present form of capitalism, ironically foreseen by Karl Marx and Friedrich Engels, poses a central dilemma between efficiency and fairness. The specialization and integration of firms make it possible to increase aggregate wealth, but the logic of pure capitalism does not favor social justice. Economic globalization has thus become a formidable cause of inequality among and within states, and the concern for global competitiveness limits the aptitude of states and other actors to address this problem.

Optimism regarding globalization rests on very fragile foundations.

Next comes cultural globalization. It stems from the technological revolution and economic globalization, which together foster the flow of cultural goods. Here the key choice is between uniformization (often termed "Americanization") and diversity. The result is both a "disenchantment of the world" (in Max Weber's words) and a reaction against uniformity. The latter takes form in a renaissance of local cultures and languages as well as assaults against Western culture, which is denounced as an arrogant bearer of a secular, revolutionary ideology and a mask for U.S. hegemony.

Finally there is political globalization, a product of the other two. It is characterized by the preponderance of the United States and its political institutions and by a vast array of international and regional organizations and transgovernmental networks (specializing in areas such as policing or migration or justice). It is also marked by private institutions that are neither governmental nor purely national—say, Doctors Without Borders or Amnesty International. But many of these agencies lack democratic accountability and are weak in scope, power, and authority. Furthermore, much uncertainty hangs over the fate of American hegemony, which faces significant resistance abroad and is affected by America's own oscillation between the temptations of domination and isolation.

The benefits of globalization are undeniable. But Friedman-like optimism rests on very fragile foundations. For one thing, globalization is neither inevitable nor irresistible. Rather, it is largely an American creation, rooted in the period after World War II and based on U.S. economic might. By extension, then, a deep and protracted economic crisis in the United States could have as devastating an effect on globalization as did the Great Depression.

Second, globalization's reach remains limited because it excludes many poor countries, and the states that it does transform react in different ways. This fact stems from the diversity of economic and social conditions at home as well as from partisan politics. The world is far away from a perfect integration of markets, services, and factors of production. Sometimes the simple existence of borders slows down and can even paralyze this integration; at other times it gives integration the flavors and colors of the dominant state (as in the case of the Internet).

Third, international civil society remains embryonic. Many nongovernmental organizations reflect only a tiny segment of the populations of their members' states. They largely represent only modernized countries, or those in which the weight of the state is not too heavy. Often, NGOs have little independence from governments.

Fourth, the individual emancipation so dear to Friedman does not quickly succeed in democratizing regimes, as one can see today in China. Nor does emancipation prevent public institutions such as the International Monetary Fund, the World Bank, or the World Trade Organization from remaining opaque in their activities and often arbitrary and unfair in their rulings.

Fifth, the attractive idea of improving the human condition through the abolition of barriers is dubious. Globalization is in fact only a sum of techniques (audio and videocassettes, the Internet, instantaneous communications) that are at the disposal of states or private actors. Self-interest and ideology, not humanitarian reasons, are what drive these actors. Their behavior is quite different from the vision of globalization as an Enlightenment-based utopia that is simultaneously scientific, rational, and universal. For many reasons—misery, injustice, humiliation, attachment to traditions, aspiration to more than just a better standard of living—this "Enlightenment" stereotype of globalization thus provokes revolt and dissatisfaction.

Another contradiction is also at work. On the one hand, international and transnational cooperation is necessary to ensure that globalization will not be undermined by the inequalities resulting from market fluctuations, weak state-sponsored protections, and the incapacity of many states to improve their fates by themselves. On the other hand, cooperation presupposes that many states and rich private players operate altruistically—which is certainly not the essence of international relations—or practice a remarkably generous conception of their long-term interests. But the fact remains that most rich states still refuse to provide sufficient development aid or to intervene in crisis situations such as the genocide in Rwanda. That reluctance compares poorly with the American enthusiasm to pursue the fight against al Qaeda and the Taliban. What is wrong here is not patriotic enthusiasm as such, but the weakness of the humanitarian impulse when the national interest in saving non-American victims is not self-evident.

IMAGINED COMMUNITIES

Among the many effects of globalization on international politics, three hold particular importance. The first concerns institutions. Contrary to realist predictions, most states are not perpetually at war with each other. Many regions and countries live in peace; in other cases, violence is internal rather than state-to-state. And since no government can do everything by itself, interstate organisms have emerged. The result, which can be termed "global society," seeks to reduce the potentially destructive effects of national regulations on the forces of integration. But it also seeks to ensure fairness in the world market and create international regulatory regimes in such areas as trade, communications, human rights, migration, and refugees. The main obstacle to this effort is the reluctance of states to accept global directives that might constrain the market or further reduce their sovereignty. Thus the UN's powers remain limited and sometimes only purely theoretical. International criminal justice is still only a spotty and contested last resort. In the world economy—where the market, not global governance, has been the main beneficiary of the state's retreat—the network of global institutions is fragmented and incomplete. Foreign investment remains ruled by bilateral agreements. Environmental protection is badly ensured, and issues such as migration and population growth are largely ignored. Institutional networks are not powerful enough to address unfettered short-term capital movements, the lack of international regulation on bankruptcy and competition, and primitive coordination among rich countries. In turn, the global "governance" that does exist is partial and weak at a time when economic globalization deprives many states of independent monetary and fiscal policies, or it obliges them to make cruel choices between economic competitiveness and the preservation of social safety nets. All the while, the United States displays an increasing impatience toward institutions that weigh on American freedom of action. Movement toward a world state looks increasingly unlikely. The more state sovereignty crumbles under the blows of globalization or such recent developments as humanitarian intervention and the fight against terrorism, the more states cling to what is left to them.

Second, globalization has not profoundly challenged the enduring national nature of citizenship. Economic life takes place on a global scale, but human identity remains national—hence the strong resistance to cultural homogenization. Over the centuries, increasingly centralized states have expanded their functions and tried to forge a sense of common identity for their subjects. But no central power in the world can do the same thing today, even in the European Union. There, a single currency and advanced economic coordination have not yet produced a unified economy or strong central institutions endowed with legal autonomy, nor have they resulted in a sense of postnational citizenship. The march from national identity to one that would be both national and European has only just begun. A world very partially unified by technology still has no collective consciousness or collective solidarity. What states are unwilling to do the world market cannot do all by itself, especially in engendering a sense of world citizenship.

Third, there is the relationship between globalization and violence. The traditional state of war, even if it is limited in scope, still persists. There are high risks of regional explosions in the Middle East and in East Asia, and these could seriously affect relations between the major powers. Because of this threat, and because modern arms are increasingly costly, the "anarchical society" of states lacks the resources to correct some of globalization's most flagrant flaws. These very costs, combined with the classic distrust among international actors who prefer to try to preserve their security alone or through traditional alliances, prevent a more satisfactory institutionalization of world politics—for example, an increase of the UN's powers. This step could happen if global society were provided with sufficient forces to prevent a conflict or restore peace—but it is not.

Globalization, far from spreading peace, thus seems to foster conflicts and resentments. The lowering of various barriers celebrated by Friedman, especially the spread of global media, makes it possible for the most deprived or oppressed to compare their fate with that of the free and well-off. These dispossessed then ask for help from others with common resentments, ethnic origin, or religious faith. Insofar as globalization enriches some and uproots many, those who are both poor and uprooted may seek revenge and self-esteem in terrorism.

GLOBALIZATION AND TERROR

TERRORISM is the poisoned fruit of several forces. It can be the weapon of the weak in a classic conflict among states or within a state, as in Kashmir or the Palestinian territories. But it can also be seen as a product of globalization. Transnational terrorism is made possible by the vast array of communication tools. Islamic terrorism, for example, is not only based on support for the Palestinian struggle and opposition to an invasive American presence. It is also fueled by a resistance to "unjust" economic globalization and to a Western culture deemed threatening to local religions and cultures.

If globalization often facilitates terrorist violence, the fight against this war without borders is potentially disastrous for both economic development and globalization. Antiterrorist measures restrict mobility and financial flows, while new terrorist attacks could lead the way for an antiglobalist reaction comparable to the chauvinistic paroxysms of the 1930s. Global terrorism is not the simple extension of war among states to nonstates. It is the subversion of traditional ways of war because it does not care about the sovereignty of either its enemies or the allies who shelter them. It provokes its victims to take measures that, in the name of legitimate defense, violate knowingly the sovereignty of those states accused of encouraging terror. (After all, it was not the Taliban's infamous domestic violations of human rights that led the United States into Afghanistan; it was the Taliban's support of Osama bin Laden.)

But all those trespasses against the sacred principles of sovereignty do not constitute progress toward global society, which has yet to agree on a common definition of terrorism or on a common policy against it. Indeed, the beneficiaries of the antiterrorist "war" have been the illiberal, poorer states that have lost so much of their sovereignty of late. Now the crackdown on terror allows them to tighten their controls on their own people, products, and money. They can give themselves new reasons to violate individual rights in the name of common defense against insecurity—and thus stop the slow, hesitant march toward international criminal justice.

Another main beneficiary will be the United States, the only actor capable of carrying the war against terrorism into all corners of the world. Despite its power, however, America cannot fully protect itself against future terrorist acts, nor can it fully overcome its ambivalence toward forms of interstate cooperation that might restrict U.S. freedom of action. Thus terrorism is a global phenomenon that ultimately reinforces the enemy—the state—at the same time as it tries to destroy it. The states that are its targets have no interest in applying the laws of war to their fight against terrorists; they have every interest in treating terrorists as outlaws and pariahs. The champions of globalization have sometimes glimpsed the "jungle" aspects of economic globalization, but few observers foresaw similar aspects in global terrorist and antiterrorist violence.

Finally, the unique position of the United States raises a serious question over the future of world affairs. In the realm of interstate problems, American behavior will determine whether the nonsuperpowers and weak states will continue to look at the United States as a friendly power (or at least a tolerable hegemon), or whether they are provoked by Washington's hubris into coalescing against American preponderance. America may be a hegemon, but combining rhetorical overkill and ill-defined designs is full of risks. Washington has yet to understand that nothing is more dangerous for a "hyperpower" than the temptation of unilateralism. It may well believe that the constraints of international agreements and organizations are not necessary, since U.S. values and power are all that is needed for world order. But in reality, those same international constraints provide far better opportunities for leadership than arrogant demonstrations of contempt for others' views, and they offer useful ways of restraining unilateralist behavior in other states. A hegemon concerned with prolonging its rule should be especially interested in using internationalist methods and institutions, for the gain in influence far exceeds the loss in freedom of action.

In the realm of global society, much will depend on whether the United States will overcome its frequent indifference to the costs that globalization imposes on poorer countries. For now, Washington is too reluctant to make resources available for economic development, and it remains hostile to agencies that monitor and regulate the global market. All too often, the right-leaning tendencies of the American political system push U.S. diplomacy toward an excessive reliance on America's greatest asset—military strength—as well as an excessive reliance on market capitalism and a "sovereigntism" that offends and alienates. That the mighty United States is so afraid of the world's imposing its "inferior" values on Americans is often a source of ridicule and indignation abroad.

ODD MAN OUT

FOR ALL THESE TENSIONS, it is still possible that the American war on terrorism will be contained by prudence, and that other governments will give priority to the many internal problems created by interstate rivalries and the flaws of globalization. But the world risks being squeezed between a new Scylla and Charybdis. The Charybdis is universal intervention, unilaterally decided by American leaders who are convinced that they have found a global mission provided by a colossal threat. Presentable as an epic contest between good and evil, this struggle offers the best way of rallying the population and overcoming domestic divisions. The Scylla is resignation to universal chaos in the form of new attacks by future bin Ladens, fresh humanitarian disasters, or regional wars that risk escalation. Only through wise judgment can the path between them be charted.

We can analyze the present, but we cannot predict the future. We live in a world where a society of uneven and often virtual states overlaps with a global society burdened by weak public institutions and underdeveloped civil society. A single power dominates, but its economy could become unmanageable or disrupted by future terrorist attacks. Thus to predict the future confidently would be highly incautious or naive. To be sure, the world has survived many crises, but it has done so at a very high price, even in times when WMD were not available.

Precisely because the future is neither decipherable nor determined, students of international relations face two missions. They must try to understand what goes on by taking an inventory of current goods and disentangling the threads of present networks. But the fear of confusing the empirical with the normative should not prevent them from writing as political philosophers at a time when many philosophers are extending their conceptions of just society to international relations. How can one make the global house more livable? The answer presupposes a political philosophy that would be both just and acceptable even to those whose values have other foundations. As the late philosopher Judith Shklar did, we can take as a point of departure and as a guiding thread the fate of the victims of violence, oppression, and misery; as a goal, we should seek material and moral emancipation. While taking into account the formidable constraints of the world as it is, it is possible to loosen them.

STANLEY HOFFMANN is Buttenwieser University Professor at Harvard University and a regular book reviewer for *Foreign Affairs*.

[The Hard Questions]

A NEW GRAND STRATEGY

For more than fifty years American foreign policy has sought to prevent the emergence of other great powers—a strategy that has proved burdensome, futile, and increasingly risky. The United States will be more secure, and the world more stable, if America now chooses to pass the buck and allow other countries to take care of themselves

BY BENJAMIN SCHWARZ
AND CHRISTOPHER LAYNE

Since the end of the Cold War, U.S. grand strategy has revolved around maintaining this country's overwhelming military, economic, and political preponderance. Until now most Americans have acquiesced in that strategy, because the costs seemed to be tolerably low. But the September 11 attacks have proved otherwise. Those assaults were neither random nor irrational. Those who undertook them acted with cool calculation to force the United States to alter specific policies—policies that largely flow from the global role America has chosen. The attacks were also a violent reaction to the very fact of America's pre-eminence.

Several tasks confront us. The most immediate is the one that rightly preoccupies the nation now: tracking down the al Qaeda terrorists and destroying their networks and their infrastructure, and waging war on the Taliban movement that harbors them. The larger task will take time, because it amounts to inventing a new American stance toward the world for the century ahead. We need to come to grips with an ironic possibility: that the very preponderance of American power may now make us not more secure but less secure. By the same token, it may actually be possible to achieve more of our ultimate foreign-policy goals by means of a diminished global presence.

Great powers have two basic strategic options: they can pursue geopolitical dominance (a "unipolar" strategy), or they can seek to maintain a rough balance of power among the strongest states in a region or around the world (a "multipolar" strategy). Since the late 1940s the United States has chosen the former course. True, even during the Cold War, when the world was essentially divided between the United States and the Soviet Union, a number of astute foreign-policy thinkers—including Walter Lippmann, George Kennan, and J. William Fulbright—argued that it was in America's interest to encourage Western Europe's and Japan's revival as independent great powers to relieve the United States of what Kennan called

the "burdens of 'bi-polarity.'" But almost all American policymakers held that the United States had to contain its allies as much as it had to contain Moscow. By providing for the security of Britain, France, and (especially) Germany and Japan—by defending their access to far-flung economic and natural resources, and by enmeshing their foreign and military policies in alliances that America dominated—Washington prevented these former and potential great powers from embarking on independent, and (from the U.S. perspective) possibly destabilizing, foreign policies. This "reassurance strategy" (to use a term currently favored by policymakers) allowed for an unprecedented level of political and economic cooperation among the states of Western Europe and East Asia.

As noted, American policy since the end of the Cold War has aimed to ensure that the United States maintains its lofty perch. Every post-Cold War Pentagon assessment of national-security needs has insisted that America maintain its globe-girdling Cold War alliances, along with its Cold War defense-spending levels, even though the threat against which those alliances and budgets were ostensibly erected has disappeared. Some critics argue that this apparent stasis is born of bureaucratic inertia, or of a defense establishment's jealous guarding of its turf (and its trough). But in fact, given the logic behind American grand strategy, this continuity is entirely justifiable. The collapse of the Soviet Union hasn't altered the conviction among many American policymakers that a stable global economic and political order depends on Washington's maintaining preponderance (or, according to the official rhetoric, "leadership") over potential great powers. This means ameliorating their security problems.

The now infamous draft of the Pentagon's Defense Planning Guidance (prepared under the direction of the current undersecretary of defense for policy, Paul Wolfowitz), which was leaked to *The New York Times* in 1992, merely stated in undiplomatic language the logic that has long informed Washington's strategy. The United States, it argued, must continue to dominate the international system and thus to "discourage" the "advanced industrial nations from challenging our leadership or... even aspiring to a larger regional or global role." To accomplish this Washington must do nothing less than "retain the preeminent responsibility for addressing... those wrongs which threaten not only our interests, but those of our allies or friends, or which could seriously unsettle international relations." In other words, America must provide its allies with what one of the document's authors (now a special assistant to the President on the National Security Council) termed "adult supervision": the United States must not only impose a military protectorate over Europe and East Asia—regions composed of wealthy and technologically sophisticated states—but also safeguard Europe's and East Asia's worldwide interests, so that they need not develop military forces capable of "global power projection." (As Gabriel Robin, a former French representative to NATO, acknowledged, the U.S.-led alliance's "real function... is to serve as the chaperon of Europe." It is, Robin said, "the means to prevent [Europe] from establishing itself as an independent fortress and perhaps one day, a rival.") Those who argue that America's national-security spending is too high, given the end of the Cold War, often fail to appreciate the task Washington has assigned itself. The adult supervision of the world is an enormously expensive and complex undertaking, which perforce means that the United States must spend more on its military than do the next nine countries together—including Russia, Japan, China, France, Britain, and Germany.

An adult-supervision strategy entails a peculiar and recondite calculation of the world situation. For instance, although most Americans might believe that a reunified, democratic Korea would indisputably be in America's interests, the former National Security Advisor Zbigniew Brzezinski, in his 1997 book *The Grand Chessboard* (probably the fullest and frankest public exposition of America's post-Cold War global strategy), repeatedly explains how this development would in fact jeopardize America's unipolar strategy: it would, he argues, reflecting views long and widely held in policymaking circles, obviate the ostensible need for U.S. troops on the peninsula, which could lead to a U.S. pullback from East Asia, which could, in turn, lead to Japan's becoming "military, more self-sufficient," which would lead to political, military, and economic rivalry among the region's states. Thus the best situation is the status quo in Korea, which allows for U.S. forces to be stationed there indefinitely.

Similar—and more urgent, given the current war on terrorism—is the thinking underlying Washington's policy in the Persian Gulf. Why is the United States so deeply embroiled in this turbulent region? Many people, echoing a comment by Secretary of State James A. Baker during the Persian Gulf War, would probably answer with one word: oil. This answer is—and was—both true and misleading. America derives most of its oil from Alaska, Canada, the continental United States, Mexico, and Venezuela. About 25 percent of U.S. petroleum imports come from the Persian Gulf. If the United States adopted a national energy strategy, it could free itself from dependence on

Persian Gulf oil. Nevertheless, Washington assumes responsibility for stabilizing the region because Western Europe and Japan are heavily dependent on its oil, and because soon China, owing to rapid economic growth, will be as well—and America wants to discourage those powers from developing the means to protect that resource for themselves. In an interview on National Public Radio early in October, Walter Russell Mead, a senior foreign-policy analyst at the Council on Foreign Relations, explained the basis of U.S. policy in the terms that NSC staffers, think-tank analysts, and State and Defense Department policy planners have used for years: "We do not get that large a percentage of our oil from the Middle East. Japan gets a lot more... And one of the reasons that we are sort of assuming this role of policeman of the Middle East, more or less, has more to do with making Japan and some other countries feel that their oil flow is assured... so that they don't then feel more need to create a great power, armed forces, and security doctrine, and you don't start getting a lot of great powers with conflicting interests sending their militaries all over the world."

Despite its sometimes esoteric logic, America's strategy of preponderance is seductive. In the abstract it makes sense that the United States should seek to amass as much power as possible. In this way the rationale behind U.S. strategy is analogous to that of a firm in an oligopolistic market, which drives its rivals out of business rather than risk its profits in a competitive environment. Theoretically, if a state can establish—and maintain—itself as the only great power in the international system, it will enjoy something very close to absolute security. But as history amply shows, when one state acquires too much power, others invariably fear that it will aggrandize itself at their expense. "Hegemonic empires," Henry Kissinger recently noted, "almost automatically elicit universal resistance, which is why all such claimants have sooner or later exhausted themselves."

More than two hundred years ago Edmund Burke warned his countrymen,

> Among precautions against ambition it may not be amiss to take one precaution against our *own*... I dread our being too much dreaded... We may say that we shall not abuse this astonishing and hitherto unheard-of power. But every nation will think we shall abuse it... Sooner or later, this state of things must produce a combination against us which may end in our ruin.

Like some optimistic Britons in the late eighteenth century, many American strategists today assert that the United States, the only superpower, is a "benevolent" hegemon, immunized from a backlash against its preponderance by what they call its "soft power"—that is, by the attractiveness of its liberal-democratic ideology and its open, syncretic culture. Washington also believes that others don't fear U.S. geopolitical pre-eminence because they know the United States will use its unprecedented power to promote the good of the international system rather than to advance its own selfish aims.

But states must always be more concerned with a predominant power's capabilities than with its intentions, and in fact well before September 11—indeed, throughout most of the past decade—other states have been profoundly anxious about the *im*balance of power in America's favor. This simmering mistrust of U.S. predominance intensified during the Clinton Administration, as other states responded to American hegemony by concerting their efforts against it. Russia and China, although long estranged, found common ground in a nascent alliance that opposed U.S. "hegemonism" and expressly aimed at re-establishing "a multipolar world." Arguing that the term "superpower" is inadequate to convey the true extent of America's economic and military pre-eminence, the French Foreign Minister Hubert Vedrine called the United States a "hyperpower." Even the Dutch Prime Minister declared that the European Union should make itself "a counterweight to the United States."

American intervention in Kosovo crystallized fears of U.S. hegemony, prompting the emergence of an anti-U.S. constellation of China, Russia, and India. Viewing the Kosovo war as a dangerous precedent establishing Washington's self-declared right to interfere in other countries' internal affairs, and asserting their support for a multipolar world, these three states increased their arms transfers and their sharing of military technology, specifically to counter American power. Also, the Kosovo conflict made apparent the disparity between America's geopolitical power and Europe's, inciting Europe to take its first serious steps toward redressing that disparity by acquiring—through the European Defense and Security Identity—the kinds of military capabilities it would need to act independent of the United States. If the European Union fulfills EDSI's longer-term goals, it will emerge as an unfettered strategic player in world politics. And that emergence will have been driven by the clear objective of investing Europe with the capability to act as a brake on America's aspirations.

Any remaining doubt that American hegemony could trigger a hostile reaction, whether reasonable or not, surely dissipated on September 11. The role the United States has assigned itself in the Persian Gulf has made it—not Japan, not the states of Western Europe, not China—vulnerable to a backlash. Iran, Iraq, and Afghanistan resent America's intrusion into regional affairs. The widespread perception within the region that the Middle East has long been a victim of "Western imperialism" of course exacerbates this animosity. Moreover, aggrieved groups throughout the Middle East contest the legitimacy of the regimes in Saudi Arabia, Kuwait, and the Gulf emirates which the United States is compelled to support, making America even more of a lightning rod for the politically disaffected. In this sense Osama bin Laden's brand of terrorism (which aims to compel the United States to remove its military forces from the Persian Gulf, and to replace America's client, the Saudi monarchy, with a fundamentalist Islamic government) dramatically illustrates U.S. vulnerability to the kind of "asymmetric warfare" of which some defense experts have warned.

The rise of new great powers is inevitable, and America's very primacy accelerates this process. If Washington continues to follow an adult-supervision strategy, which treats its "allies" as irresponsible adolescents and China and Russia as future enemies to be suppressed, its relations with these emerging great powers will be increasingly dangerous, as they coalesce against what they perceive as an American threat. But that is not even the worst conceivable outcome. What if a sullen and resentful China were to align itself with Islamic fundamentalist groups? Such a situation is hardly beyond the realm of possibility; partners form alliances not because they are friends, or because they have common values, but because they fear someone else more than they fear each other.

A strategy of preponderance is burdensome, Sisyphean, and profoundly risky. It is therefore time for U.S. policymakers to adopt a very different grand strategy: one that might be called offshore balancing. Rather than fear multipolarity, this strategy embraces it. It recognizes that instability—caused by the rise and fall of great powers, great-power rivalries, and messy regional conflicts—is a geopolitical fact of life. Offshore balancing accepts that the United States cannot prevent the rise of new great powers, either within the present American sphere (the European Union, Germany, Japan) or outside it (China, a resurgent Russia). Instead of exhausting its resources and drawing criticism or worse by keeping these entities weak, the United States would allow them to develop their militaries to provide for their own national and regional security. Among themselves, then, these states would maintain power balances, check the rise of overly ambitious global and regional powers, and stabilize Europe, East Asia, and the Persian Gulf. It would naturally be in their interests to do so.

It's always safest and cheapest to get others to stabilize the turbulent regions of the globe. Historically, however, this has seldom been an option, because if one lives in a dangerous neighborhood, one must be prepared to protect oneself from troublemakers rather than relying on someone else to do so. In fact, the only two great powers in modern history that successfully devolved onto others the responsibility for maintaining regional stability are Britain during its great-power heyday (1700–1914) and the United States (until 1945). They were able to do so because they had moats—a narrow one for England, and two very big ones for the United States—that kept predatory Eurasian great powers at bay.

As offshore balancers, Britain and the United States reaped enormous strategic dividends. While they were shielded from threatening states by geography, London and Washington could afford to maintain militaries smaller than those of Continental powers, and concentrate instead on getting rich. Often they could stay out of Europe's turmoil entirely, gaining in strength as other great powers fought debilitating wars. And even in wartime offshore balancers have enjoyed advantages that Continental powers have not. Instead of sending big armies to fight costly Continental wars, Britain, for instance, relied on its navy to blockade those states bidding for mastery of Europe and on its financial power to underwrite coalitions against them, and stuck its allies with the greater part of the blood price of defeating those powers that aspired to dominate the Continent.

The United States, of course, followed a similar strategy during World War II. From 1940 to 1944 it confined its role in the European war to providing economic assistance and munitions to the Soviet Union and Britain and—after entering the war, in December of 1941—to relatively low-cost strategic air bombardment of Germany, and peripheral land campaigns in North Africa and Italy. The United States was more than happy to delay the invasion of Europe until June of 1944. By then the Red Army—which inflicted about 88 percent of the Wehrmacht's casualties throughout the war—had mortally weakened Germany, but at a staggering cost.

Taken together, the experiences of Britain and America highlight the central feature of the offshore balancing strategy: it allows for burden *shifting*, rather than burden sharing. Offshore balancers can afford to be bystanders in the opening stages of conflict. Because the security of others is most immediately at risk, an offshore balancer can be confident that those others will attempt to defend themselves. Often they will do so expeditiously, obviating the offshore balancer's intervention. If, on the other hand, a predominant power seems to be winning, an offshore balancer can intervene decisively to forestall its victory (as Britain did against Philip II, Louis XIV, and Napoleon). And if the offshore balancer must intervene, the state aspiring to dominance will already have been at least somewhat bloodied, and thus not as formidable as it was for those who had the geopolitical misfortune to constitute the first line of defense.

The same dynamics apply—or would, if the United States gave them a chance—in regional conflicts, although not quite as dramatically. Great powers that border restive neighbors, or that are economically dependent on unstable regions, have a much larger interest than does the United States in policing those areas. Most regional power balances (the relative positions of, say, Hungary and Romania, or of one sub-Saharan state and another) need not concern the United States. America must intervene only to prevent a single power from dominating a strategically crucial area—and then only if the efforts of great powers with a larger stake in that region have failed to redress the imbalance. So for an offshore balancing strategy to work, the world must be multipolar—that is, there must be several other great powers, and major regional powers as well, onto which the United States can shift the burden of maintaining stability in various parts of the world.

For America the most important grand-strategic issue is what relations it will have with these new great powers. In fostering a multipolar world—in which the foreign and national-security policies of the emerging great powers will be largely devoted to their rivalries with one another and to quelling and containing regional instability—an offshore balancing strategy is, of course, opportunistic and self-serving. But it also exercises restraint and shows geopolitical respect. By abandoning the "preponderance" strategy's extravagant objectives, the United States can minimize the risks of open confrontation with the new great powers.

Although jockeying for advantage is a fact of life for great powers, coexistence, and even cooperation between and among them, is not unusual. Offshore balancing seeks to promote America's relative power and security, but it also aims to maximize the opportunity for the United States to be on decent terms with the other great powers. In this sense the strategy has

much in common with Richard Nixon and Henry Kissinger's vision of détente. That policy was a significant departure from previous Cold War approaches, in that the United States explicitly recognized the Soviet Union as a collaborator in, rather than a challenger to, the effort to maintain the stability of the international system. To understand this dramatic shift, contrast the inaugural address of John F. Kennedy, in which that paragon of Cold War liberalism advanced the stirring but rather dangerous notion that in the struggle with communism the United States would "pay any price, bear any burden," with Nixon's first inaugural address, which promulgated the realistic but conciliatory message that "we cannot expect to make everyone our friend, but we can try to make no one our enemy." This was détente's animating sentiment.

Détente was based on the assumption (hardly contested at the time) that the USSR wouldn't go away. Because the superpower rivalry could not be resolved without destroying humanity, there was, as Kissinger declared, "no alternative to coexistence." Détente, then, was a strategy for managing a permanent relationship. In what Nixon and Kissinger hoped would evolve into a mature relationship, Moscow and Washington would acknowledge each other's legitimate interests and try not to allow disagreements to poison accommodations. Détente was a shift in style with substance—or, rather, a shift in style with substantive consequences. The Soviets and the Chinese were to be approached not as alien ideologues but as intelligent adults with whom the United States should find a substantial area of common interest.

Similarly, an offshore balancing strategy would dictate that in order to coexist with the emerging great powers, or even to enjoy cooperative ties with them (in efforts to combat Islamic terrorism, for instance), the United States must start treating such powers like fellow adults. This would mean both accepting them as peers and acknowledging the legitimacy of their national interests. In concrete terms, here is how an offshore balancing strategy would apply to particular cases.

- Today, not for the first time in its history, Russia is down and out as a great power. But it has come back before and probably will do so again. Moreover, for the United States, Russia is crucial as a potential ally in three regions: Europe (vis-à-vis a European "superstate," or Germany if the EU project fails), East Asia (vis-à-vis China), and the Persian Gulf and Central Asia. As an offshore balancer, the United States would abandon plans for NATO expansion (which Russia regards as a strategic threat), and if Washington decides to undertake a national missile-defense program, it would follow through on the recent agreement to make deep cuts in its strategic nuclear arsenal to reassure Russia and China that it doesn't seek to gain a first-strike advantage. It would allow Russia to supervise its legitimate sphere of influence—in Chechnya and in Central Asia, where it is combating Islamic fundamentalists, as well as in parts of Eastern Europe and in the states that formerly composed the Soviet Union. America's direct sphere of influence embraces the area from the Canadian Arctic to Tierra del Fuego and from Greenland to

Guam. Surely we can tolerate other great powers' enjoying spheres of influence in their own parts of the world.

- With respect to China, the United States would recognize that the Taiwan issue is an internal Chinese matter. Taiwan's unresolved status is a legacy of the civil war that ended on the mainland in 1949. It is worth recalling that before the outbreak of the Korean War, Secretary of State Dean Acheson advised that the United States should extricate itself from the unfinished business of the Chinese civil war and leave Taiwan to its fate. A half century later it is time for the United States finally to do so. Washington would also fundamentally re-examine its notion of what constitutes a "China threat." That China, the largest and potentially most powerful state in East Asia, would seek a more assertive political, economic, and military role in the region—and would even want to end America's current strategic superiority there—hardly meets that threshold (although it is no doubt alarming to China's immediate neighbors). The United States should also mute its criticism of China's human-rights policy. Washington simply can't transform China into a liberal, free-market democracy, and U.S. pressure only exacerbates the friction in Sino-American relations. Generally, an offshore balancing strategy would hold that fatalism should replace idealism in America's attitude toward what used to be called the internal arrangements of other countries. It would also hold that our pragmatic policy choices, born of self-interest, should be embraced as such, and not clothed in altruism or idealism. Seeking to engender changes in other nations' fundamental values so that they resemble America's is an unreasonable goal of foreign policy. An offshore balancing strategy would accept other nations for what they are, or what their history has made them.

- With respect to Europe, the United States would endorse the EU's efforts—which Washington now opposes—to acquire the military capabilities it needs to defend its interests independent of the United States. At the same time, the United States would begin a phased withdrawal from its European security commitments. To be sure, many U.S. policymakers have argued that the Europeans have demonstrated their incapacity (during the Balkan crises, for instance) to act effectively without U.S. "leadership." But these protests are hypocritical—who can blame the Europeans for their inability to assert themselves in security affairs when Washington has for decades repeatedly squelched European initiatives that would have made that assertion possible? An offshore balancing strategy would hold that America's strategic interest in Europe does not demand that Washington insure against every untoward event there. Disorder in the Balkans and other places on Europe's fringes should be a matter for Europeans, who have the wherewithal to combat it, quarantine it, or, if they choose, ignore it. The United States would follow a similar policy with respect to Japan. Washington would announce to Tokyo its intention to terminate the Mutual Cooperation and Security Treaty within a specified time period (say, five years), at the end of which Japan, for more than fifty years a politically stable state, would have developed whatever military means it believes necessary to func-

tion as an independent great power. An offshore balancing strategy would turn on a simple truth: other states have at least as much interest as the United States does in secure sea-lanes, access to resources, and regional stability. The less America does, and the less others expect it to do, the more other states will do to help themselves.

Recognizing the legitimacy of other great powers' spheres of influence offers the United States a further strategic advantage. The Persian Gulf and Central Asia show why. Russia and China both are profoundly concerned about the spread of Islamic fundamentalism on their peripheries. In Chechnya, in Central Asia (where Russian troops help to defend the former Soviet republics of Uzbekistan and Tajikistan), and in the Caucasus, Moscow has fought major military campaigns to protect its southern flank against militant Islam. China, too, is combating terrorism fomented by Islamic separatists, in the Xinjiang province. Last June, Beijing and Moscow entered into a security relationship, the Shanghai Cooperation Organization (which also embraces three of the former Soviet republics in Central Asia), to coordinate efforts to combat the common threat to their security posed by these Islamic fundamentalist terrorist groups—groups linked to the Taliban and Osama bin Laden. Similarly, India, a possible future great power, has been battling Islamic terrorists who are waging a proxy war on Pakistan's behalf to wrest the disputed province of Kashmir away from New Delhi. Simply put, for reasons of security and access to oil, Russia, China, India, Western Europe, and Japan have strong reasons—stronger than America's—to pacify Central Asia and the Persian Gulf. By adopting an offshore balancing strategy, the United States will compel them to do so.

Passing the buck would help the United States out of the impasse that securing Afghanistan promises to be. The political and military challenges the war poses underscore how difficult and costly will be the effort to restore order in the country and the region when the fighting stops. When the United States has achieved its military goals in Afghanistan, it should announce a phased withdrawal from its security commitments in the region, shifting to others the hard job of stabilizing it.

The complexities involved in that job are numerous. Washington's very strategy of primacy, and America's concomitant military presence in the region, are in themselves a source of instability, especially for the regimes on which the United States relies. The regimes in Saudi Arabia and Pakistan, for instance, face doubtful prospects precisely because their close connection to Washington intensifies radical nationalist and Islamic fundamentalist opposition within those countries. For this reason none of the regional regimes in the current coalition can be especially dependable allies. Only with enormous pressure did a few of them even allow American forces to conduct offensive strikes on Afghanistan from bases on their territory. And fearing that popular anger at the U.S. military campaign will trigger domestic political explosions, many of these states pressed Washington to bring an early end to the war.

If America remains in the region indefinitely, it will have to prop up these unpopular or failing regimes. In Saudi Arabia the

United States could easily find itself militarily involved if internal upheaval threatens the monarchy's hold on power. To forestall economic collapse in Pakistan, Washington will have to donate billions of dollars in direct and indirect assistance. Finally, if the United States continues to play the role of regional gendarme, it will assume the thankless—and probably hopeless—burden of trying to put Afghanistan together again. Divided along ethnic, linguistic, and clan fault lines, the various factions inside Afghanistan cannot agree on that country's future political organization. (The forces making up the anti-Taliban contingent seem only to agree that they resent U.S. bombing of their country.) That the outside powers have conflicting goals for Afghanistan's future further complicates any sorting out of Afghanistan's political structure. If ever there was a place where America should devolve security responsibilities to others, it is the Persian Gulf and Southwest Asia region. Again, Western Europe, Japan, Russia, China, and India all have greater security and economic interests in the region than does the United States, and if America pulls out, they will police it because they must.

Rather than attempt to impose a Pax Americana on this endemically turbulent area, the United States should devote the resources it currently spends on this costly and dangerous job to rendering the region economically and strategically irrelevant. That is, America should pursue a national energy policy that would develop alternative sources of energy for the United States and, more important, the rest of the industrialized world. This colossal scientific and industrial effort should be our highest national-security priority (see "Mideast Oil Forever?," by Joseph J. Romm and Charles B. Curtis, April, 1996, *Atlantic*). If the United States shifts responsibility for stabilizing the region to the other great powers, the real price of Persian Gulf oil will become extremely high for them. It would then be in their interests to pool resources and expertise with America in what would amount to an international Manhattan Project to obviate the need for that oil—thus dramatically reducing the revenue streams to the regimes in Iran, Iraq, and Saudi Arabia. Doing so is surely a common international interest. If Washington were to spend the approximately $106 billion that—according to Earl Ravenal, a former Pentagon analyst—it is devoting this year to defending the Persian Gulf region, and if Western Europe, Japan, China, and Russia were to kick in what they would otherwise spend on policing the region, it's hard to imagine that this goal couldn't be achieved.

Some will assert, correctly, that if it abjures a strategy of preponderance, America will sacrifice some of the awe with which it is viewed by the world. But less awe and less influence will bring the United States more security. Some will object that the policy we advocate shuns the inspiring role of America as "the indispensable nation." But such a grandiose vision, while pleasing to our image of ourselves, is the antithesis of statecraft, which must be guided by discrimination on the basis of power, interest, and circumstance. Historically, the most imaginative statesmen and policies have hardly been visionary. For centuries, with flexibility and subtlety, British diplomats pursued a grand strategy that aimed at nothing more inspiring than ensuring a balance of power among the states of Western Europe. This was really just tactical fine-tuning on a grand scale, and so aroused the consternation of idealists of every stripe. For their part, America's nineteenth-century statesmen could not have been less idealistic or more pragmatic as they, by adroitly exploiting European great-power rivalries, maneuvered the British, the French, and the Spanish out of North America and established American predominance in the Western Hemisphere—probably the most stunning diplomatic achievement of modern history, and the very model of a successful multipolar strategy.

The policy we advocate is informed by the conviction that history is "just one damned thing after another"; we see no end to power politics. And we hold that the purpose of grand strategy isn't the pursuit of new world orders but simply making the best of bad choices—to use the political philosopher Michael Oakeshott's metaphor, keeping afloat in "a boundless and bottomless sea; there is neither harbor for shelter nor floor for anchorage, neither starting point nor appointed destination." Ours is a grand strategy for the long haul—and so, by the lights of visionaries who see foreign policy as a means of pursuing millennialist goals, not a very grand one. But the grander its foreign-policy vision, the more a state is trapped in the tyranny of its own construct: although recent administrations display an odd compulsion to devise and promulgate such visions of America's role in the world, those visions are in fact incompatible with the push and pull of strategy.

Finally, although some might characterize an offshore balancing strategy as isolationist, it emphatically is not. Rather, its guiding principle is a clear-eyed realism. It is a workaday policy—pragmatic, flexible, and opportunistic. But it will also bring America into a more respectful and natural relationship with the other great powers, as the United States forsakes the temptations of hegemony. "A mature great power will make measured and limited use of its power," Walter Lippmann wrote in 1965.

> It will eschew the theory of a global and universal duty, which not only commits it to unending wars of intervention, but intoxicates its thinking with the illusion that it is a crusader for righteousness… I am in favor of learning to behave like a great power, of getting rid of the globalism, which would not only entangle us everywhere, but is based on the totally vain notion that if we do not set the world in order, no matter what the price, we cannot live in the world safely… In the real world, we shall have to learn to live as a great power which defends itself and makes its way among other great powers.

From *The Atlantic Monthly*, January 2002, pp. 36-42. © 2002 by Benjamin Schwarz. Reprinted by permission.

MR. ORDER
Meets
MR. CHAOS

We live in an era of unprecedented prosperity, but when the financial bubble bursts we'll plunge into a world depression. Nations no longer go to war, but civil wars are booming. Humanity has embraced the idea of environmental interdependence, but the global ecosystem is in terminal crisis. Depending on your perspective, we stand either on the verge of a golden age or at the brink of disaster. Robert Wright and Robert Kaplan, two of the United States' most perceptive observers of world affairs and the human condition, met recently in Washington, D.C., to offer conflicting views of the path of history.

My Minivan and World Peace

By Robert Wright

Anyone who knows me would be surprised to find me cast as an optimist, but when you're juxtaposed with Robert Kaplan, it's not hard to come off looking pretty chipper and upbeat about the world.

What is the basis for my relative optimism? My prescription and diagnosis are built upon the notion of the non-zero-sum game, which is a reference to game theory. A zero-sum game is what you see in an athletic event like tennis: Every point in the match is good for one player and bad for the other. So the fates of the players are inversely correlated. In a non-zero-sum game, the fortunes can be positively correlated; the outcome can be win-win or lose-lose, depending on how competitors play the game. And, in fact, in a tennis doubles match the players on the same team have a highly non-zero-sum relationship because they'll both win or they'll both lose.

Nowadays we're all embedded in lots of non-zero-sum relationships that we really don't even think about. For instance, when I bought my Honda minivan I was in a non-zero-sum relationship with workers in various countries. The deal was I paid a tiny hit of their wages and they built me a car. It is characteristic of globalization that it embeds us in these non-zero-sum relationships. It makes our fates more correlated with the fates of people at great distances. It's a subtle process that we usually don't think about, but every once in a while this correlation of fortunes becomes glaringly evident, as was the case with the Asian crisis when we realized that a financial downturn can instantly spread around the world; or when a virus spreads across the Internet and you realize that computer users on different continents are all vulnerable, their fates are correlated.

In theory, as globalization makes relations among nations more and more non-zero-sum, you would expect to see more in the way of institutionalized cooperation to address these problems. That is not a pathbreaking insight. For some time now, political scientists have been talking about the growing interdependence of nations and the growing logic behind cooperation. But I believe that this process is now moving so fast that, much sooner than most people expect, we're going to reach a system of institutionalized cooperation among nations that is so thorough it qualifies as world governance. I don't mean world government, a single centralized authority. I imagine a looser mix of global and regional organizations. But still I'm imagining some very significant sacrifices of national sovereignty to supranational bodies. We've already seen a little of this surrender of national sovereignty with the World Trade Organization, and I would argue there was a little bit of surrender (a well-advised surrender) when 174 nations signed the Chemical Weapons Convention.

Wright: "History is not just one damn thing after another, it's a process with a direction; it has an arrow."

I fully expect this trend toward global governance to continue, although I'm much more confident about it happening in the long run than in the short run. The zone of non-zero-sumness has been expanding for a very long time: You can go back to the Stone Age when the most complex polity on earth was a hunter-gatherer village and chart the evolution to the level of the chiefdom—a multivillage polity—and then to the level of the ancient state, and then to the system of modern nation-states, and so on. The key element that has driven the evolution of social complexity and of governance to higher levels is technology. Sometimes it is information technology, as when the invention of writing often accompanied the evolution of the first ancient states. Sometimes it is transportation technology and sometimes, ironically enough, it is weapons technology. Weapons technologies can make relations much more non-zero-sum—certainly nuclear weapons make war a very non-zero-sum endeavor in the sense of making it a lose-lose game, wherein the object of the game is never to play. Nuclear weapons thus strengthen the argument for a system of collective security pursued through some supranational institution such as the United Nations.

We don't know in detail what the future of technological evolution will be, but we have a pretty good idea. Information technologies will continue to evolve and enmesh people in webs of transactions, interactions, and interdependence. Weapons technologies will evolve, but perhaps more important, the information about how to build very lethal weapons of mass destruction will likely be accessible to more and more people. Thus, almost all nations share a common interest in controlling the development and use of these weapons. Technological evolution will continue doing what it has done for the broad sweep of history, which is expanding the realm of non-zero-sumness, making the fates of peoples and nations more correlated, and in the process driving governance to a higher level, to the global level.

Kaplan: "You're right... but neither is it on a direct, predetermined course... The course of history [is] just a gradual improvement, punctuated with a lot of ups and downs."

That the fates of the world's people have grown more and more correlated over time is not by itself especially good news. As you may have noticed, many examples of non-zero-sum dynamics are actually negative-sum games, lose-lose games, where the object of the game is to break even. Global warming is an example of such a negative-sum game—where we just want to fend off the bad outcome—that I think calls for institutionalized cooperation and some real, if small, sacrifice of national sovereignty.

So when I argue that history features more and more of this non-zero-sumness, that statement isn't by itself good or bad, it just is. It's just something we have to reckon with. But there is one feature of the direction of human history that is at least mildly upbeat, in some ways redeeming. It's what I call the expanding moral compass. Philosopher Peter Singer has written about this. If you go back to ancient Greece, there was a time when members of one Greek city-state considered members of another Greek city-state literally subhuman. They would slaughter and pillage without any compunction whatsoever. Then the Greeks underwent a process of enlightenment and they decided that actually other Greeks are humans, too. It's just the Persians who aren't humans. (Okay, it was limited progress, but it was progress.) And today I think we've made more progress, especially in economically developed nations. I think almost everyone in such countries would say that people everywhere, regardless of race, creed, or color, deserve at least minimal respect.

If you ask why that has happened, I argue that it gets back to this basic dynamic of history, this growth of non-zero-sumness. If you look at Greece at the time of their limited enlightenment, relations were growing more non-zero-sum among Greek city-states because they were fighting a war together against the Persians. They needed each other more, they were in the same boat, and to cooperate they had to accord each other at least minimal respect. And if you ask why an ethos of moral universalism now prevails in economically advanced, globally integrated nations, I would say it's the same answer. If you ask me why don't I think it's a good idea to bomb the Japanese, I'd say, "For one thing, because they built my mini-van." I'm proud to say I have some more high-minded reasons as well, but I do think this basic, concrete interdependence forces people to accord one another at least minimal respect, to think a little about the welfare of people halfway around the world. I expect this dynamic to grow and persist in the future because in a world where disease can spread across borders in no time at all, it's in the interest of Americans to worry about the health of people in Africa or Asia. In a world where terrorists can wield unprecedentedly lethal technologies, it's in the interests of Americans to worry about political grievances before they fester to the point of terrorism. One feature of a globalized society is that disaster can happen at the global level, so we're now in this process where either we grasp the moral and political implications of this increasingly shared fate we have with other people or very bad things will happen.

The modern world is in many ways a disoriented and disturbing place. Things are changing very fast, but I think if you look at the broad sweep of the past it offers a way to orient ourselves. History is not just one damn thing after another, it's a process with a direction; it has an arrow. And I think if we use that arrow to orient ourselves then I would predict that the coming decades will not be characterized by chaos.

Robert Wright is author of Nonzero: The Logic of Human Destiny *(New York: Pantheon Books, 2000) and a visiting scholar at the University of Pennsylvania.*

Hope for the Best, Expect the Worst

Robert Kaplan

Well, Bob, while you've been looking ahead to discern the broad, cosmic sweep of history, I've been looking ahead just 10 or 15 years in terms of foreign policy—which is often most effective when it's conceived of in light of worst-case scenarios, in the hope that those scenarios don't occur. I should remind you that constructive pessimism is profoundly in the American tradition. It's the basis for the U.S. Constitution. If you read *The Federalist Papers*, you can see that Americans have become a country of optimists over 225 years precisely because we've had the good fortune of having our systems of government founded by pessimists. The French Revolution conversely was founded on optimism, on the belief that elites could engineer positive results from above, and it devolved into the guillotine and Napoleon's dictatorship. Alexander Hamilton, whom I consider the greatest of the Founding Fathers, said don't think there will be fewer wars in the world simply because there will be more democracies. In *Federalist Number Six* he said there are as many wars from commercial motives as from territorial aggrandizement. So it is in that spirit of *The Federalist Papers* that I'm going to present a scenario about what worries me over the next 10 or 15 years.

I wrote in 1994 that even as part of the globe was moving toward economic prosperity, another part—containing much of the population—was marching in another direction due to issues such as demography, resource scarcity and disease. So let me tell you how I see things now, seven years later. The European colonialists did a lot of terrible things, but they did bring a certain degree of order to much of sub-Saharan Africa, South Asia, and Central Asia. That colonial grid work of states started dissolving in the 1990s when we saw the weakening or outright collapse of several marginal places. I use the term "marginal" not because their well-being wasn't important, but because they had low populations, their economies were small, and they didn't really affect the region around them all that much. Somalia, Sierra Leone, Tajikistan, Haiti, and Rwanda were not core regional states in any sense, but look at how they disrupted the international community.

Wright: "We're going to reach a system of institutionalized cooperation among nations that is so thorough it qualifies as world governance."

I believe that, for a number of reasons, we're going to see the weakening, dilution, and perhaps even crackup of larger, more complex, modern societies in the next 10 or 15 years in places such as Nigeria, Ivory Coast, and Pakistan. And we're going to see severe crises in countries like Brazil and India. This dissolution of the colonial grid work is going to create the kind of crises where there will be no intervention scenarios, or the intervention scenarios will be far worse than they were in Bosnia or Sierra Leone. The problem is not that these places have particularly bad governments. They're coping as best as any could. The reasons are far more complex and intractable.

First of all, these societies are modernizing. Although history teaches us that modern democratic institutions provide stability, history also reveals that the process of creating and developing modern democratic institutions is very destabilizing. As free-market democracies develop, more and more people are brought into the political process. And all of these people are full of yearning, ambitions, and demands that governing institutions very often cannot keep pace with. So things start to break down here and there. It is economic growth that typically fuels political upheavals, not poverty.

The other challenge to the stability of the nation-state is demography. You hear a lot about how the world population is aging, but that's over the long term and throughout the world as a whole [see "The Population Implosion," FOREIGN POLICY, March/April 2001]. But when you look ahead at just 20 or 30 countries over the next 10 or 20 years, you see dramatic rises in the youth population (what demographers call "youth bulges"). When you watch your television and you see unrest or rioting in Indonesia, Ivory Coast, Gaza, and the West Bank, what's similar about all of them? All of the violence is typically conducted by young men, ages 15 to 29, who are unemployed and frustrated. The sector of the young male population within this age group is going to grow dramatically in the countries that already have tremendous unrest and are already on the edge. In other words, the places that will have a population pyramid that is bottom-heavy with the youngest members of society are the ones that can least afford it.

And if that isn't enough, you've got urbanization. The 21st century is going to be the first century in world history when more than half of humanity will live in cities. Even sub-Saharan Africa is almost 50 percent urban. Urban societies are much more challenging to govern than rural societies. In rural societies people can grow their own food, so they are less susceptible to price increases for basic commodities. Rural societies don't require the complex infrastructure of sewage, potable water, electricity, and other things that urban societies have. Urbanization widens the scope of error for leaders in the developing world while simultaneously narrowing the scope for success. It is harder to satisfy an urban population than a rural population, especially when that population is growing in such leaps and bounds that governing institutions simply cannot keep pace.

Then you have resource scarcity, particularly water. I spent the summer in a small village in Portugal where we only had running water about eight hours a week. We had to drive about half a mile to a local fountain to fill pitchers of water. Anyone who has not gone without water has no idea what it's like not to be able to flush your toilet or take a bath. There's been a drought for the last four years across a swath of South Asia from Afghanistan, Pakistan, and into India. Dams are low, so there is not enough water for drinking or generating electricity. So in these hot cities of the subcontinent you have less and less air conditioning in the summer. This kind of stuff doesn't necessarily cause political crises, but it's all part of the background noise that aggravates existing crises. This frustration worsens ethnic tension and makes social divides harder to resolve. In short, people get angry. There was a spate of riots in Karachi, Pakistan, not long ago that was preceded by an extended period when there was very little electricity due to water shortages.

Kaplan: "We're going to see the weakening, dilution, and perhaps even crackup of larger, more complex, modern societies in the next 10 or 15 years..."

Then there's the issue of climate change. Let's just say for the sake of argument that this whole global warming issue has been exaggerated, that it really doesn't exist, that it's not going to be a problem. Well, even if you factor out global warming, the normal climatic variations of the earth during the next few decades will still ensure devastating floods and other upheavals because, for the first time in world history, you have hundreds of millions of human beings living in environmentally fragile terrain— where perhaps human beings were never meant to live at all. So even without global warming you're going to have natural events that can spark political upheaval.

And finally, the other factor that's going to spark serious institutional crises in a lot of states is democracy. Everyone wants to be democratic, no use denying it. But democracy tends to emerge best when it emerges last. It should be the capstone to all other types of development, when you already have middle classes that pay income taxes, when you already have institutions run by literate bureaucrats, when the major issues of a society (such as territorial borders) are all resolved and you already have a functioning polity. Then, and only then, can a society cope with weak minority governments. Then, and only then, can democracy unleash a nation's full potential. Right now, we're seeing democracy evolve in many places around the earth accompanied by unemployment and inflation rates every bit as dire as Germany in the 1930s, when Hitler emerged under democratic conditions, and in Italy, when Mussolini came to power in the early 1920s. I'm not arguing against democracy, but I believe democracy will be another destabilizing factor.

If it seems like I'm deliberately cultivating a sense of the tragic it's because that's how you avoid tragedy in the first place. Remember that Klemens von Metternich was so brilliant in creating a post-Napoleonic order that Europe saw decades of peace and prosperity—so much so that politicians in France and England lost their sense of the tragic. All they saw ahead were optimistic scenarios and, as such, they stumbled and miscalculated their way into World War I. Take my concern in that spirit.

Robert Kaplan is author of The Coming Anarchy: Shattering the Dreams of the Post Cold War *(New York: Vintage Books, 2000) and a senior fellow at the New America Foundation.*

In the Long Run, We're All Interdependent

Robert Wright responds.

Well Bob, I'm actually something of a fan of pessimism myself. I think it focuses us on the problems that need our attention. I find it particularly heartening that your books have a sizable American readership, since that suggests that Americans increasingly realize their fates are intertwined with the fates of people around the world. But I don't want to overdo the pessimism. And in particular I don't want to make it sound like globalization and its attendant technological fluctuations are part of some kind of uniformly bad force. I'm actually something of a cheerleader for globalization. It has problems, but I think on balance it's a good thing.

You said that the world was increasingly dividing into two parts, echoing the common refrain that globalization exacerbates income inequality worldwide. But that conclusion actually depends on how you examine the data. If you look at the number of rich versus poor nations, then you can certainly make that argument. But if you look at the total number of people in the world, ignoring where the borders fall, then what's happening in absolute terms is that there are fewer poor people than there used to be. And even in relative terms, it's far from clear that income inequality is growing, and a number of people have argued that the income gap is actually shrinking worldwide. It turns out that many of the world's poor people are concentrated in a few very large countries (like China and India) that have seen more progress than some of the smaller countries (notably those in Africa). But even in Africa, globalization has seen a kind of vindication: The countries that have seen the most economic advancement are the ones that are most open to trade and investment.

Another virtue of globalization is that it is basically an antiwar activity. I think as peoples and nations become more economically intertwined, war becomes more of a lose-lose kind of non-zero-sum game that it doesn't make sense to play. There still are wars in the world, but there is a very interesting feature of the modern world that is insufficiently noted: We increasingly think of wars between nations as something that poor countries do. Nobody expects any of the most economically advanced nations to go to war with one another, which represents a real shift of mind-set. If you look back at most of history it was really standard procedure for the most powerful polities to go to war with one another. Nowadays, most interstate fighting breaks out in parts of the world that could be termed "underglobalized" areas. I don't mean that pejoratively. It's not their fault that they're underglobalized. There are various quirks of history or geographical circumstance that explain why some parts of the world have advanced faster economically than others. But the fact is that wars are mostly a threat in the poorest parts of the world.

Now, when you get to subnational conflict, war within nations, I agree, Bob, that's a problem that may grow more serious. You argue that conflict is often exacerbated by economic development. I'd add another way in which modernization has given rise to intranational conflict, and that is through the propagation of information technology. As I suggested earlier, information technology has certain globalizing effects, but it also has fragmenting effects because whenever you lower the cost of communication you make it easier for small groups with meager resources to organize. It's no coincidence that the Protestant Reformation roughly coincided with the invention of the printing press. After Martin Luther had tacked up his 95 Theses, printers took it upon themselves to start printing them in various cities. That is how Luther first organized the masses, because printing was suddenly so cheap.

You're seeing the same thing in the modern world thanks to the Internet. Inevitably, information technology is going to empower separatist groups such as Muslims in the west of China and Basques in Spain. But, in the long run, you can imagine this secessionist frenzy working itself out, because as some of these subnational groups choose to drop out of nations they can at the same time cement themselves into supranational bodies. In fact, the Quebec separatists have said they plan to join the North American Free Trade Agreement (NAFTA) as soon as they get out of Canada, and I would expect that European separatist groups would be strongly tempted to join the European Union. So, I certainly agree that globalization presents us with all kinds of short-term difficulties, but I do still think it's a process that is fundamentally beneficial and will lead to a new equilibrium in the long run.

Passion Play

Robert Kaplan responds.

Bob, let me draw some distinctions here, just in the spirit of argument. You tend to put a lot of emphasis on the ability of people to make good, rational choices. But if you think that people are always going to behave accord-

The Clash of Interpretations

Samuel Huntington, cofounder of FOREIGN POLICY *and one of the world's most influential political scientists, incited widespread debate during the 1990S with his "Clash of Civilizations" thesis, which maintained that cultural fault lines would dominate the post-Cold War world. How do Huntington's views fare when examined through the prisms of order and chaos?*

Robert Wright:
I thought Dr. Huntington somewhat overdid the fissures between civilizations. The world is full of examples where nations with very different cultural heritages are on very good and stable terms, the United States and Japan to name just one example. At the same time, I realize that cultural history matters and it can be a source of tension. But to the extent that he was right about that, what bothered me was his prescription that we should be true to our cultural spirit. He said, for instance, that Australia is not really an Asian nation, so it should not be in a trade bloc with Asian nations, it should be part of an expanded NAFTA. It seems to me that the moral of the story is exactly the opposite: If indeed fissures among civilizations are deep and threatening (and sometimes they are), then the effort should be to bridge them with supranational organizations like trade blocs.

Robert Kaplan:
I thought Sam Huntington's thesis was brilliant, which is proven by the fact that it got everybody angry. Just look at the three countries that were brought into NATO: Hungary, Poland, and the Czech Republic were all part of Western Christendom, and countries that were left out (Bulgaria and Romania) were part of the Eastern Orthodox world. When you look at the current borders of NATO, it's basically a variation of the Holy Roman Empire in the 11th century. If you travel through the Middle East you see increasing tensions in villages between Christians and Muslims. All the corporate mergers are between American and European companies. It's true they have different corporate cultures, but they're similar compared to corporate cultures in other parts of the world. And our main foreign-policy challenge of the next 20 years will be managing a relationship with China. Probably never before have we had a major adversary where there is such a great chance for cultural misunderstanding.

But it's important to recall that Huntington's scenario was a paradigm. It was just a big abstract argument that you judge on the basis of whether it is better than any other generalizing abstract argument. I think on that basis he succeeded very well, but I would agree with Bob that the solution is maybe to bring Romania into NATO, to bridge some of these gaps so that NATO doesn't evolve into a bastion of Western Christendom.

ing to their best, rational self-interest, read *Mein Kampf.* As Hamilton said, "the passions of men will not conform to the dictates of reason and justice, without constraint." The U.S. Constitution was established to slyly organize and control our passions. I'm not convinced that we're going to act any more rationally than we have in the past. It is true that there is a movement toward world governance, but a single, unifying thread is not necessarily a good thing. For instance, the European Union could readily devolve into a benign bureaucratic despotism that will ignore the interests of the lower middle classes. I think the nationalist movements popping up throughout Europe are already a reaction to this benign bureaucratic despotism from Brussels. If there is to be world governance, it has to be a kind that doesn't only appeal to the elites.

And those who feel marginalized have resources at their disposal that go way beyond the Internet. The Industrial Revolution was about bigness—big aircraft carriers, tanks, and railway grids—so that only large states could take advantage of the power the Industrial Revolution had to offer. But when you're talking about cyberwarfare, biological weapons, and this whole new gamut of weaponry in the post-Industrial Revolution, when you live in a world where just a telephone jack and a petri dish give you power, then it's not just large nations that can benefit. Nonstate actors who feel shut out can also magnify their power through this new technology. Power relationships are going to be more complex than ever. You were right when you said that technology drives history, but it doesn't necessarily do so in an orderly manner.

As for the moral universalism that you mention, I think we have to be a bit careful because the West is now using the term "global community" in the way we used to use the term "free world." We're trying to define the whole world in terms of our own moral outlook and what we want. There may be other powers and other cultures that have different views of how the world should be organized, so we have to be careful not to sound triumphalist.

And although inequality might be decreasing, I believe the most significant form of inequality is not what we see between the United States and sub-Saharan Africa, but the income gap you see between the wealthy coastal community and poor interior of a place such as Ghana. The biggest divides are between these globalized communities within the poorest countries—with their own electricity generators, their own water wells, and their own private security guards—that are hooked up to the world economy and surrounded by people with whom they have less and less in common.

You're right Bob, history is not one damn thing after another, but neither is it on a direct, predetermined course à la Karl Marx. The philosopher who captures it all best is Charles de Montesquieu who, in *The Spirit of the Laws*, sees the course of history as just a gradual improvement, punctuated with a lot of ups and downs.

But, lest I sound too contrarian, allow me to point out that I've been concentrating on the zero-sum games that occur within your vast non-zero-sum game. So, in that sense, there is no contradiction between us.

Coffee, Tea, or Apocalypse

Robert Wright responds.

I would hate to let stand the accusation that I think people behave rationally. I've long argued that people are really quite spotty on this particular front, and the saving grace of history has been that whenever people screwed up in one part of the world, there were people in another part of the world who picked up the torch. So when the Roman emperors messed up and began exhibiting the sort of increasingly autocratic rule that people are prone to given the opportunity, and the barbarian hordes did us the service of dismantling the Roman Empire, there were other empires that could continue to grow and thrive.

But I agree that once we reach the global level of organization, we face exactly the threat you describe in that there is no longer a plurality of experiments going on around the world. Increasingly it's one big experiment, and if it collapses that is bad news on a very large scale. That is one reason why I always say that history almost reads better than any novel. The protagonist of this story, that is to say the human species, has been driven more or less inexorably to a moment of fundamental moral and political choice. Making the right decision really depends on our level of moral enlightenment, an understanding of the commonality of all human beings. And if we fail to make the right moral choice, then apocalypse could well ensue. We live at a time of great drama and I absolutely acknowledge that chaos is one of the prospects we face.

A Whiff of Medievalism

Robert Kaplan responds.

Bob, I think chaos is more than a prospect. As divisions within societies and nation-states become greater, chaos might very well be inevitable. A new global community is taking root, but it is doing so at the top. Right now that community is so small that it's still a Potemkin village, but it's nor going to be that way forever. The middle and upper middle classes, what I call the "nouvelle cuisine classes," are merging together at the pinnacle. If you've ever been to the annual Davos conference, you are struck by how medieval it has become: You see the world's elites gathered together, just like the aristocrats of Germany, France, and England 200 or 300 years ago, who had more in common with each other than with their own peasants at home. Our sense of identity is being driven more by what economic class we belong to than what country we live in. The world is moving very slowly and inexorably out of the nation-state phase. That may ultimately lead to something better, but the process of leading to something better is very chaotic.

24

Want to Know More?

This debate is based on a dialogue between Robert Wright and Robert Kaplan that took place at the **Meridian International Center** in Washington, D.C., on March 13, 2001. Ambassador Walter Cutler moderated the event.

Robert Wright Recommends:

The great historian William McNeill's books *The Rise of the West: A History of the Human Community* (Chicago: University of Chicago Press, 1999) and *A World History* (New York: Oxford University Press,, 1999) convey the directional nature of history without sacrificing nuance. But seeing direction in history is so unpopular among historians that the task is often left to nonhistorians, such as political scientist Francis Fukuyama. His justly famous *The End of History and the Last Man* (New York: Free Press, 1992) differs from my *Nonzero: The Logic of Human Destiny* (New York: Pantheon Books, 2000) in that mine is a more materialist (in a somewhat Marxist sense) account of history's driving force. Peter Singer's *The Expanding Circle: Ethics and Sociobiology* (Oxford: Oxford University Press, 1981) acutely noted history's moral direction—the erratic movement over the last few millenniums toward moral universalism (though, again, my explanation is more materialist than his). Thomas Friedman's bestseller *The Lexus and the Olive Tree: Understanding Globalization* (New York: Farrar, Straus & Giroux, 1999) gives due weight to globalization's positive side without being Panglossian. His term the "superempowered angry man" captures the growing ability of small groups to wreak massive damage. For an abstract but fascinating account of how non-zero-sumness can beget more non-zero-sumness as the growth of cooperation strengthens the logic of further cooperation, see Robert Axelrod's classic *The Evolution of Cooperation* (New York: Penguin Books, 1990).

Robert Kaplan Recommends:

Paul Kennedy's *Preparing for the 21st Century*. (New York: Random House, 1993) is the best primer available on how demography, the environment, technology, and other related factors will test humankind's political skills in the future. Martin van Creveld's *The Transformation of War* (New York: Free Press, 1991) is a brilliant work of sustained, abstract thinking (perhaps the greatest since Carl von Clausewitz's *On War*) that shows how technology will be used toward primitive ends in future wars. In *The Clash of Civilizations and the Remaking of World Order* (New York: Simon & Schuster, 1996) Harvard professor Samuel Huntington—who, since the 1950s, has been a controversial but uncannily accurate forecaster—offers a prognosis of a future where tangible cultural divides will replace artificial, ideological ones. My own *The Coming Anarchy: Shattering the Dreams of the Post Cold War* (New York: Vintage Books, 2000) identifies the demographic, environmental, and political problems ahead while finding hope in ancient and modern philosophy.

Among the greatest works of philosophy that demonstrate how and why order must precede democracy—and how the search for order stems from the human need to be protected from others—is Thomas Hobbes' *Leviathan* (New York: W. W. Norton & Company, 1997). Arnold J. Toynbee's *A Study of History* (Oxford: Oxford University Press, 1987) presents an epic pageant explaining how culturally based states have arisen out of the ashes of preceding ones.

• For links to relevant Web sites, as well as a comprehensive index of related FOREIGN POLICY articles, access **www.foreignpolicy.com**.

UNIT 2

Population and Food Production

Unit Selections

5. **The Big Crunch**, Jeffrey Kluger
6. **SARS Has Lots of Company as Emerging Disease**, Seth Borenstein
7. **Bittersweet Harvest: The Debate Over Genetically Modified Crops**, Honor Hsin

Key Points to Consider

- What are the basic characteristics and trends of the world's population? How many people are there? How long do people typically live?

- How fast is the world's population growing? What are the reasons for this growth? How do population dynamics vary from one region to the next?

- How does rapid population growth affect the quality of the environment, social structures, and the ways in which humanity views itself?

- What are some of the threats to the human population which might cause it to decline or collapse?

- There is a growing debate about genetically modified food. What are the differing perspectives on this debate and what issues are likely to be contested in the near future?

- How can economic and social policies be changed in order to reduce the impact of population growth on environmental quality?

- In an era of global interdependence, how much impact can individual governments have on demographic changes?

 Links: www.dushkin.com/online/
These sites are annotated in the World Wide Web pages.

The Hunger Project
http://www.thp.org
Penn Library: Resources by Subject
http://www.library.upenn.edu/cgi-bin/res/sr.cgi
World Health Organization
http://www.who.int
WWW Virtual Library: Demography & Population Studies
http://demography.anu.edu.au/VirtualLibrary/

After World War II, the world's population reached an estimated 2 billion people. It had taken 250 years to triple to that level. In the 60 years since the end of World War II, the population has tripled again to 6 billion. When the typical reader of this book reaches the age of 50, experts estimate that the global population will have reached 8 1/2 billion! By 2050, or about 100 years after World War II, some experts forecast that 10 to 12 billion people may populate the world. A person born in 1946 (a so-called baby boomer) who lives to be 100 could see a six-fold increase in population.

Nothing like this has ever occurred before. To state this in a different way: In the next 50 years there will have to be twice as much food grown, twice as many schools and hospitals available, and twice as much of everything else just to maintain the current and rather uneven standard of living. We live in an unprecedented time in human history.

One of the most interesting aspects of this population growth is that there is little agreement about whether this situation is good or bad. The government of China, for example, has a policy that encourages couples to have only one child. In contrast, there are a few governments that use various financial incentives to promote large families.

Some experts view population growth as the major problem facing the world, while others see it as secondary to social, economic, and political problems. The theme of conflicting views, in short, has been carried forward from the introductory unit of this book to the more specific discussion of population.

In the first decade of the new millennium, there are many population issues that transcend numerical or economic considerations. The disappearance of indigenous cultures is a good example of the pressures of population growth on people who live on the margins of modern society. Finally, while demographers develop various scenarios forecasting population growth, it is important to remember that there are circumstances that could lead not to growth but to a significant decline in global population. The spread of AIDS and other infectious diseases reveals that confidence in modern medicine's ability to control these scourges may be premature. Nature has its own checks and balances to the population dynamic that are not policy instruments of some international organization. This factor is often overlooked in an age of technological optimism.

The lead article in this section provides an overview of the general demographic trends in the contemporary world. The unit continues with an overview of the evolution of new strains of infectious diseases and the threats they pose to populations.

There is, of course, no greater check on population growth than the ability to produce an adequate food supply. Some experts question whether current technologies are sustainable over the long run. How much food are we going to need in the decades to come, and how are farmers and fishermen going to produce it? The debate about genetically modified crops is the lightening rod for this critical topic.

Making predictions about the future of the world's population is a complicated task, for there are a variety of forces at work and considerable variation from region to region. The danger of oversimplification must be overcome if governments and international organizations are going to respond with meaningful policies. Perhaps one could say that there is not a global population problem but rather many population challenges that vary from country to country and region to region.

THE BIG CRUNCH

Birthrates are falling, but it may be a half-century before the number of people—and their impact—reaches a peak

By Jeffrey Kluger

ODDS ARE YOU'LL NEVER MEET ANY OF THE ESTIMATED 247 HUMAN BEINGS WHO WERE BORN IN THE PAST MINUTE. IN A POPULATION OF 6 BILLION, 247 IS A DEMO-GRAPHIC HICCUP. IN THE MINUTE BEFORE LAST, HOWEVER, THERE WERE ANOTHER 247. IN THE MINUTES TO COME THERE WILL be another, then another, then another. By next year at this time, all those minutes will have produced nearly 130 million newcomers to the great human mosh pit. That kind of crowd is awfully hard to miss.

For folks inclined to fret that the earth is heading for the environmental abyss, the population problem has always been one of the biggest causes for worry—and with good reason. The last time humanity cele-brated a new century there were 1.6 billion people here for the party—or a quarter as many as this time. In 1900 the average life expectancy was, in some places, as low as 23 years; now it's 65, meaning the extra billions are staying around longer and de-manding more from the planet. The 130 million or so births registered annually—

even after subtracting the 52 million deaths—is still the equivalent of adding nearly one new Germany to the world's population each year.

But things may not be as bleak as they seem. Lately demographers have come to the conclusion that the population locomo-tive—while still cannonballing ahead—may be chugging toward a stop. In country after country, birthrates are easing, and the population growth rate is falling.

To be sure, this kind of success is un-even. For every region in the world that has brought its population under control, there's another where things are still ex-ploding. For every country that has figured out the art of sustainable agriculture, there are others that have worked their land to exhaustion. The population bomb may yet go off before governments can snuff the fuse, but for now, the news is better than it's been in a long time. "We could have an end in sight to population growth in the next century," says Carl Haub, a demogra-pher with the nonprofit Population Re-search Bureau. "That's a major change."

Cheering as the population reports are becoming today, for much of the past 50 years, demographers were bearers of mostly bad tidings. In census after census, they reported that humanity was not just settling the planet but smothering it. It was not until the century was nearly two-thirds over that scientists and governments fi-nally bestirred themselves to do something about it. The first great brake on popula-tion growth came in the early 1960s, with the development of the birth-control pill, a magic pharmacological bullet that made contraception easier—not to mention ti-dier—than it had ever been before. In 1969 the United Nations got in on the population game, creating the U.N. Population Fund, a global organization dedicated to bringing family-planning techniques to women who would not otherwise have them. In the de-cades that followed, the U.N. increased its commitment, sponsoring numerous global symposiums to address the population problem further. The most significant was the 1994 Cairo conference, where attend-ees pledged $5.7 billion to reduce birthrates in the developing world and acknowledged that giving women more education and reproductive freedom was the key to accomplishing that goal. Even a global calamity like AIDS has yielded unexpected dividends, with international campaigns to promote condom use and abstinence helping to prevent not only disease transmission but also conception.

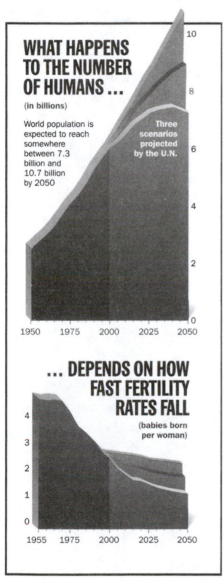

WHAT HAPPENS TO THE NUMBER OF HUMANS ...

(in billions)

World population is expected to reach somewhere between 7.3 billion and 10.7 billion by 2050

Three scenarios projected by the U.N.

1950 1975 2000 2025 2050

... DEPENDS ON HOW FAST FERTILITY RATES FALL

(babies born per woman)

1955 1975 2000 2025 2050

Source: United Nations

Such efforts have paid off in a big way. According to U.N. head counters, the average number of children produced per couple in the developing world—a figure that reached a whopping 4.9 earlier this century—has plunged to just 2.7. In many countries, including Spain, Slovenia, Greece and Germany, the fertility rate is well below 1.5, meaning parents are pro-

ducing 25% fewer offspring than would be needed to replace themselves—in effect, throwing the census into reverse. A little more than 30 years ago, global population growth was 2.04% a year, the highest in human history. Today it's just 1.3%. "It was a remarkable century," says Joseph Chamie of the U.N. Population Division. "We quadrupled the population in 100 years, but that's not going to happen again."

Sunny as the global averages look, however, things get a lot darker when you break them down by region. Even the best family-planning programs do no good if there is neither the money nor governmental expertise to carry them out, and in less-developed countries—which currently account for a staggering 96% of the annual population increase—both are sorely lacking. In parts of the Middle East and Africa, the fertility rate exceeds seven babies per woman. In India, nearly 16 million births are registered each year, for a growth rate of 1.8%. While Europe's population was three times that of Africa in 1950, today the two continents have about the same count. At the current rate, Africa will triple Europe in another 50 years.

Many of the countries in the deepest demographic trouble have imposed aggressive family-planning programs, only to see them go badly—even criminally—awry. In the 1970s, Indian Prime Minister Indira Gandhi tried to reduce the national birthrate by offering men cash and transistor radios if they would undergo vasectomies. In the communities in which those sweeteners failed, the government resorted to coercion, putting millions of males—from teenage boys to elderly men—on the operating table. Amid the popular backlash that followed, Gandhi's government was turned out of office, and the public rejected family planning.

China's similarly notorious one-child policy has done a better job of slowing population growth but not without problems. In a country that values boys over girls, one-child rules have led to abandonments, abortions and infanticides, as couples limited to a single offspring keep spinning the reproductive wheel until it comes up male. "We've learned that there is no such thing as 'population control,'" says Alex Marshall of the U.N. Population Fund. "You don't control it. You allow people to make up their own mind."

That strategy has worked in many countries that once had runaway population growth. Mexico, one of Latin America's population success stories, has

made government-subsidized contraception widely available and at the same time launched public-information campaigns to teach people the value of using it. A recent series of ads aimed at men makes the powerful point that there is more machismo in clothing and feeding offspring than in conceiving and leaving them. In the past 30 years, the average number of children born to a Mexican woman has plunged from seven to just 2.5. Many developing nations are starting to recognize the importance of educating women and letting them—not just their husbands—have a say in how many children they will have.

But bringing down birthrates loses some of its effectiveness as mortality rates also fall. At the same time Mexico reduced its children-per-mother figure, for example, it also boosted its average life expectancy from 50 years to 72—a wonderful accomplishment, but one that offsets part of the gain achieved by reducing the number of births.

When people live longer, populations grow not just bigger but also older and frailer. In the U.S. there has been no end of hand wringing over what will happen when baby boomers—who owe their very existence to the procreative free-for-all that followed World War II—retire, leaving themselves to be supported by the much smaller generation they produced. In Germany there are currently four workers for every retired person. Before long that ratio will be down to just 2 to 1.

STATE OF THE PLANET

Humans already use 54% of Earth's rainfall, says the U.N. report, and 70% of that goes to agriculture

---

For now, the only answer may be to tough things out for a while, waiting for the billions of people born during the great population booms to live out their long life, while at the same time continuing to reduce birthrates further so that things don't get thrown so far out of kilter again. But there's no telling if the earth—already worked to exhaustion feeding the 6 billion people currently here—can take much more. People in the richest countries consume a disproportionate share of the world's resources, and as poorer nations push to catch up, pressure on the planet will keep growing. "An ecologist looks at the population size relative to the carrying capacity of Earth," says Lester Brown, president of the Worldwatch Institute. "Looking at it that way, things are much worse than we expected them to be 20 years ago."

How much better they'll get will be decided in the next half-century (*see chart*). According to three scenarios published by the U.N., the global population in the year 2050 will be somewhere between 7.3 billion and 10.7 billion, depending on how fast the fertility rate falls. The difference between the high scenario and the low scenario? Just one child per couple. With the species poised on that kind of demographic knife edge, it pays for those couples to make their choices carefully.

—Reported by William Dowell/New York, Meenakshi Ganguly/New Delhi and Dick Thompson/Washington

SARS has lots of Company as Emerging Disease

SETH BORENSTEIN

WASHINGTON—Get used to SARS, West Nile, hantavirus, Ebola, Nipah, Hendra, AIDS and other new or recent nasty infectious diseases. Health experts say we're living in a new age of infections.

And we have mostly ourselves to blame.

The nation's top scientists say environmental, economic, social and scientific changes have helped trigger an unprecedented explosion of more than 35 new infectious diseases that have burst upon the world in the past 30 years. The U.S. death rate from infectious disease, which dropped in the first part of the 20th century and then stablized, is now double what it was in 1980.

SARS is only the latest of these new germs.

"The kinds of things we are doing for SARS we can anticipate we are going to do again and again," Dr. Julie Gerberding, the Centers for Disease Control and Prevention director, told the U.S. Senate's Health, Education, Labor and Pensions Committee last week.

Even as most of the world struggles to cope with the threat of severe acute respiratory syndrome, or SARS, infectious-disease specialists are warily examining an outbreak of a bird influenza in the Netherlands. Only one person has died so far, but forms of flu that are new to humans, as this one is, can be very deadly.

"Things are closing in on all sides," said Dr. W. Paul Glezen, an epidemiologist at the Influenza Research Center at Baylor College of Medicine in Houston.

"Today's outlook with regard to microbial threats to health is bleak" concludes a 396-page March report from the prestigious Institute of Medicine, a re-

search arm of the federal government. "Microbial threats present us with new surprises every year."

Every hour, 1,500 people worldwide fie of an old or new infectious disease, and more than half of those are children under 5, according to that report.

Many of the infectious diseases or their pathogens that now scem common-food-borne E. coli, water-borne Cryptosporidium, air-borne Legionnaire's disease, blood-torne hepatitis C and sexually transmitted AIDS—first surfaced in roughly the past 30 years.

"This period from the 1970s is without precedent in the history of the annals of medicine," said Dr. Paul Epstein, associate director or Harvard Medical School's Center for Health and Global Environment.

To figure out why this is happening, the Institute of Medicine convened a panel of top U.S. researchers. They attributed the surge in new diseases to 13 specific changes in the world and the way we live.

Those 13 factors are microbial adaptation and change; human susceptibility to infection; climate and weather; changing ecosystems; human demographics and behavior; economic development and land use; international travel and commerce; technology and industry; breakdown of public health measures; poverty and social inequality; war and famine; lack of political will; and bioterrorism.

Of the more than 35 emerging diseases since the 1970s, "a substantial proportion relate to man's manipulation of ecology," Dr. Anthonly Fauci, director of the National Institute of Allergy and

Infecticious Diseases, said in an interview.

"It's a confluence of many factors," said Dr. Fred Sparling, a medical and microbiology professor at the University of North Carolina at Chapel Hill and a co-author of the Institute of Medicine report. "We have more people in megacities, which increases the risk for dissemination, more people moving into habitat that is used by animals as vectors for (disease) transmission, more international travel and commerce, which disseminates (germs) once they have arisen. And more poverty and social inequalities, which clearly add to it."

Most of these diseases live in animals; the epidemic begins when they jump from animal humans, typically because of close contact.

That often follows the clearing of land for housing or industry—in effect putting people closer to the ticks, mosquitoes and rodents that carry new diseases.

Lyme disease soared in the United States after land clearing in the Northeast drew whitetailed deer closer to humans, said Stanford University professor of medicine Dr. Stan Deresinski, editor of Infectious Disease Alert. Lyme-carrying ticks, which lived on those deer, could more easily reach humans and infect them.

New dams cause changes in water flow, often increasing mosquito populations. In 1977 and 1978, for example, standing water behind Egypt's Aswan dam provided a spectacular ground for the mosquito that carried Rift Valley Fever. North Carolina's Sparling said. The

New and re-emerging diseases

These are some of the more than 35 infectious illnesses that have begun spreading or have flared up recently.

Map locates origin of disease or serious outbreak

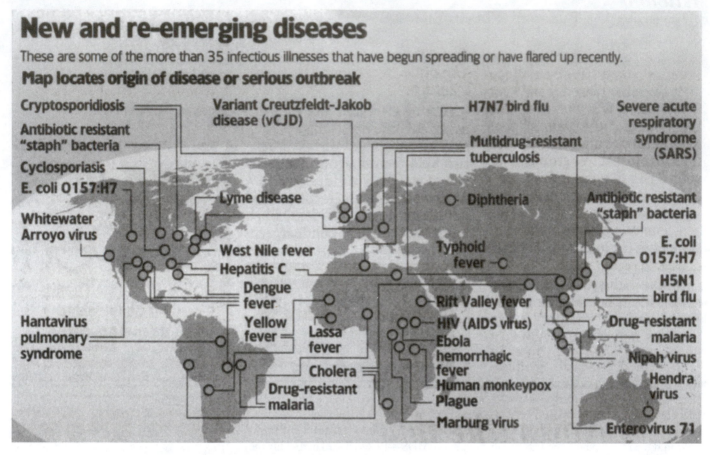

Source: Infectious disease Society of America

disease infected 200,000 people and killed 598.

Dense urban populations, people living longer, and the increased number of immune-compromised people also help disease spread faster, the IOM report said.

When countries hide new outbreaks, as China did with SARS, or lack public health systems capable of detecting outbreaks, diseases spread faster, said Dr. Howard Markel, a professor of infectious diseases at the University of Michigan. Markel, who also teaches the history of medicine, is author of the forthcoming book "When Germs Travel."

Weather changes can be a factor in emerging diseases, too, experts said. West Nile and hantavirus struck in the United States after severe droughts, said Epstein, of Harvard Medical School. That's because transfers of mosquito-borne diseases are easier when large numbers of animals congregate around small pools of water.

Another human factor: Many new diseases are strains of old ones that have become resistant to antibiotics, Sparling said.

Once a disease takes hold these days, it tends to be globalized quickly by travel and trade.

West Nile, for example, is thought to have reached New York from its traditional home in the Middle East on an infected bird carried by a ship or plane. With SARS, doctors have been able to pinpoint tourists, businessmen and doctors who have taken the virus from Hong Kong to Hanoi, Singapore and Toronto. It set a record for speed of continent-to-continent transmission.

"This is the first (widespread) emerging disease that's airborne, and that's what has everybody afraid," Epstein said. Although SARS isn't as contagious as measles or the common flu, airborne transmission makes it easier to spread than diseases requiring a mosquito bite such as West Nile.

"It's amazing, it's scary," said Dr. Jim Diaz, professor of public health and preventive medicine at the Louisiana Sate University School of Medicine. "It just shows you that international travel can move things so fast."

Fortunately, the world's health systems are moving faster, too.

The CDC's Gerberding noted that within about a month scientists identified the SARS virus, something never seen before, and developed a crude diagnostic test for the disease.

The main goal is to keep infections from getting out of control.

"We don't conquer germs," said Michigan's Markel. "We wrestle them to a draw."

Bittersweet Harvest

The Debate Over Genetically Modified Crops

Honor Hsin

In 1982 scientists on the 4th floor of the Monsanto Company U Building successfully introduced a foreign gene into a plant cell for the first time in history. These plants were genetically modified: they continued to express the new gene while exhibiting normal plant physiology and producing normal offspring. This breakthrough spawned the field of genetically modified (GM) crop production. Since the discovery, however, the international response to GM crops has been mixed. Along with the tremendous potential that lies vested in this technology, there are many risks and uncertainties involved as well. Arguments have centered on the health implications and environmental impact of cultivating GM crops and have raised disputes over national interests, global policy, and corporate agendas. Although there are many sides to this debate, discussions on GM crop regulation should be held within the context of scientific evidence, coupled with a careful weighing of present and future agricultural prospects.

Benefits and Costs

The possibility of environmental benefits first spurred the development of GM crops. The environmental issues at stake can be illustrated by one example of a potent genetic modification, the introduction of an endotoxin gene from *Bacillus thuringiensis* (Bt), a soil microorganism used for decades by organic growers as an insecticide, into soybeans, corn, and cotton. These GM crops promise to reduce the need to spray large amounts of chemicals into a field's ecosystem since the toxins are produced by the plants themselves. The Bt crops pose environmental risks, however, and could possibly harm other organisms. Bt corn was shown to harm monarch butterfly caterpillars in the laboratory, although later studies performed with more realistic farming conditions found this result conclusively only with Syngenta Company's Bt maize, which expressed up to 40 percent more toxin than other brands. Another pertinent environmental issue is the possible evolution of Bt resistance in pests. Since the Bt toxin expressed by the crops is ubiquitous in the field, there is positive selection for resistance against it, which would quickly make Bt's effect obsolete. Experimentation has begun, however, that involves regulating the percentage of Bt crops in a field so that a balance can be achieved between high yields and survival of Bt-sensitive pests. Although there are still multiple layers of ecosystem complexity that need to be considered, careful scientific research can begin to address these questions.

Another potential area of risk that needs to be analyzed is the effect of GM crops on human health. A possible consequence of Bt expression in crops is the development of allergic reactions in farmers since the toxin is more highly concentrated in the crops than in the field. Furthermore, the method used to insert foreign genes into GM crops always risks manipulation of unknown genes in the plant, resulting in unforeseen consequences. The effects of GM crops on humans therefore must be tested rigorously. Fortunately, no solid evidence yet exists for adverse physiological reactions to GM crops in humans, and some scientists argue that these same genetic-modification techniques are also currently being used in the development of pharmaceutical and industrial products.

A prevailing theme in the GM debate is that when discrepancies between scientific consensus and government policy result in unwanted consequences, the blame is often placed directly on GM crop technology itself. In 2000 about 300,000 acres of StarLink corn, a Bt crop produced by Aventis CropScience, were being cultivated in the United States. Since the US Environmental Protection Agency had declared its uncertainty over the allergenic potential of StarLink, the crops were grown with the understanding that they would be used solely as animal feed. Later that year news broke that StarLink corn had found its way into numerous taco food products around the world. This incident received wide press coverage and brought instant attention to the debate over GM crop safety. More at issue, though, were the United States' lax policies of GM crop approval and regulation. For nearly a decade, the US government made no distinction between GM crops and organically grown crops, and allergenicity safety tests were not mandatory. Only recently has the US Food and Drug Administration begun to reconsider its policies.

Canada is another leading producer of GM crops, with regulatory policies similar to those of the United States. Recent controversy surrounding Canada's cultivation of GM rapeseed, or canola, brought attention to another major environmental risk of GM crops. Unlike wheat and soybeans, which can self-pollinate to reproduce, the pollen of rapeseed plants spreads up to 800 meters beyond the field. There have been concerns in Ottawa over the government's refusal to reveal the location of ongoing GM wheat testing by Monsanto, resulting in fear of unwanted pollen spreading. This issue demonstrates one of the most potent risks of GM crops: uncontrolled breeding and the introduction of foreign genes into the natural ecosystem. An example of such an incident is Mexico's discovery of transgenic genes in non-GM strains of maize, although this result is still under scrutiny. More measures must be tested to restrain these possibilities. Current research on introducing the foreign genes into chloroplasts, which are only carried in the maternal line and not in pollen, offers a promising example.

Unfortunately, activist organizations rarely cite credible scientific evidence in their positions and have won much public sympathy by exploiting popular fears and misconceptions about genetic-engineering technology.

Europe's policy toward GM crops lies on the opposite end of the spectrum. In 1996 Europe approved the import of Monsanto's Roundup Ready soybeans and in 1997 authorized the cultivation of GM corn from Novartis. At around this time, however, there were rising concerns in Britain over BSE (bovine spongiform encephalopathy), or mad cow disease, which was thought to have killed more than two dozen people and cost the country the equivalent of billions of US dollars. The public was enraged over what it believed was a failure of government regulation, and in 1998 the European Commission voted to ban the import and cultivation of new GM crops. Besides the disappointment of private GM corporations like Monsanto, the United States claims to have lost US$600 million in corn exports to the

European Union. Recently, several European countries have considered lifting the ban contingent on the establishment of adequate labeling practices. The United States has complained to EU officials that labeling requirements discriminate against its agricultural exports, bringing the GM debate into the midst of a world trade dispute. In late January 2000, a tentative agreement was reached on the Montreal Biosafety Protocol in which the United States, Canada, Australia, Argentina, Uruguay, and Chile agreed to preliminary labeling of international exports and a precautionary principle allowing EU countries to reject imports if a scientific risk assessment of the imported crop is provided. This agreement, however, does not override decisions made by the World Trade Organization.

Corporate Control

The European public's anti-GM crop stance stems primarily from the success of environmental advocacy groups such as Greenpeace and Friends of the Earth. Numerous demonstrations have occurred throughout Britain, France, and other EU countries where GM crops have been uprooted and destroyed. Unfortunately, activist organizations rarely cite credible scientific evidence in their positions and have won much public sympathy by exploiting popular fears and misconceptions about genetic-engineering technology.

One issue they highlight that might prove significant, however, is the role of corporate interests in the GM-crop debate. A few years ago, Monsanto's attempt to acquire the "terminator" technology sparked tremendous controversy. This patent consisted of an elaborate genetically engineered control system designed to inhibit the generation of fertile seeds from crops. In essence, it was developed so that farmers would need to purchase new GM seeds each year, although arguments were raised that this technology could help prevent uncontrolled GM crop breeding. After much pressure from the nonprofit advocacy group Rural Advancement Foundation International, however, Monsanto announced in late 1999 that it would not market the "terminator" technology.

The "terminator" ordeal attracted so much attention because it placed Monsanto's corporate interest directly against the strongest argument in favor of genetic-engineering technology: potential cost savings and nutritional value of GM crops to developing countries. The UN Development Programme recently affirmed that GM crops could be the key to alleviating global hunger. Although the United Nations has expressed concern over precautionary testing of crops (through agencies like the World Health Organization), some contend that Western opposition to this technology ignores concerns of sub-Saharan and South Asian countries where malnutrition and poverty are widespread.

India is among those nations that could benefit from GM-crop technology. India's population has been growing by 1.8 percent annually; by 2025 India will need to produce 30 percent more grain per year to feed the twenty million new mouths added to its population. The need for higher food productivity is highlighted by incidents of poor farmers in Warangal and Punjab who have committed suicide when faced with devastated crops and huge debts on pesticides. The Indian government has approved several GM crops for commercial production, and testing has also commenced on transgenic cotton, rice, maize, tomato, and cauliflower, crops that would reduce the need for pesticides. A recent furor erupted over the discovery of around 11,000 hectares of illegal Bt cotton in Gujarat. The Gujarat administration responded immediately by ordering the fields stripped, the crops burned, and the seeds destroyed. There is still uncertainty over who will repay the farmers, who claim that Mahyco, a Monsanto subsidiary, is attempting to monopolize the distribution of Bt crops in India, and that the Indian government is also yielding to pressure from pesticide manufacturers. Corporate battles still abound in a nation where many farmers appear to be in need of agricultural change.

Feed the World

Many opponents of GM crops argue that the technology is not needed to help solve the problem of world hunger, with 800 million people who do not have enough to eat. They often argue that the world produces enough food to feed nine billion people while there are only six billion people today, implying that global hunger is simply a matter of distribution and not food productivity. Unfortunately, fixing the distribution problem is a complex issue. Purchasing power would need to increase in developing countries, coupled with increased food production in both developing and developed countries so that crops can be marketed at a price the underprivileged can afford. Since land for farming is limited, the remaining option for increasing crop productivity is to increase yield. While GM-crop technology is not the only method that can be used to achieve this end, it can contribute greatly toward it.

On Dr. Shiva's argument for supporting local knowledge in agricultural practices, Dr. Prakash argues that, from experience, "[local knowledge] is losing one third of your children before they hit the age of three. Is that the local knowledge that you want to keep reinforcing and keep perpetuating?"

Some consider GM crops part of a series of corporate attempts to control markets in developing countries and thus they brand GM technology another globalization "evil." Dr. Vandana Shiva of the Research Foundation for Science, Technology, and Ecology argues that globalization has pressured farmers in developing countries to grow monocultures—single-crop farming—instead of fostering sustainable agricultural diversity. Genetic engineering, in this view, is the next industrialization effort after chemical pesticides, and would also bear

no greater benefit than indigenous poly-cultural farming. The Food and Agricultural Organization of the United Nations also notes the leaning of research investment toward monocultures, spurred on by the profit potential of GM crops.

On the other hand, GM-crop technology serves to increase crop yield on land already in use for agricultural purposes, thereby preserving biodiversity in unused land. In the words of Dr. C. S. Prakash "using genetics helped [to] save so much valuable land from being under the plow." On Shiva's argument for supporting local knowledge in agricultural practices, Dr. Prakash argues that, from experience, "[local knowledge] is losing one third of your children before they hit the age of three. Is that the local knowledge that you want to keep reinforcing and keep perpetuating?"

Continuing along these lines and bringing GM technology in developing countries into the broader context of morality, leaders including Per Pinstrup-Andersen, director of the International Food Policy Research Institute, and Hassan Adamu, Nigeria's minister of agriculture, emphasize the importance of providing freedom of access, education, and choice in GM technology to the individual farmer himself. In Africa, for example, many local farmers have benefited from hybrid seeds obtained from multinational corporations. On a larger scale, however, Africa's agricultural production per unit area is among the lowest in the world, and great potential lies in utilizing GM crops to help combat pestilence and drought problems. On the issue of local knowledge, Dr. Florence Wambugu of the International Service for the Acquisition of Agribiotech Applications in Kenya (ISAAA) asserts that GM crops consist of "packaged technology in the seed" that can yield benefits without a change in local agricultural customs.

On another front of the world hunger debate, a promising benefit that GM-crop technology brings to developing countries is the introduction or enhancement of nutrients in crops. The first prod-

uct to address this was "golden rice," an engineered form of rice that expresses high levels of beta-carotene, a precursor of Vitamin A, which could be used to combat Vitamin A deficiency found in over 120 million children worldwide. Although many advocacy groups claim that the increased levels of Vitamin A from a golden rice diet are not high enough to fully meet recommended doses of Vitamin A, studies suggest that a less-than-full dose can still make a difference in an individual whose Vitamin A intake is already deficiently low. Currently the International Rice Research Institute is evaluating environmental and health concerns. After such tests are completed, however, there remains one final hurdle in the marketing process that advocates on both sides of the GM debate do agree on: multilateral access and sharing between public and private sectors. The International Undertaking on Plant Genetic Resources was established to foster such relationships for the world's key crops, but more discussions will have to take place on the intellectual-property rights of GM-crop patents.

Science First

Monsanto recently drafted a pledge of Five Commitments: Respect, Transparency, Dialogue, Sharing, and Benefits. These are qualities that all multinational organizations should bring to the debate over GM crops. In the meantime, the technology of genetic engineering has already emerged and bears promising potential. On the question of world hunger, GM crops are not the full solution, but they can play a part in one. There are possible risks which must be examined and compared to the risks associated with current agricultural conditions, and progress must not be sought too hastily. It is important to base considerations of the benefits and risks of GM crops on careful scientific research, rather than corporate interest or public fears.

HONOR HSIN, Staff Writer, *Harvard International Review*

UNIT 3

The Global Environment and Natural Resources Utilization

Unit Selections

8. **The Challenges We Face**, Jeffrey Kluger and Andrea Dorfman
9. **Shifting the Pain: World's Resources Feed California's Growing Appetite**, Tom Knudson
10. **Water Scarcity Could Overwhelm the Next Generation**, Janet L. Sawin
11. **We *Can* Build a Sustainable Economy**, Lester R. Brown

Key Points to Consider

- What are the basic environmental challenges that confront both governments and individual consumers?

- What are the dimensions to the debate about global warming and the use of fossil fuel?

- What transformations will societies that are heavy users of fossil fuels undergo in order to meet future energy needs?

- Has the international community adequately responded to problems of pollution and threats to our common natural heritage? Why or why not?

- What is the natural resource picture going to look like 30 years from now?

- How is society, in general, likely to respond to the conflicts between lifestyle and resource conservation?

- Can a sustainable economy be organized and what changes in behavior and values are necessary to accomplish this?

 Links: www.dushkin.com/online/
These sites are annotated in the World Wide Web pages.

National Geographic Society
http://www.nationalgeographic.com

National Oceanic and Atmospheric Administration (NOAA)
http://www.noaa.gov

Public Utilities Commission of Ohio (PUCO)
http://www.puc.state.oh.us/consumer/gcc/index.html

SocioSite: Sociological Subject Areas
http://www.pscw.uva.nl/sociosite/TOPICS/

United Nations Environment Programme (UNEP)
http://www.unep.ch

Beginning in the eighteenth century, the modern nation-state was conceived, and over many generations it has evolved to the point where it is now difficult to imagine a world without national governments. These legal entities have been viewed as separate, self-contained units that independently pursue their "national interests." Scholars often described the world as a political community of independent units that interact with each other (a concept that has been described as a billiard ball model).

This perspective of the international community as comprised of self-contained and self-directed units has undergone major rethinking in the past 30 years. One of the reasons for this is the international dimensions of the growing demands being placed on natural resources. The Middle East, for example, contains a majority of the world's oil reserves. The United States, Western Europe and Japan are very dependent on this vital source of energy. This unbalanced supply and demand equation has created an unprecedented lack of self-sufficiency for the world's major economic powers.

The increased interdependence of countries is further illustrated by the fact that air and water pollution often do not respect political boundaries. One country's smoke is often another country's acid rain. The concept that independent political units control their own destiny, in short, makes less sense than it may have 100 years ago. In order to more fully understand why this is so, one must first look at how Earth's natural resources are being utilized and how this may be affecting the global environment.

The initial articles in the unit examine the broad dimensions of the uses and abuses of natural resources. The central theme in these articles is whether or not human activity is in fact bringing about fundamental changes in the functioning of Earth's self-regulating ecological systems. In many cases an unsustainable rate of usage is under way, and, as a consequence, an alarming decline in the quality of the natural resource base is taking place.

An important conclusion resulting from this analysis is that contemporary methods of resource utilization often create problems that transcend national boundaries. Global climate changes, for example, will affect everyone, and if these changes are to be successfully addressed, international collaboration will be required. The consequences of basic human activities such as growing and cooking food are profound when multiplied billions of times every day. A single country or even a few countries working together cannot have a significant impact on redressing these problems. Solutions will have to be conceived that are truly global in scope. Just as there are shortages of natural resources, there are also shortages of new ideas for solving many of these problems.

Unit 3 continues by examining specific case studies that explore in greater detail the issues raised in the first chapter. The unit concludes with a discussion of the issues involved in moving from a perspective of the environment as simply an economic resource to be consumed to a perspective that has been defined as "sustainable development." This change is easily called for, but in fact it goes to the core of social values and basic economic activities. Developing sustainable practices, therefore, is a challenge of unprecedented magnitude.

Nature is not some object "out there" to be visited at a national park. It is the food we eat and the energy we consume. Human beings are joined in the most intimate of relationships with the natural world in order to survive from one day to the next. It is ironic how little time is spent thinking about this relationship. This lack of attention, however, is not likely to continue, for rapidly growing numbers of people and the increased use of energy-consuming technologies are placing unprecedented pressures on Earth's carrying capacity.

The Challenges We Face

By JEFFREY KLUGER and ANDREA DORFMAN

For starters, let's be clear about what we mean by "saving the earth." The globe doesn't need to be saved by us, and we couldn't kill it if we tried. What we do need to save—and what we have done a fair job of bollixing up so far—is the earth as we like it, with its climate, air, water and biomass all in that destructible balance that best supports life as we have come to know it. Muck that up, and the planet will simply shake us off, as it's shaken off countless species before us. In the end, then, it's us we're trying to save—and while the job is doable, it won't be easy.

The 1992 Earth Summit in Rio de Janeiro was the last time world leaders assembled to look at how to heal the ailing environment. Now, 10 years later, Presidents and Prime Ministers are convening at the World Summit on Sustainable Development in Johannesburg next week to reassess the planet's condition and talk about where to go from here. In many ways, things haven't changed: the air is just as grimy in many places, the oceans just as stressed, and most treaties designed to do something about it lie in incomplete states of ratification or implementation. Yet we're oddly smarter than we were in Rio. If years of environmental false starts have taught us anything, it's that it's time to quit seeing the job of cleaning up the world as a zero-sum game between industrial progress on the one hand and a healthy planet on the other. The fact is, it's development—well-planned, well-executed sustainable development—that may be what saves our bacon before it's too late.

Food
As we try to nourish 6 billion people, both bioengineering and organic farming will help

As the summiteers gather in Johannesburg, TIME is looking ahead to what the unfolding century—a green century—could be like. In this special report, we will examine several avenues to a healthier future, including green industry, green architecture, green energy, green transportation and even a greener approach to wilderness preservation. All of them have been explored before, but never so urgently as now. What gives such endeavors their new credibility is the hope and notion of sustainable development, a concept that can be hard to implement but wonderfully simple to understand.

Population
The tide of people may not ebb until the head count hits the 11 billion mark

With 6.1 billion people relying on the resources of the same small planet, we're coming to realize that we're drawing from a finite account. The amount of crops, animals and other biomatter we extract from the earth each year exceeds what the planet can replace by an estimated 20%, meaning it takes 14.4 months to replenish what we use in 12—deficit spending of the worst kind. Sustainable development works to reverse that, to expand the resource base and adjust how we use it so we're living off biological interest without ever touching principal. "The old environmental movement had a reputation of elitism," says Mark Malloch Brown, administrator of the United Nations Development Program (UNDP). "The key now is to put people first and the environment second, but also to remember that when you exhaust resources, you destroy people." With that in mind, the summiteers will wrestle with a host of difficult issues that affect both people and the environment. Among them:

• POPULATION AND HEALTH: While the number of people on earth is still rising rapidly, especially in the developing countries of Asia, the good news is that the growth rate is slowing. World population increased 48% from 1975 to 2000, compared with 64% from 1950 to 1975. As this gradual deceleration continues, the population is expected to level off eventually, perhaps at 11 billion sometime in the last half of this century.

Though it's not easy to see it from the well-fed West, a third of the world goes hungry

Economic-development and family-planning programs have helped slow the tide of people, but in some places, population growth is moderating for all the wrong reasons. In the poorest parts of the world, most notably Africa, infectious diseases such as AIDs, malaria, cholera and tuberculosis are having a Malthusian effect. Rural-land degradation is pushing people into cities, where crowded, polluted living conditions create the perfect breeding grounds for sickness. Worldwide, at least 68 million are expected to die of AIDs by 2020, including 55 million in sub-Saharan Africa. While any factor that eases population pressures may help the environment, the situation would be far less tragic if rich nations did more to help the developing world reduce birth rates and slow the spread of disease.

Efforts to provide greater access to family planning and health care have proved effective. Though women in the poorest countries still have the most children, their collective fertility rate is 50% lower than it was in 1969 and is expected to decline more by 2050. Other programs targeted at women include basic education and job training. Educated mothers not only have a stepladder out of poverty, but they also choose to have fewer babies.

Rapid development will require good health care for the young since there are more than 1 billion people ages 15 to 24. Getting programs in place to keep this youth bubble healthy could make it the most productive generation ever conceived. Says Thoraya Obaid, executive director of the U.N. Population Fund: "It's a window of opportunity to build the economy and prepare for the future."

• FOOD: Though it's not always easy to see it from the well-fed West, up to a third of the world is in danger of starving. Two billion people lack reliable access to safe, nutritious food, and 800 million of them—including 300 million children—are chronically malnourished.

Agricultural policies now in place define the very idea of unsustainable development. Just 15 cash crops such as corn, wheat and rice provide 90% of the world's food, but planting and replanting the same crops strips fields of nutrients and makes them more vulnerable to pests. Slash-and-burn planting techniques and overreliance on pesticides further degrade the soil.

Solving the problem is difficult, mostly because of the ferocious debate over how to do it. Biotech partisans say the answer lies in genetically modified crops—foods engineered for vitamins, yield and robust growth. Environmentalists worry that fooling about with genes is a recipe for Frankensteinian di-

saster. There is no reason, however, that both camps can't make a contribution.

Better crop rotation and irrigation can help protect fields from exhaustion and erosion. Old-fashioned cross-breeding can yield plant strains that are heartier and more pest-resistant. But in a world that needs action fast, genetic engineering must still have a role—provided it produces suitable crops. Increasingly, those crops are being created not just by giant biotech firms but also by home-grown groups that know best what local consumers need.

The National Agricultural Research Organization of Uganda has developed corn varieties that are more resistant to disease and thrive in soil that is poor in nitrogen. Agronomists in Kenya are developing a sweet potato that wards off viruses. Also in the works are drought-tolerant, disease-defeating and vitamin-fortified forms of such crops as sorghum and cassava—hardly staples in the West, but essentials elsewhere in the world. The key, explains economist Jeffrey Sachs, head of Columbia University's Earth Institute, is not to dictate food policy from the West but to help the developing world build its own biotech infrastructure so it can produce the things it needs the most. "We can't presume that our technologies will bail out poor people in Malawi," he says. "They need their own improved varieties of sorghum and millet, not our genetically improved varieties of wheat and soybeans."

Water
In 25 years two-thirds of humanity may live in nations running short of life's elixir

• WATER: For a world that is 70% water, things are drying up fast. Only 2.5% of water is fresh, and only a fraction of that is accessible. Meanwhile, each of us requires about 50 quarts per day for drinking, bathing, cooking and other basic needs. At present, 1.1 billion people lack access to clean drinking water and more than 2.4 billion lack adequate sanitation. "Unless we take swift and decisive action," says U.N. Secretary-General Kofi Annan, "by 2025, two-thirds of the world's population may be living in countries that face serious water shortages."

Only 2.5% of water is fresh, and only a fraction of that is accessible

The answer is to get smart about how we use water. Agriculture accounts for about two-thirds of the fresh water consumed. A report prepared for the summit thus endorses the "more crop per drop" approach, which calls for more efficient irrigation techniques, planting of drought- and salt-tolerant crop varieties that require less water and better monitoring of growing conditions, such as soil humidity levels. Improving water-delivery systems would also help, reducing the amount that is lost en route to the people who use it.

One program winning quick support is dubbed WASH—for Water, Sanitation and Hygiene for All—a global effort that aims to provide water services and hygiene training to everyone who lacks them by 2015. Already, the U.N., 28 governments and many nongovernmental organizations (NGOs) have signed on.

Climate
Car exhaust is a major source of the heat-trapping gases that produce global warming

• ENERGY AND CLIMATE: In the U.S., people think of rural electrification as a long-ago legacy of the New Deal. In many parts of the world, it hasn't even happened yet. About 2.5 billion people have no access to modern energy services, and the power demands of developing economies are expected to grow 2.5% per year. But if those demands are met by burning fossil fuels such as oil, coal and gas, more and more carbon dioxide and other greenhouse gases will hit the atmosphere. That, scientists tell us, will promote global warming, which could lead to rising seas, fiercer storms, severe droughts and other climatic disruptions.

Of more immediate concern is the heavy air pollution caused in many places by combustion of wood and fossil fuels. A new U.N. Environment Program report warns of the effects of a haze across all southern Asia. Dubbed the "Asian brown cloud" and estimated to be 2 miles thick, it may be responsible for hundreds of thousands of deaths a year from respiratory diseases.

The better way to meet the world's energy needs is to develop cheaper, cleaner sources. Pre-Johannesburg proposals call for eliminating taxation and pricing systems that encourage oil use and replacing them with policies that provide incentives for alternative energy. In India there has been a boom in wind power because the government has made it easier for entrepreneurs to get their hands on the necessary technology and has then required the national power grid to purchase the juice that wind systems produce.

Other technologies can work their own little miracles. Micro-hydroelectric plants are already operating in numerous nations, including Kenya, Sri Lanka and Nepal. The systems divert water from streams and rivers and use it to run turbines without complex dams or catchment areas. Each plant can produce as much as 200 kilowatts—enough to electrify 200 to 500 homes and businesses—and lasts 20 years. One plant in Kenya was built by 200 villagers, all of whom own shares in the cooperative that sells the power.

The Global Village Energy Partnership, which involves the World Bank, the UNDP and various donors, wants to provide energy to 300 million people, as well as schools, hospitals and clinics in 50,000 communities worldwide over 10 years. The key will be to match the right energy source to the right users. For example, solar panels that convert sunlight into electricity might be cost-effective in remote areas, while extending the power grid might be better in Third World cities.

Biodiversity
Unless we guard wilderness, as many as half of all species could vanish in this century

• BIODIVERSITY: More than 11,000 species of animals and plants are known to be threatened with extinction, about a third of all coral reefs are expected to vanish in the next 30 years and about 36 million acres of forest are being razed annually. In his new book, *The Future of Life*, Harvard biologist Edward O. Wilson writes of his worry that unless we change our ways half of all species could disappear by the end of this century.

Once you tear out swaths of ecosystem, you harm areas you didn't want to touch

The damage being done is more than aesthetic. Many vanishing species provide humans with both food and medicine. What's more, once you start tearing out swaths of ecosystem, you upset the existing balance in ways that harm even areas you didn't intend to touch. Environmentalists have said this for decades, and now that many of them have tempered ecological absolutism with developmental realism, more people are listening.

The Equator Initiative, a public-private group, is publicizing examples of sustainable development in the equatorial belt. Among the projects already cited are one to help restore marine fisheries in Fiji and another that promotes beekeeping as a source of supplementary income in rural Kenya. The Global Conservation Trust hopes to raise $260 million to help conserve genetic material from plants for use by local agricultural programs. "When you approach sustainable development from an environmental view, the problems are global," says the U.N.'s Malloch Brown. "But from a development view, the front line is local, local, local."

If that's the message environmental groups and industry want to get out, they appear to be doing a good job of it. Increasingly, local folks act whether world political bodies do or not. California Governor Gray Davis signed a law last month requiring automakers to cut their cars' carbon emissions by 2009.

Many countries are similarly proactive. Chile is encouraging sustainable use of water and electricity; Japan is dangling financial incentives before consumers who buy environmentally sound cars; and tiny Mauritius is promoting solar cells and discouraging use of plastics and other disposables.

Business is getting right with the environment too. The Center for Environmental Leadership in Business, based in Washington, is working with auto and oil giants including Ford, Chevron, Texaco and Shell to draft guidelines for incorporating biodiversity conservation into oil and gas exploration. And the center has helped Starbucks develop purchasing guidelines that reward coffee growers whose methods have the least impact on the environment. Says Nitin Desai, secretary-general of the Johannesburg summit: "We're hoping that partnerships—involving governments, corporations, philanthropies and NGOs—will increase the credibility of the commitment to sustainable development."

Will that happen? In 1992 the big, global measures of the Rio summit seemed like the answer to what ails the world. In 2002 that illness is—in many respects—worse. But if Rio's goal was to stamp out the disease of environmental degradation, Johannesburg's appears to be subtler—and perhaps better: treating the patient a bit at a time, until the planet as a whole at last gets well.

From *Time*, August 26, 2002, pp. A6–A12. © 2002 by Time Inc. Reprinted by permission.

Shifting the pain

World's resources feed California's growing appetite

Tom Knudson

Half a hemisphere separates the headwaters of the Amazon River and the frostbitten northern latitudes of Canada.

But the two landscapes have one thing in common.

You can see it along a muddy rain-forest road in Ecuador, in the silver glint of a pipeline snaking through the grass. North of Edmonton, Alberta, a different sight catches your eye: an old-growth forest of spruce, pine and aspen shredded by a dusty maze of logging roads.

That oil pipeline and those logging roads are linked, via quiet rivers of commerce, to the largest concentration of consumers in North America, to a culture that proudly protects its own coastline and forests from exploitation while using more gasoline, wood and paper than any other state in America: California.

With 34 million people and the world's fifth-largest economy, California has long consumed more than it produces. But today, its passion for protecting natural resources at home while importing them in record quantities from afar is backfiring on the world's environment.

It is exporting the pain of producing natural resources—polluted water, pipeline accidents, piecemeal forests and human conflicts—to the far corners of the planet, to places out of sight and out of mind. California is the state of denial.

"There is a disconnect going on," said William Libby, a professor emeritus of forestry at the University of California, Berkeley, who lectures and consults on forest issues around the globe. "We consume like mad. And we preserve like mad."

Since the days of John Muir—the California naturalist whose writings and ramblings helped kindle the conservation movement just over a century ago—concern for the environment has been a cornerstone of California life.

And seldom has conservation touched California so deeply as during the past 10 years. Since 1992, environmental rules have eliminated or sharply reduced logging on 10 million acres of national forest land in the state—an area 13 times larger than Yosemite National Park. In the Mojave and Great Basin deserts, 3.5 million acres were declared wilderness in 1994—an expanse half again the size of Yellowstone National Park.

And while that conservation legacy will enrich Californians—and California ecosystems—for generations to come, its reach also extends far beyond the Golden State.

Libby was one of the first to notice, while on sabbatical in New Zealand in 1992. As the volume of wood cut from California forests dropped due to regulations to protect spotted owls, the demand for logs in New Zealand soared—making loggers there happy.

"Prices were insane," Libby said. "The New Zealanders wanted me to get them a dead spotted owl so they could stuff it, put it in the lobby and genuflect to it."

He soon discovered logging was on the rise in other places, too, and has since published several articles that link preservation of California forests with species extinctions elsewhere.

"We Californians are really not very good conservationists—we're very good preservationists," he said. "Conservation means you use resources well and responsibly. Preservation means you are rich enough to set aside things you want and buy them from someone else."

A half-century ago, California was self-sufficient in wood. Today, the state imports 80 percent of what it uses. Follow some of that wood back to its source and you find yourself in the northern boreal forest, where Canada allows trees to be cut in ways not permitted in California.

On average, nine of every 10 acres logged in Canada are clear-cut—the contentious practice of leveling large patches of the forest. And more than two-thirds of Canadian logging takes place in stands that have never been nicked by a chain saw—virgin forests that in California would be regarded as sanctuaries.

"Many Americans believe Canada is this incredible wilderness, but it's not true," said Richard Thomas, an Edmonton consultant and author of a 1998 provincial study critical of logging practices in Alberta. "We are very much like a Third World country when it comes to our resources. We just let other countries have at it."

Six thousand miles south, a wave of development for another resource crucial to California—crude oil—is inflicting similarly serious wounds across Ecuador's Amazon. Rain forests that were home to kaleidoscopic displays of plant and animal life in the 1970s and '80s now are showcases of pollution and poverty.

Every day, an average of 235,000 barrels of oil is pumped from the region for export to world markets. The largest portion—65,000 to 85,000 barrels a day—is shipped to refineries in Los Angeles and San Francisco.

The discovery of more reserves in the Amazon is setting off a new wave of controversy and threatening the cultural survival of semi-nomadic rain-forest tribes. Still, the country's new president, Lucio Gutiérrez, assured financiers in New York earlier this year that he supports more drilling because Ecuador is deeply in debt and needs foreign investment.

"The historical challenges for my government are very clear," Gutiérrez said at the time.

California's hunger for the planet's natural resources need not stir up trouble, if a system were in place to prevent it. You can find such a safeguard in the storm-tossed North Pacific, where Canadian fishermen, working under a federal plan that gives them an ownership stake in fish, are harvesting millions of pounds of rockfish every year for California without hurting the environment.

"Everybody is quite conscientious," said Jim Harris, a Canadian trawler. "We've got a fishery that is going to be here for the duration."

The clash of conservation and consumption in California may be large, but it is not unique in this country.

"We're the largest consuming nation basically of everything," said James Bowyer, a professor at the University of Minnesota who specializes in conservation policy and natural resource consumption.

"Yet we find every reason in the world why we shouldn't mine steel, why we shouldn't drill for oil," Bowyer said. "It's ironic because we are transferring the impacts to someplace else. And then we are telling ourselves what we are doing is good for the environment.

"And not only are we transferring those impacts, we are magnifying them by turning to nations that don't have the stringent environmental controls that we do."

No government agency maps the global impact of California consumers. But a small Oakland think tank, Redefining Progress, has assembled estimates of the mountain of resources, from wood to fossil fuel, fresh water to seafood, consumed by 146 nations—and some California counties—a yardstick it calls an "ecological footprint."

The United States, a world leader in the conservation of natural resources, has a larger footprint (24 acres per person) than all nations except the United Arab Emirates (with 25 acres). Do the math and you find America's 291 million people draw upon a 7 billion-acre chunk of the planet—an area roughly three times the size of the United States.

An assessment for Marin County—the pricey, conservation-minded San Francisco suburb—found citizens there eat, drink, spend and drive their way through even more of the planet's natural wealth: 27 acres per person a year—the largest ecological footprint ever calculated.

The group's footprints have attracted attention from scientists and policy-makers around the world. And although some people criticize the methods as imprecise, none denies the basic premise.

"The idea is right," said Libby, the forestry professor.

Last year, Libby found some Californians are not eager to hear about the global consequences of conservation and consumption.

At a conference on Sierra Nevada forest management, held in Nevada City, Libby asked the 250 people attending how many of them lived in houses made of wood.

Almost everyone did. Then he asked how many had houses built with alternatives such as used tires and straw bales. Only two or three people responded.

A few moments later, Libby recalled, he asked, "How many people are comfortable with species going extinct somewhere else because we're not going to cut any wood on the Tahoe National Forest?"

At that point, Libby said, "Somebody in the audience shouted: 'We don't like your question.'"

Water scarcity could overwhelm the next generation

Janet L. Sawin

Up to 7 billion people in 60 countries—more than the whole present population of the world—will face water scarcity within the next half-century, according to the United Nations' *World Water Development Report* released in March. The UN report is the most complete appraisal of water resources to date. Other reports released the same month offer dire projections of the impacts of water scarcity on human health, the environment, and global political stability. [1]

Demand for fresh water has tripled in the past 50 years, and is continuing to rise as a result of population growth and economic development. About 70 percent of the demand is for agriculture. But increasing water use is not just a function of the greater number of people needing to eat and drink; it also results from pollution and misuse of available water supplies, both directly via dumping or runoff of effluents into water, and indirectly via pollution of air and soil. The damage is accelerated by wetlands destruction and other abuses—including human-caused global warming. According to the UN report, "recent estimates suggest that climate change will account for about 20 percent of the increase in global water scarcity" in coming decades.

Water mismanagement has become a crisis of governance that will impact heavily on public health and the environment, while heightening tensions and conflicts over declining resources. Worldwide, the greatest impacts will be on the poor, who are most vulnerable to water-borne illness—which further perpetuates their poverty. At present, 1.1 billion people lack access to clean water and 2.4 billion lack adequate sanitation. In 2000, an estimated 2.2 million people, most of them infants or children under five, died from water-related diseases.

It is not only the poor who are at risk, however. Around the same time the UN report was issued, two reports published in New Jersey revealed that the state's drinking water supply contains hundreds of chemicals, including steroids, detergents, pesticides, preservatives, and prescription medications. While all were found only in low concentrations, researchers cannot rule out possible health risks of ingesting low levels of multiple chemicals over long time periods. The water bodies studied contained antibiotics that could breed drug-resistant bacteria, and endocrine disruptors believed to cause neurological and reproductive birth defects. Other studies in the U.S. and Europe have yielded comparable results, with hundreds of lakes, streams, and rivers testing positive for traces of similar chemicals.

The UN report concludes that a more positive outcome is possible, *if* the global community improves infrastructure, takes full advantage of conservation and efficiency technologies, and enacts appropriate water pricing plans and water treaties. In the global North, upgrading water treatment facilities to deal with contaminants will be costly, but necessary to safeguard human health. One implication of the report is that attempting to rely only on water treatment is a losing battle and that it is becoming increasingly urgent, in all countries, to reduce or eliminate toxins at their source.

Notes

1. The report estimates that humans need 20 to 50 liters of water free from harmful contaminants daily to meet basic needs. For monitoring purposes, the World Health Organixation/UN Children's Fund defines reasonable access to water as at least 20 liters per day, from an improved source within one kilometer of the user's dwelling.

From *World Watch*, Vol. 16 No. 4. © 2003 by Worldwatch Institute, www.worldwatch.org.

We *Can* Build a Sustainable Economy

The keys to securing the planet's future lie in stabilizing both human population and climate. The challenges are great, but several trends look promising.

by Lester R. Brown

The world economy is growing faster than ever, but the benefits of this rapid growth have not been evenly distributed. As population has doubled since mid-century and the global economy has nearly quintupled, the demand for natural resources has grown at a phenomenal rate.

Since 1950, the need for grain has nearly tripled. Consumption of seafood has increased more than four times. Water use has tripled. Demand for beef and mutton has tripled. Firewood demand has tripled, lumber demand has more than doubled, and paper demand has gone up sixfold. The burning of fossil fuels has increased nearly fourfold, and carbon emissions have risen accordingly.

These spiraling human demands for resources are beginning to outgrow the earth's natural systems. As this happens, the global economy is damaging the foundation on which it rests.

To build an environmentally sustainable global economy, there are many obstacles, but there are also several promising trends and factors in our favor. One is that we know what an environmentally sustainable economy would look like. In a sustainable economy:

• Human births and deaths are in balance.
• Soil erosion does not exceed the natural rate of new soil formation.
• Tree cutting does not exceed tree planting.
• The fish catch does not exceed the sustainable yield of fisheries.
• The number of cattle on a range does not exceed the range's carrying capacity.
• Water pumping does not exceed aquifer recharge.
• Carbon emissions and carbon fixation are in balance.
• The number of plant and animal species lost does not exceed the rate at which new species evolve.

We know how to build an economic system that will meet our needs without jeopardizing prospects for future generations. And with some trends already headed in the right direction, we have the cornerstones on which to build such an economy.

Stabilizing Population

With population, the challenge is to complete the demographic transition, to reestablish the balance between births and deaths that characterizes a sustainable society. Since populations are rarely ever precisely stable, a stable population is defined here as one with a growth rate below 0.3%. Populations are effectively stable if they fluctuate narrowly around zero.

Thirty countries now have stable populations, including most of those in Europe plus Japan. They provide the solid base for building a world population stabilization effort. Included in the 30 are all the larger industrialized countries of Europe—France, Germany, Italy, Russia, and the United Kingdom. Collectively, these 30 countries contain 819 million people or 14% of humanity. For this goal, one-seventh of humanity is already there.

The challenge is for the countries with the remaining 86% of the world's people to reach stability. The two large nations that could make the biggest difference in this effort are China and the United States. In both, population growth is now roughly 1% per year. If the global food situation becomes desperate, both could reach stability in a decade or two if they decided it were important to do so.

The world rate of population growth, which peaked around 2% in 1970, dropped below 1.6% in 1995. Although the rate is declining, the annual addition is still close to 90 million people per year. Unless populations can be stabilized with demand below the sustainable yield of local ecosystems, these systems will be destroyed. Slowing growth may delay the eventual collapse of ecosystems, but it will not save them.

The European Union, consisting of some 15 countries and containing 360 million people, provides a model for the rest of the world of an environmentally sustainable food/population balance. At the same time that the region has reached zero population growth, movement up the food chain has come to a halt as diets have become saturated with livestock products. The result is that Europe's grain consumption has been stable for close to two decades at just under 160 million tons—a level that is within the region's carrying capacity. Indeed, there is a potential for a small but sustainable export surplus of grain that can help countries where the demand for food has surpassed the carrying capacity of their croplands.

World Fertilizer and Grainland
(Per Person, 1950-94)

SOURCES: USDA, FAO, IFA

As other countries realize that continuing on their current population trajectory will prevent them from achieving a similar food/ population balance, more and more may decide to do what China has done—launch an all-out campaign to stabilize population. Like China, other governments will have to carefully balance the reproductive rights of the current generation with the survival rights of the next generation.

Very few of the group of 30 countries with stable populations had stability as an explicit policy goal. In those that reached population stability first, such as Belgium, Germany, Sweden, and the United Kingdom, it came with rising living standards and expanding employment opportunities for women. In some of the countries where population has stabilized more recently, such as Russia and other former Soviet republics, the deep economic depression accompanying economic reform has substantially lowered birth rates, much as the Great Depression did in the United States. In addition, with the rising number of infants born with birth defects and deformities since Chernobyl, many women are simply afraid to bear children. The natural decrease of population (excluding migration) in Russia of 0.6% a year—leading to an annual population loss of 890,000—is the most rapid on record.

Not all countries are achieving population stability for the right reasons. This is true today and it may well be true in the future. As food deficits in densely populated countries expand, governments may find that there is not enough food available to import. Between fiscal year 1993 and 1996, food aid dropped from an all-time high of 15.2 million tons of grain to 7.6 million tons. This cut of exactly half in three years reflects primarily fiscal stringencies in donor countries, but also, to a lesser degree, higher grain prices in fiscal 1996. If governments fail to establish a humane balance between their people and food supplies, hunger and malnutrition may raise death rates, eventually slowing population growth.

Some developing countries are beginning to adopt social policies that will encourage smaller families. Iran, facing both land hunger and water scarcity, now limits public subsidies for housing, health care, and insurance to three children per family. In Peru, President Alberto Fujimori, who was elected overwhelmingly to his second five-year term in a predominantly Catholic country, said in his inaugural address in August 1995 that he wanted to provide better access to family-planning services for poor women. "It is only fair," he said, "to disseminate thoroughly the methods of family planning to everyone."

Stabilizing Climate

With climate, as with population, there is disagreement on the need to stabilize. Evidence that atmospheric carbon-dioxide levels are rising is clear-cut. So, too, is the greenhouse effect that these gases produce in the atmosphere. That is a matter of basic physics. What is debatable is the rate at which global temperatures will rise and what the precise local effects will be. Nonetheless, the consensus of the mainstream scientific community is that there is no alternative to reducing carbon emissions.

How would we phase out fossil fuels? There is now a highly successful "phase out" model in the case of chlorofluorocarbons (CFCs). After two British scientists discovered the "hole" in the ozone layer over Antarctica and published their findings in *Nature* in May 1985, the international community convened a conference in Montreal to draft an agreement designed to reduce CFC production sharply. Subsequent meetings in London in 1990 and Copenhagen in 1992 further advanced the goals set in Montreal. After peaking in 1988 at 1.26 million tons, the manufacture of CFCs dropped to an estimated 295,000 tons in 1994—a decline of 77% in just six years.

As public understanding of the costs associated with global warming increases, and as evidence of the effects of higher temperatures accumulates, support for reducing dependence on fossil fuels is building. At the March 1995 U.N. Climate Convention in Berlin, environmental groups were joined in lobbying for a reduction in carbon emissions by a group of 36 island communities and insurance industry representatives.

The island nations are beginning to realize that rising sea levels would, at a minimum, reduce their land area and displace people. For some low-lying island countries, it could actually threaten their survival. And the insurance industry is beginning to realize that increasing storm intensity can threaten the survival of insurance companies as well. When Hurricane Andrew tore through Florida in 1992, it took down not only thousands of buildings, but also eight insurance firms.

In September 1995, the U.S. Department of Agriculture reported a sharp drop in the estimated world grain harvest because of crop-withering heat waves in the northern tier of industrial countries. Intense late-summer heat had damaged harvests in Canada and the United States, across Europe, and in Russia. If farmers begin to see that the productivity of their land is threatened by global warming, they, too, may begin to press for a shift to renewable sources of energy.

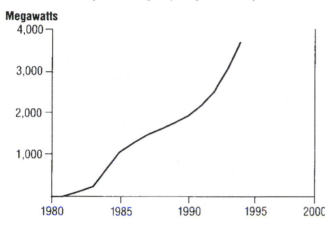

World Wind Energy
(Generating Capacity, 1980-94)

SOURCES: Gipe and Asociates, BTM Consulting

As with CFCs, there are alternatives to fossil fuels that do not alter climate. Several solar-based energy sources, including wind power, solar cells, and solar thermal power plants, are advancing rapidly in technological sophistication, resulting in steadily falling costs. The cost of photovoltaic cells has fallen precipitously over the last few decades. In some villages in developing countries where a central grid does not yet exist, it is now cheaper to install an array of photovoltaic cells than to build a centralized power plant plus the grid needed to deliver the power.

Wind power, using the new, highly efficient wind turbines to convert wind into electricity, is poised for explosive growth in the years ahead. In California, wind farms already supply enough electricity to meet the equivalent of San Francisco's residential needs.

The potential for wind energy is enormous, dwarfing that of hydropower, which provides a fifth of the world's electricity. In the United States, the harnessable wind potential in North Dakota, South Dakota, and Texas could easily meet national electricity needs. In Europe, wind power could theoretically satisfy all the continent's electricity needs. With scores of national governments planning to tap this vast resource, rapid growth in the years ahead appears inevitable.

A Bicycle Economy

Another trend to build on is the growing production of bicycles. Human mobility can be increased by investing in public transportation, bicycles, and automobiles. Of these, the first two are by far the most promising environmentally. Although China has announced plans to move toward an automobile-centered transportation system, and car production in India is expected to double by the end of the decade, there simply may not be enough land in these countries to support such a system and to meet the food needs of their expanding populations.

Against this backdrop, the creation of bicycle-friendly transportation systems, particularly in cities, shows great promise. Market forces alone have pushed bicycle production to an estimated 111 million in 1994, three times the level of automobile production. It is in the interest of societies everywhere to foster the use of bicycles and public transportation—to accelerate the growth in bicycle manufacturing while restricting that of automobiles. Not only will this help save cropland, but this technology can greatly increase human mobility without destabilizing climate. If food becomes increasingly scarce in the years ahead, as now seems likely, the land-saving, climate-stabilizing nature of bicycles will further tip the scales in their favor and away from automobiles.

The stabilization of population in some 30 countries, the stabilization of food/people balance in Europe, the reduction in CFC production, the dramatic growth in the world's wind power generating capacity, and the extraordinary growth in bicycle use are all trends for the world to build on. These cornerstones of an environmentally sustainable global economy provide glimpses of a sustainable future.

Regaining Control of Our Destiny

Avoiding catastrophe is going to take a far greater effort than is now being contemplated by the world's political leaders. We know what needs to be done, but politically we are unable to do it because of inertia and the investment of powerful interests in the status quo. Securing food supplies for the next generation depends on an all-out effort to stabilize population and climate, but we resist changing our reproductive behavior, and we refrain from converting our climate-destabilizing, fossil-fuel-based economy to a solar/hydrogen-based one.

As we move to the end of this century and beyond, food security may well come to dominate international affairs, national economic policy making, and—for much of humanity —personal concerns about survival. There is now evidence from enough countries that the old formula of substituting fertilizer for land is no longer working, so we need to search urgently for alternative formulas for humanly balancing our numbers with available food supplies.

Unfortunately, most national political leaders do not even seem to be aware of the fundamental shifts occurring in the world food economy, largely because the official projections by the World Bank and the U.N. Food and Agriculture Organization are essentially extrapolations of past trends.

If we are to understand the challenges facing us, the teams of economists responsible for world food supply-and-demand projections at these two organizations need to be replaced with an interdisciplinary team of analysts, including, for example, an agronomist, hydrologist, biologist, and meteorologist, along with an economist. Such a team could assess and incorporate into projections such things as the effect of soil erosion on land productivity, the effects of aquifer depletion on future irrigation water supplies, and the effect of increasingly intense heat waves on future harvests.

The World Bank team of economists argues that, because the past is the only guide we have to the future, simple extrapolations of past trends are the only reasonable way to make projections. But the past is also filled with a body of scientific literature on growth in finite environments, and it shows that biological growth trends typically conform to an S-shaped curve over time.

The risk of relying on these extrapolative projections is that they are essentially "no problem" projections. For example, the most recent World Bank projections, which use 1990 as a base and which were published in late 1993, are departing further and further from reality with each passing year. They show the world grain harvest climbing from 1.78 billion tons in 1990 to 1.97 billion tons in the year 2000. But instead of the projected gain of nearly 100 million tons since 1990, world grain production has not grown at all. Indeed, the 1995 harvest, at 1.69 billion tons, is 90 million tons below the 1990 harvest.

One of the most obvious needs today is for a set of country-by-country carrying-capacity assessments. Assessments using an interdisciplinary team can help provide information needed to face the new realities and formulate policies to respond to them.

Setting Priorities

The world today is faced with an enormous need for change in a period of time that is all too short. Human behavior and values, and the national priorities that reflect them, change in response to either new information or new experiences. The effort now needed to reverse the environmental degradation of the planet and ensure a sustainable future for the next generation will require mobilization on a scale comparable to World War II.

Regaining control of our destiny depends on stabilizing population as well as climate. These are both key to the achievement of a wide array of social goals ranging from the restoration of a rise in food consumption per person to protection of the diversity of plant and animal species. And neither will be easy. The first depends on a revolution in human reproductive behavior; the second, on a restructuring of the global energy system.

Serving as a catalyst for these gargantuan efforts is the knowledge that if we fail our future will spiral out of control as the acceleration of history overwhelms political institutions. It will almost guarantee a future of starvation, economic insecurity, and political instability. It will bring political conflict between societies and among ethnic and religious groups within societies. As these forces are unleashed, they will leave social disintegration in their wake.

Offsetting the dimensions of this challenge, including the opposition to change that is coming from vested interests and the momentum of trends now headed in the wrong direction, are some valuable assets. These include a well-developed global communications network, a growing body of scientific knowledge, and the possibility of using fiscal policy—a potentially powerful instrument for change—to build an environmentally sustainable economy.

Bicycles vs. Cars
(Worldwide Production, 1950-94)

SOURCES: U.N. Interbike Directory

Policies for Progress

Satisfying the conditions of sustainability—whether it be reversing the deforestation of the planet, converting a throwaway economy into a reuse-recycle one, or stabilizing climate—will require new investment. Probably the single most useful instrument for converting an unsustainable world economy into one that is sustainable is fiscal policy. Here are a few proposals:

• **Eliminate subsidies for unsustainable activities**. At present, governments subsidize many of the very activities that threaten the sustainability of the economy. They support fishing fleets to the extent of some $54 billion a year, for example, even though existing fishing capacity already greatly exceeds the sustainable yield of oceanic fisheries. In Germany, coal production is subsidized even though the country's scientific community has been outspoken in its calls for reducing carbon emissions.

• **Institute a carbon tax**. With alternative sources of energy such as wind power, photovoltaics, and solar thermal power plants becoming competitive or nearly so, a carbon tax that would reflect the cost to society of burning fossil fuels—the costs, that is, of air pollution, acid rain, and global warming—could quickly tip the scales away from further investment in fossil fuel production to investment in wind and solar energy.

Today's fossil-fuel-based energy economy can be replaced with a solar/hydrogen economy that can meet all the energy needs of a modern industrial society without causing disruptive temperature rises.

• **Replace income taxes with environmental taxes**. Income taxes discourage work and savings, which are both positive activities that should be encouraged. Taxing environmentally destructive activities instead would help steer the global economy in an environmentally sustainable direction. Among the activities to be taxed are the use of pesticides, the generation of toxic wastes, the use of virgin raw materials, the conversion of cropland to nonfarm uses, and carbon emissions.

The time may have come also to limit tax deductions for children to two per couple: It may not make sense to subsidize childbearing beyond replacement level when the most pressing need facing humanity is to stabilize population.

The challenge for humanity is a profound one. We have the information, the technology, and the knowledge of what needs to be done. The question is, Can we do it? Can a species that is capable of formulating a theory that explains the birth of the universe now implement a strategy to build an environmentally sustainable economic system?

About the Author Lester R. Brown is president of the Worldwatch Institute, 1776 Massachusetts Avenue, N.W., Washington, D.C. 20036. Telephone 202/452-1999; fax 202/296-7365.

From *The Futurist*, July/August 1996, pp. 8–12. © 1996 by The World Future Society, Bethesda, MD; http://www.wfs.org/wfs. Reprinted by permission.

UNIT 4
Political Economy

Unit Selections

Key Points to Consider

- Are those who argue that there is in fact a process of globalization overly optimistic? Why or why not?

- What are some of the impediments to a truly global political economy?

- How has globalization accelerated the growth of criminal behavior?

- How are the political economies of traditional societies different from those of the consumer-oriented societies?

- What are some of the barriers that make it difficult for non-industrial countries to develop?

- How are China and other emerging countries trying to alter their ways of doing business in order to meet the challenges of globalization? Are they likely to succeed?

- What economic challenges do countries like Japan and the U.S. face in the years to come?

 Links: www.dushkin.com/online/
These sites are annotated in the World Wide Web pages.

Belfer Center for Science and International Affairs (BCSIA)
http://ksgwww.harvard.edu/csia/

U.S. Agency for International Development
http://www.info.usaid.gov

Virtual Seminar in Global Political Economy/Global Cities & Social Movements
http://csf.colorado.edu/gpe/gpe95b/resources.html

The World Bank Group
http://www.worldbank.org

A defining characteristic of the twentieth century's was the intense struggle between proponents of two economic ideologies. At the heart of the conflict was the question of what role government should play in the management of a country's economy. For some the dominant capitalist economic system appeared to be organized primarily for the benefit of a few wealthy people. From their perspective, the masses were trapped in poverty, supplying cheap labor to further enrich the privileged elite. These critics argued that the capitalist system could be changed only by gaining control of the political system and radically changing the ownership of the means of production. In striking contrast to this perspective, others argued that the best way to create wealth and eliminate poverty was through the profit motive, which encouraged entrepreneurs to create new products and businesses. An open and competitive marketplace, from this point of view, minimized government interference and was the best system for making decisions about production, wages, and the distribution of goods and services.

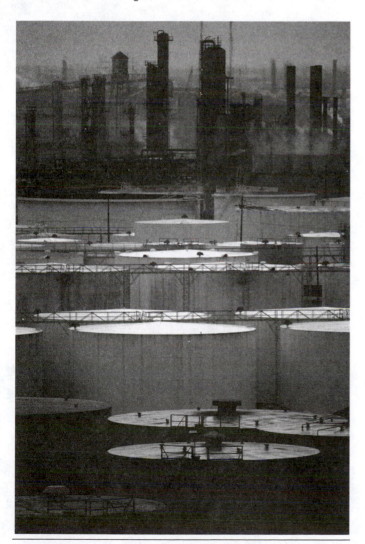

Photo Credit: Kent Knudson/PhotoLink/Getty Images

Both academic discourse and pragmatic and often violent conflict characterized the contest of capitalism versus socialism/communism. The Russian and Chinese revolutions overthrew the old social order and created radical changes in the political and economic systems in these two important countries. The political structures that were created to support new systems of agricultural and industrial production (along with the centralized planning of virtually all aspects of economic activity) eliminated most private ownership of property. These two revolutions were, in short, unparalleled experiments in social engineering.

The collapse of the Soviet Union and the dramatic reforms that have taken place in China have recast the debate about how to best structure contemporary economic systems. Some believe, that with the end of communism and the resulting participation of hundreds of millions of new consumers in the global market, an unprecedented new era has been entered. Many have noted that this process of "globalization" is being accelerated by a revolution in communication and computer technologies. Proponents of this view argue that a new global economy is emerging that will ultimately eliminate national economic systems.

Others are less optimistic about the prospects of globalization. They argue that the creation of a single economic system where there are no boundaries to impede the flow of capital and goods and services does not mean a closing of the gap between the world's rich and poor. Rather, they argue that giant corporations will have fewer legal constraints on their behavior, and this will lead to greater exploitation of workers and the accelerated destruction of the environment. Further, these critics point out that the unintended globalization of drug trafficking and other criminal behaviors is developing more rapidly than remedies can be developed.

The use of the term "political economy" for the title of this unit is recognition that economic and political systems are not separate. All economic systems have some type of marketplace where goods and services are bought and sold. Government (either national or international) regulates these transactions to some degree; that is, government sets the rules that regulate the marketplace.

One of the most important concepts in assessing the contemporary political economy is "development." For the purposes of this unit, the term "development" is defined as an improvement in the basic aspects of life: lower infant mortality rates, longer life expectancy, lower disease rates, higher rates of literacy, healthier diets, and improved sanitation. Judged by these standards, some countries are more "developed" than others. A fundamental question that a thoughtful reader must consider is whether globalization is resulting in increased development not only for a few people but also for all of those participating in the global political economy.

The unit is organized into two sections. The first is a general discussion of the concept of globalization. How is the concept defined, and what are the differing perspectives on it? For example, is the idea of a global economy merely wishful thinking by those who sit on top of the power hierarchy, self-deluded into believing that globalization is an inexorable force that will evolve in its own way, following its own rules? Or will there continue to be the traditional tensions of power politics, that is, between the powerful and those who are either ascending or descending in power?

The second subsection is a selection of case studies that focus on specific countries or economic sectors. These case studies have been selected to challenge the reader to develop his or her own conclusions about the positive and negative consequences of the globalization process. Does the contemporary global political economy result in an age-old system of winners and losers, or can everyone positively benefit from its system of wealth creation and distribution?

The Complexities and Contradictions of Globalization

Globalization, we are told, is what every business should be pursuing, and what every nation should welcome. But what, exactly, is it? James Rosenau offers a nuanced understanding of a process that is much more real, and transforming, than the language of the marketplace expresses.

JAMES N. ROSENAU

The mall at Singapore's airport has a food court with 15 food outlets, all but one of which offering menus that cater to local tastes; the lone standout, McDonald's, is also the only one crowded with customers. In New York City, experts in *feng shui*, an ancient Chinese craft aimed at harmonizing the placement of man-made structures in nature, are sought after by real estate developers in order to attract a growing influx of Asian buyers who would not be interested in purchasing buildings unless their structures were properly harmonized.

Most people confronted with these examples would probably not be surprised by them. They might even view them as commonplace features of day-to-day life late in the twentieth century, instances in which local practices have spread to new and distant sites. In the first case the spread is from West to East and in the second it is from East to West, but both share a process in which practices spread and become established in profoundly different cultures. And what immediately comes to mind when contemplating this process? The answer can be summed up in one word: globalization, a label that is presently in vogue to account for peoples, activities, norms, ideas, goods, services, and currencies that are decreasingly confined to a particular geographic space and its local and established practices.

Indeed, some might contend that "globalization" is the latest buzzword to which observers resort when things seem different and they cannot otherwise readily account for them. That is why, it is reasoned, a great variety of activities are labeled as globalization, with the result that no widely accepted formulation of the concept has evolved. Different observers use it to describe different phenomena, and often there is little overlap among the various usages. Even worse, the elusiveness of the concept of globalization is seen as underlying the use of a variety of other, similar terms—world society, interdependence, centralizing tendencies, world system, globalism, universalism, internationalization, globality—that come into play when efforts are made to grasp why public affairs today seem significantly different from those of the past.

Such reasoning is misleading. The proliferation of diverse and loose definitions of globalization as well as the readiness to use a variety of seemingly comparable labels are not so much a reflection of evasive confusion as they are an early stage in a profound ontological shift, a restless search for new ways of understanding unfamiliar phenomena. The lack of precise formulations may suggest the presence of buzzwords for the inexplicable, but a more convincing interpretation is that such words are voiced in so many different contexts because of a shared sense that the human condition is presently undergoing profound transformations in all of its aspects.

WHAT IS GLOBALIZATION?

Let us first make clear where globalization fits among the many buzzwords that indicate something new in world affairs that is moving important activities and concerns beyond the national seats of power that have long served as the foundations of economic, political, and social life. While all the buzzwords seem to cluster around the same dimension of the present human condition, useful distinctions can be drawn among them.

Most notably, if it is presumed that the prime characteristic of this dimension is change—a transformation of practices and norms—then the term "globalization" seems appropriate to denote the "something" that is changing humankind's preoccupation with territoriality and the traditional arrangements of the state system. It is a term that directly implies change, and thus differentiates the phenomenon as a process rather than as a prevailing condition or a desirable end state.

Conceived as an underlying process, in other words, globalization is not the same as globalism, which points to aspirations for a state of affairs where values are shared by or pertinent to all the world's more than 5 billion people, their environment, and their role as citizens, consumers, or producers with an interest in collective action to solve common problems. And it can also be distinguished from universalism, which refers to those values that embrace all of humanity (such as the values that science or religion draws on), at any time or place. Nor is it coterminous with complex interdependence, which signifies structures that link people and communities in various parts of the world.

Although related to these other concepts, the idea of globalization developed here is narrower in scope. It refers neither to values nor to structures, but to sequences that unfold either in the mind or in behavior, to processes that evolve as people and organizations go about their daily tasks and seek to realize their particular goals. What distinguishes globalizing processes is that they are not hindered or prevented by territorial or jurisdictional barriers. As indicated by the two examples presented at the outset, such processes can readily spread in many directions across national boundaries, and are capable of reaching into any community anywhere in the world. They consist of all those forces that impel individuals, groups, and institutions to engage in similar forms of behavior or to participate in more encompassing and coherent processes, organizations, or systems.

Contrariwise, localization derives from all those pressures that lead individuals, groups, and institutions to narrow their horizons, participate in dissimilar forms of behavior, and withdraw to less encompassing processes, organizations, or systems. In other words, any technological, psychological, social, economic, or political developments that foster the expansion of interests and practices beyond established boundaries are both sources and expressions of the processes of globalization, just as any developments in these realms that limit or reduce interests are both sources and expressions of localizing processes.

Note that the processes of globalization are conceived as only capable of being worldwide in scale. In fact, the activities of no group, government, society, or company have never been planetary in magnitude, and few cascading sequences actually encircle and encompass the entire globe. Televised events such as civil wars and famines in Africa or protests against governments in Eastern Europe may sustain a spread that is worldwide in scope, but such a scope is not viewed as a prerequisite of globalizing dynamics. As long as it has the potential of an unlimited spread that can readily transgress national jurisdictions, any interaction sequence is considered to reflect the operation of globalization.

Obviously, the differences between globalizing and localizing forces give rise to contrary conceptions of territoriality. Globalization is rendering boundaries and identity with the land less salient while localization, being driven by pressures to narrow and withdraw, is highlighting borders and intensifying the deep attachments to land that can dominate emotion and reasoning.

In short, globalization is boundary-broadening and localization is boundary-heightening. The former allows people, goods, information, norms, practices, and institutions to move about oblivious to despite boundaries. The boundary-heightening processes of localization are designed to inhibit or prevent the movement of people, goods, information, norms, practices, and institutions. Efforts along this line, however, can be only partially successful. Community and state boundaries can be heightened to a considerable extent, but they cannot be rendered impervious. Authoritarian governments try to make them so, but their policies are bound to be undermined in a shrinking world with increasingly interdependent economies and communications technologies that are not easily monitored. Thus it is hardly surprising that some of the world's most durable tensions flow from the fact that no geographic borders can be made so airtight to prevent the infiltration of ideas and goods. Stated more emphatically, some globalizing dynamics are bound, at least in the long run, to prevail.

The boundary-expanding dynamics of globalization have become highly salient precisely because recent decades have witnessed a mushrooming of the facilities, interests, and markets through which a potential for worldwide spread can be realized. Likewise, the boundary-contracting dynamics of localization have also become increasingly significant, not least because some people and cultures feel threatened by the incursions of globalization. Their jobs, their icons, their belief systems, and their communities seem at risk as the boundaries that have sealed them off from the outside world in the past no longer assure protection. And there is, of course, a basis of truth in these fears. Globalization does intrude; its processes do shift jobs elsewhere; its norms do undermine traditional mores. Responses to these threats can vary considerably. At one extreme are adaptations that accept the boundary-broadening processes and make the best of them by integrating them into local customs and practices. At the other extreme are responses intended to ward off the globalizing processes by resort to ideological purities, closed borders, and economic isolation.

THE DYNAMICS OF FRAGMEGRATION

The core of world affairs today thus consists of tensions between the dynamics of globalization and localization. Moreover, the two sets of dynamics are causally linked, almost as if every increment of globalization gives rise to an increment of localization, and vice versa. To account for these tensions I have long used the term "fragmegration," an awkward and perhaps even grating label that has the virtue of capturing the pervasive interactions between the fragmenting forces of localization and the integrative forces of globalization.1 One can readily observe the unfolding of fragmegrative dynamics in the struggle of the European Union to cope with proposals for monetary unification or in the electoral campaigns and successes of Jean-Marie Le Pen in France, Patrick Buchanan in the United States, and Pauline Hanson in Australia—to mention only three examples.

It is important to keep in mind that fragmegration is not a single dynamic. Both globalization and localization are clusters of forces that, as they interact in different ways and through different channels, contribute to more encompassing processes in the case of globalization and to less encompassing processes in the case of localization. These various dynamics, moreover, operate in all realms of human activity, from the cultural and social to the economic and political.

In the political realm, globalizing dynamics underlie any developments that facilitate the expansion of authority, policies, and interests beyond existing socially constructed territorial boundaries, whereas the politics of localization involves any trends in which the scope of authority and policies undergoes contraction and reverts to concerns, issues, groups, and institutions that are less extensive than the prevailing socially constructed territorial boundaries. In the economic realm, globalization encompasses the expansion of production, trade, and investments beyond their prior locales, while localizing dynamics are at work when the activities of producers and consumers are constricted to narrower boundaries. In the social and cultural realms, globalization operates to extend ideas, norms, and practices beyond the settings in which they originated, while localization highlights or compresses the original settings and thereby inhibits the inroad of new ideas, norms, and practices.

It must be stressed that the dynamics unfolding in all these realms are long-term processes. They involve fundamental human needs and thus span all of human history. Globalizing dynamics derive from peoples' need to enlarge the scope of their self-created orders so as to increase the goods, services, and ideas available for their well-being. The agricultural revolution, followed by the industrial and postindustrial transformations, are among the major sources that have sustained globalization. Yet even as these forces have been operating, so have contrary tendencies toward contraction been continuously at work. Localizing dynamics derive from people's need for the psychic comforts of close-at-hand, reliable support—for the family and neighborhood, for local cultural practices, for a sense of "us" that is distinguished from "them." Put differently, globalizing dynamics have long fostered large-scale order, whereas localizing dynamics have long created pressure for small-scale order. Fragmegration, in short, has always been an integral part of the human condition.

GLOBALIZATION'S EVENTUAL PREDOMINANCE

Notwithstanding the complexities inherent in the emergent structures of world affairs, observers have not hesitated to anticipate what lies beyond fragmegration as global history unfolds. All agree that while the contest between globalizing and localizing dynamics is bound to be marked by fluctuating surges in both directions, the underlying tendency is for the former to prevail over the latter. Eventually, that is, the dynamics of globalization are expected to serve as the bases around which the course of events is organized.

Consensus along these lines breaks down, however, over whether the predominance of globalization is likely to have desirable or noxious consequences. Those who welcome globalizing processes stress the power of economic variables. In this view the globalization of national economies through the diffusion of technology and consumer products, the rapid transfer of financial resources, and the efforts of transnational companies to extend their market shares is seen as so forceful and durable as to withstand and eventually surmount any and all pressures toward fragmentation. This line acknowledges that the diffusion that sustains the processes of globalization is a centuries-old dynamic, but the difference is that the present era has achieved a level of economic development in which it is possible for innovations occurring in any sector of any country's economy to be instantaneously transferred to and adapted in any other country or sector. As a consequence,

> when this process of diffusion collides with cultural or political protectionism, it is culture and protectionism that wind up in the shop for repairs. Innovation accelerates. Productivity increases. Standards of living improve. There are setbacks, of course. The newspaper headlines are full of them. But we believe that the time required to override these setbacks has shortened dramatically in the developed world. Indeed, recent experience suggests that, in most cases, economic factors prevail in less than a generation....
>
> Thus understood, globalization—the spread of economic innovations around the world and the political and cultural adjustments that accompany this diffusion—cannot be stopped.... As history teaches, the political organizations and ideologies that yield superior economic performance survive, flourish, and replace those that are less productive.[2]

While it is surely the case that robust economic incentives sustain and quicken the processes of globalization, this line of theorizing nevertheless suffers from not allowing for its own negation. The theory offers no alternative interpretations as to how the interaction of economic, political, and social dynamics will play out. One cannot demonstrate the falsity—if falsity it is—of the theory because any contrary evidence is seen merely as "setbacks," as expectable but temporary deviations from the predicted course. The day may come, of course, when event so perfectly conform to the predicted patterns of globalization that one is inclined to conclude that the theory has been affirmed. But in the absence of alternative scenarios, the theory offers little guidance as to how to interpret intervening events, especially those that highlight the tendencies toward fragmentation. Viewed in this way, it is less a theory and more an article of faith to which one can cling.

Other observers are much less sanguine about the future development of fragmegration. They highlight a litany of noxious consequences that they see as following from the eventual predominance of globalization: "its economism; its economic reductionism; its technological determinism; its political cynicism, defeatism, and immobilism; its de-socialization of the subject and resocialization of risk; its teleological subtext of inexorable global 'logic' driven exclusively by capital accumulation and the market; and its ritual exclusion of factors, causes, or goals other than capital accumulation and the market from the priority of values to be pursued by social action."[3]

Still another approach, allowing for either desirable or noxious outcomes, has been developed by Michael Zurn. He identifies a mismatch between the rapid extension of boundary-crossing activities and the scope of effective governance. Consequently, states are undergoing what is labeled "uneven denationalization," a primary process in which "the rise of international governance is still remarkable, but not accompanied by mechanisms for… democratic control; people, in addition, become alienated from the remote political process…. The democratic state in the Western world is confronted with a situation in which it is undermined by the process of globalization and overarched by the rise of international institutions."[4]

There is no inherent contradiction between localizing and globalizing tendencies.

While readily acknowledging the difficulties of anticipating where the process of uneven denationalization is driving the world, Zurn is able to derive two scenarios that may unfold: "Whereas the pessimistic scenario points to instances of fragmentation and emphasizes the disruption caused by the transition, the optimistic scenario predicts, at least in the long run, the triumph of centralization." The latter scenario rests on the presumption that the increased interdependence of societies will propel them to develop ever more effective democratic controls over the very complex arrangements on which international institutions must be founded.

UNEVEN FRAGMEGRATION

My own approach to theorizing about the fragmegrative process builds on these other perspectives and a key presumption of my own—that there is no inherent contradiction between localizing and globalizing tendencies—to develop an overall hypothesis that anticipates fragmegrative outcomes and that allows for its own negation: the more pervasive globalizing tendencies become, the less resistant localizing reactions will be to further globalization. In other words, globalization and localization will coexist, but the former will continue to set the context for the latter. Since the degree of coexistence will vary from situation to situation (depending on the salience of the global economy and the extent to which ethnic and other noneconomic factors actively contribute to localization), I refer, borrowing from Zurn, to the processes depicted by the hypothesis as uneven fragmegration. The hypothesis allows for continuing pockets of antagonism between globalizing and localizing tendencies even as increasingly (but unevenly) the two accommodate each other. It does not deny the pessimistic scenario wherein fragmentation disrupts globalizing tendencies; rather it treats fragmentation as more and more confined to particular situations that may eventually be led by the opportunities and requirements of greater interdependence to conform to globalization.

For globalizing and localizing tendencies to accommodate each other, individuals have to come to appreciate that they can achieve psychic comfort in collectivities through multiple memberships and multiple loyalties, that they can advance both local and global values without either detracting from the other. The hypothesis of uneven fragmegration anticipates a growing appreciation along these lines because the contrary premise, that psychic comfort can only be realized by having a highest loyalty, is becoming increasingly antiquated. To be sure, people have long been accustomed to presuming that, in order to derive the psychic comfort they need through collective identities, they had to have a hierarchy of loyalties and that, consequently, they had to have a highest loyalty that could only be attached to a single collectivity. Such reasoning, however, is a legacy of the state system, of centuries of crises that made people feel they had to place nation-state loyalties above all others. It is a logic that long served to reinforce the predominance of the state as the "natural" unit of political organization and that probably reached new heights during the intense years of the cold war.

But if it is the case, as the foregoing analysis stresses, that conceptions of territoriality are in flux and that the failure of states to solve pressing problems has led to a decline in their capabilities and a loss of legitimacy, it follows that the notion that people must have a "highest loyalty" will also decline and give way to the development of multiple loyalties and an understanding that local, national, and transnational affiliations need not be mutually exclusive. For the reality is that human affairs are organized at all these levels for good reasons; people have needs that can only be filled by close-at-hand organizations and other needs that are best served by distant entities at the national or transnational level.

In addition, not only is an appreciation of the reality that allows for multiple loyalties and memberships likely to grow as the effectiveness of states and the salience of national loyalties diminish, but it also seems likely to widen as the benefits of the global economy expand and people become increasingly aware of the extent to which their well-being is dependent on events and trends elsewhere in the world. At the same time, the distant economic processes serving their needs are impersonal and hardly capable of advancing the need to share with others in a collective affiliation. This need was long served by the nation-state, but with fragmegrative dynamics having undermined the national level as a source of psychic comfort and with transnational entities seeming too distant to provide the psychic benefits of affiliation, the satisfactions to be gained through more close-at-hand affiliations are likely to seem ever more attractive.

THE STAKES

It seems clear that fragmegration has become an enduring feature of global life; it is also evident that globalization is not merely a buzzword, that it encompasses pervasive complexities and contradictions that have the potential both to enlarge and to degrade our humanity. In order to ensure that the enlargement is more prevalent than the degradation, it is important that people and their institutions become accustomed to the multiple dimensions and nuances as our world undergoes profound and enduring transformations. To deny the complexities and contradictions in order to cling to a singular conception of what globalization involves is to risk the many dangers that accompany oversimplification.

NOTES

1. For an extensive discussion of the dynamics of fragmegration, see James N. Rosenau, *Along the Domestic-Foreign Frontier: Exploring Governance in a Turbulent World* (Cambridge: Cambridge University Press, 1997), ch. 6.
2. William W. Lewis and Marvin Harris, "Why Globalization Must Prevail," *The McKinsey Quarterly*, no. 2 (1992), p. 115.
3. Barry K. Gills, "Editorial: 'Globalization' and the 'Politics of Resistance,'" *New Political Economy*, vol. 2 (March 1997), p. 12.
4. Michael Zurn, "What Has Changed in Europe? The Challenge of Globalization and Individualization," paper presented at a meeting on What Has Changed? Competing Perspectives on World Order (Copenhagen, May 14–16, 1993), p. 40.

JAMES N. ROSENAU *is University Professor of International Affairs at George Washington University. His latest book is* Along the Domestic-Foreign Frontier: Exploring Governance in a Turbulent World *(Cambridge: Cambridge University Press, 1997). This article draws on the author's "New Dimensions of Security: The Interaction of Globalizing and Localizing Dynamics,"* Security Dialogue, *September 1994, and "The Dynamics of Globalization: Toward an Operational Formulation,"* Security Dialogue, *September 1996).*

Reprinted with permission from *Current History* magazine, November 1997, pp. 360–364. © 1997 by Current History, Inc.

THE
Five Wars
OF GLOBALIZATION

The illegal trade in drugs, arms, intellectual property, people, and money is booming. Like the war on terrorism, the fight to control these illicit markets pits governments against agile, stateless, and resourceful networks empowered by globalization. Governments will continue to lose these wars until they adopt new strategies to deal with a larger, unprecedented struggle that now shapes the world as much as confrontations between nation–states once did.

By Moisés Naím

The persistence of al Qaeda underscores how hard it is for governments to stamp out stateless, decentralized networks that move freely, quickly, and stealthily across national borders to engage in terror. The intense media coverage devoted to the war on terrorism, however, obscures five other similar global wars that pit governments against agile, well-financed networks of highly dedicated individuals. These are the fights against the illegal international trade in drugs, arms, intellectual property, people, and money. Religious zeal or political goals drive terrorists, but the promise of enormous financial gain motivates those who battle governments in these five wars. Tragically, profit is no less a motivator for murder, mayhem, and global insecurity than religious fanaticism.

In one form or another, governments have been fighting these five wars for centuries. And losing them. Indeed, thanks to the changes spurred by globalization over the last decade, their losing streak has become even more pronounced. To be sure, nation-states have benefited from the information revolution, stronger political and economic linkages, and the shrinking importance of geographic distance. Unfortunately, criminal networks have benefited even more. Never fettered by the niceties of sovereignty, they are now increasingly free of geographic constraints. Moreover, globalization has not only expanded illegal markets and boosted the size and the resources of criminal networks, it has also imposed more burdens on governments: Tighter public budgets, decentralization, privatization, deregulation, and a more open environment for international trade and investment all make the task of fighting global criminals more difficult. Governments are made up of cumbersome bureaucracies that generally cooperate with difficulty, but drug traffickers, arms dealers, alien smugglers, counterfeiters, and money launderers have refined networking to a high science, entering into complex and improbable strategic alliances that span cultures and continents.

Defeating these foes may prove impossible. But the first steps to reversing their recent dramatic gains must be to recognize the fundamental similarities among the five wars and to treat these conflicts not as law enforcement problems but as a new global trend that shapes the world as much as confrontations between nation-states did in the past. Customs officials, police officers, lawyers, and judges alone will never win these wars. Governments must recruit and deploy more spies, soldiers, diplomats, and economists who understand how to use incentives and regulations to steer markets away from bad social outcomes. But changing the skill set of government combatants alone will not end these wars. Their doctrines and institutions also need a major overhaul.

THE FIVE WARS

Pick up any newspaper anywhere in the world, any day, and you will find news about illegal migrants, drug busts, smuggled weapons, laundered money, or counterfeit goods. The global nature of these five wars was unimaginable just a decade ago. The resources—financial, human, institutional, technological—deployed by the combatants have reached unfathomable orders of magnitude. So have the numbers of victims. The tactics and tricks of both sides boggle the mind. Yet if you cut through the fog of daily headlines and orchestrated photo ops, one inescapable truth emerges: The world's governments are fighting a qualitatively new phenomenon with obsolete tools, inadequate laws, inefficient bureaucratic arrangements, and ineffective strategies. Not surprisingly, the evidence shows that governments are losing.

Drugs

The best known of the five wars is, of course, the war on drugs. In 1999, the United Nations' "Human Development Report" calculated the annual trade in illicit drugs at $400 billion, roughly the size of the Spanish economy and about 8 percent of world trade. Many countries are reporting an increase in drug use. Feeding this habit is a global supply chain that uses everything from passenger jets that can carry shipments of cocaine worth $500 million in a single trip to custom-built submarines that ply the waters between Colombia and Puerto Rico. To foil eavesdroppers, drug smugglers use "cloned" cell phones and broadband radio receivers while also relying on complex financial structures that blend legitimate and illegitimate enterprises with elaborate fronts and structures of cross-ownership.

The United States spends between $35 billion and $40 billion each year on the war on drugs; most of this money is spent on interdiction and intelligence. But the creativity and boldness of drug cartels has routinely outstripped steady increases in government resources. Responding to tighter security at the U.S.-Mexican border, drug smugglers built a tunnel to move tons of drugs and billions of dollars in cash until authorities discovered it in March 2002. Over the last decade, the success of the Bolivian and Peruvian governments in eradicating coca plantations has shifted production to Colombia. Now, the U.S.-supported Plan Colombia is displacing coca production and processing labs back to other Andean countries. Despite the heroic efforts of these Andean countries and the massive financial and technical support of the United States, the total acreage of coca plantations in Peru, Colombia, and Bolivia has increased in the last decade from 206,200 hectares in 1991 to 210,939 in 2001. Between 1990 and 2000, according to economist Jeff

DeSimone, the median price of a gram of cocaine in the United States fell from $152 to $112.

Even when top leaders of drug cartels are captured or killed, former rivals take their place. Authorities have acknowledged, for example, that the recent arrest of Benjamin Arellano Felix, accused of running Mexico's most ruthless drug cartel, has done little to stop the flow of drugs to the United States. As Arellano said in a recent interview from jail, "They talk about a war against the Arellano brothers. They haven't won. I'm here, and nothing has changed."

Arms Trafficking

Drugs and arms often go together. In 1999, the Peruvian military parachuted 10,000 AK-47s to the Revolutionary Armed Forces of Colombia, a guerrilla group closely allied to drug growers and traffickers. The group purchased the weapons in Jordan. Most of the roughly 80 million AK-47s in circulation today are in the wrong hands. According to the United Nations, only 18 million (or about 3 percent) of the 550 million small arms and light weapons in circulation today are used by government, military, or police forces. Illicit trade accounts for almost 20 percent of the total small arms trade and generates more than $1 billion a year. Small arms helped fuel 46 of the 49 largest conflicts of the last decade and in 2001 were estimated to be responsible for 1,000 deaths a day; more than 80 percent of those victims were women and children.

Small arms are just a small part of the problem. The illegal market for munitions encompasses top-of-the-line tanks, radar systems that detect Stealth aircraft, and the makings of the deadliest weapons of mass destruction. The International Atomic Energy Agency has confirmed more than a dozen cases of smuggled nuclear-weapons-usable material, and hundreds more cases have been reported and investigated over the last decade. The actual supply of stolen nuclear-, biological-, or chemical-weapons materials and technology may still be small. But the potential demand is strong and growing from both would-be nuclear powers and terrorists. Constrained supply and increasing demand cause prices to rise and create enormous incentives for illegal activities. More than one fifth of the 120,000 workers in Russia's former "nuclear cities"—where more than half of all employees earn less than $50 a month—say they would be willing to work in the military complex of another country.

Governments have been largely ineffective in curbing either supply or demand. In recent years, two countries, Pakistan and India, joined the declared nuclear power club. A U.N. arms embargo failed to prevent the reported sale to Iraq of jet fighter engine parts from Yugoslavia and the Kolchuga anti-Stealth radar system from Ukraine. Multilateral efforts to curb the manufacture and distribution of weapons are faltering, not least because some powers are unwilling to accept curbs on their own activities. In 2001, for example, the United States blocked a legally binding global treaty to control small arms in part because it worried about restrictions on its own citizens' rights to own guns. In the absence of effective international legislation and enforcement, the laws of economics dictate the sale of more weapons at cheaper prices: In 1986, an AK-47 in Kolowa, Kenya, cost 15 cows. Today, it costs just four.

Other Fronts

Drugs, arms, intellectual property, people, and money are not the only commodities traded illegally for huge profits by international networks. They also trade in human organs, endangered species, stolen art, and toxic waste. The illegal global trades in all these goods share several fundamental characteristics: Technological innovations and political changes open new markets, globalization is increasing both the geographical reach and the profit opportunities for criminal networks, and governments are on the losing end of the fight to stop them. Some examples:

Human organs: Corneas, kidneys, and livers are the most commonly traded human parts in a market that has boomed thanks to technology, which has improved preservation techniques and made transplants less risky. In the United States, 70,000 patients are on the waiting list for major organ transplants while only 20,000 of them succeed in getting the organ they need. Unscrupulous "organ brokers" partly meet this demand by providing, for a fee, organs and transplant services.

Some of the donors, especially of kidneys, are desperately poor. In India, an estimated 2,000 people a year sell their organs. Many organs, however, come from nonconsenting donors forced to undergo operations or from cadavers in police morgues. For example, medical centers in Germany and Austria were recently found to have used human heart valves taken without consent from the cadavers of poor South Africans.

Endangered species: From sturgeon for caviar in gourmet delicatessens to tigers or elephants for private zoos, the trade in endangered animals and plants is worth billions of dollars and includes hundreds of millions of plant and animal types. This trade ranges from live animals and plants to all kinds of wildlife products derived from them, including food products, exotic leather goods, wooden musical instruments, timber, tourist curiosities, and medicines.

Stolen art: Paintings and sculptures taken from museums, galleries, and private homes, from Holocaust victims, or from "cultural artifacts" poached from archeological digs and other ancient ruins are also illegally traded internationally in a market worth an estimated $2 billion to $6 billion each year. The growing use of art-based transactions in money laundering has spurred demand over the last decade. The supply has boomed because the Soviet Union's collapse flooded the world's market with art that had been under state control. The Czech Republic, Poland, and Russia are three of the five countries most affected by art crime worldwide.

Toxic waste: Innovations in maritime transport, tighter environmental regulations in industrialized countries coupled with increased integration of poor countries to the global economy and better telecommunications have created a market where waste is traded internationally. Greenpeace estimates that during the 20 years prior to 1989, just 3.6 million tons of hazardous waste were exported; in the five years after 1989, the trade soared to about 6.7 billion tons. The environmental organization also reckons that 86 to 90 percent of all hazardous waste shipments destined for developing countries—purportedly for recycling, reuse, recovery, or humanitarian uses—are toxic waste.

—*M.N.*

Intellectual Property

In 2001, two days after recording the voice track of a movie in Hollywood, actor Dennis Hopper was in Shanghai where a street vendor sold him an excellent pirated copy of the movie with his voice already on it. "I don't know how they got my voice into the country before I got here," he wondered. Hopper's experience is one tiny slice of an illicit trade that cost the United States an estimated $9.4 billion in 2001. The piracy rate of business software in Japan and France is 40 percent, in Greece and South Korea it is about 60 percent, and in Germany and Britain it hovers around 30 percent. Forty percent of Procter & Gamble shampoos and 60 percent of Honda motorbikes sold in China in 2001 were pirated. Up to 50 percent of medical drugs in Nigeria and Thailand are bootleg copies. This problem is not limited to consumer products: Italian makers of industrial valves worry that their $2 billion a year export market is eroded by counterfeit Chinese valves sold in world markets at prices that are 40 percent cheaper.

The drivers of this bootlegging boom are complex. Technology is obviously boosting both the demand and the supply of illegally copied products. Users of Napster, the now defunct Internet company that allowed anyone, anywhere to download and reproduce copyrighted music for free, grew from zero to 20 million in just one year. Some 500,000 film files are traded daily through file-sharing services such as Kazaa and Morpheus; and in late 2002, some 900 million music files could be downloaded for free on the Internet—that is, almost two and a half times more files than those available when Napster reached its peak in February 2001.

Global marketing and branding are also playing a part, as more people are attracted to products bearing a well-known brand like Prada or Cartier. And thanks to the rapid growth and integration into the global economy of countries, such as China, with weak central governments and ineffective laws, producing and exporting near perfect knockoffs are both less expensive and less risky. In the words of the CEO of one of the best known Swiss watchmakers: "We now compete with a product manufactured by Chinese prisoners. The business is run by the Chinese military, their families and friends, using roughly the same machines we have, which they purchased at the same industrial fairs we go to. The way we have rationalized this problem is by assuming that their customers and ours are different. The person that buys a pirated copy of one of our $5,000 watches for less than $100 is not a client we are losing. Perhaps it is a future client that some day will want to own the real thing instead of a fake. We may be wrong and we do spend money to fight the piracy of our products. But given that our efforts do not seem to protect us much, we close our eyes and hope for the better." This posture stands in contrast to that of companies that sell cheaper products such as garments, music, or videos, whose revenues are directly affected by piracy.

Governments have attempted to protect intellectual property rights through various means, most notably the World Trade Organization's Agreement on Trade-Related Aspects of Intellectual Property Rights (TRIPS). Several other organizations such as the World Intellectual Property Organization, the World Customs Union, and Interpol are also involved. Yet the large and growing volume of this trade, or a simple stroll in the streets of Manhattan or Madrid, show that governments are far from winning this fight.

Alien Smuggling

The man or woman who sells a bogus Hermes scarf or a Rolex watch in the streets of Milan is likely to be an illegal alien. Just as likely, he or she was transported across several continents by a trafficking network allied with another network that specializes in the illegal copying, manufacturing, and distributing of high-end, brand-name products.

Alien smuggling is a $7 billion a year enterprise and according to the United Nations is the fastest growing business of organized crime. Roughly 500,000 people enter the United States illegally each year—about the same number as illegally enter the European Union, and part of the approximately 150 million who live outside their countries of origin. Many of these backdoor travelers are voluntary migrants who pay smugglers up to $35,000, the top-dollar fee for passage from China to New York. Others, instead, are trafficked—that is, bought and sold internationally—as commodities. The U.S. Congressional Research Service reckons that each year between 1 million and 2 million people are trafficked across borders, the majority of whom are women and children. A woman can be "bought" in Timisoara, Romania, for between $50 and $200 and "resold" in Western Europe for 10 times that price. The United Nations Children's Fund estimates that cross-border smugglers in Central and Western Africa enslave 200,000 children a year. Traffickers initially tempt victims with job offers or, in the case of children, with offers of adoption in wealthier countries, and then keep the victims in subservience through physical violence, debt bondage, passport confiscation, and threats of arrest, deportation, or violence against their families back home.

Governments everywhere are enacting tougher immigration laws and devoting more time, money, and technology to fight the flow of illegal aliens. But the plight of the United Kingdom's government illustrates how tough that fight is. The British government throws money at the problem, plans to use the Royal Navy and Royal Air Force to intercept illegal immigrants, and imposes large fines on truck drivers who (generally unwittingly) transport stowaways. Still, 42,000 of the 50,000 refugees who have passed through the Sangatte camp (a main entry point for illegal immigration to the United Kingdom) over the last three years have made it to Britain. At current rates, it will take 43 years for Britain to clear its asylum backlog. And that country is an island. Continental nations such as Spain, Italy, or the United States face an even greater challenge as immigration pressures overwhelm their ability to control the inflow of illegal aliens.

Money Laundering

The Cayman Islands has a population of 36,000. It also has more than 2,200 mutual funds, 500 insurance companies, 60,000 businesses, and 600 banks and trust companies with almost $800 billion in assets. Not surprisingly, it figures prominently in any discussion of money laundering. So does the United States, several of whose major banks have been caught up in investigations of money laundering, tax evasion, and fraud. Few, if any, countries can claim to be free of the practice of helping individuals and companies hide funds from governments, creditors, business partners, or even family members, including the proceeds of tax evasion, gambling, and other crimes. Estimates of the volume of global money laundering range between 2 and 5 percent of the world's annual gross national product, or between $800 billion and $2 trillion.

Smuggling money, gold coins, and other valuables is an ancient trade. Yet in the last two decades, new political and economic trends coincided with technological changes to make this ancient trade easier, cheaper, and less risky. Political changes led to the deregulation of financial markets that now facilitate cross-border money transfers, and technological changes made distance less of a factor and money less "physical." Suitcases full of banknotes are still a key tool for money launderers, but computers, the Internet, and complex financial schemes that combine legal and illegal practices and institutions are more common. The sophistication of technology, the complex web of financial institutions that crisscross the globe, and the ease with which "dirty" funds can be electronically morphed into legitimate assets make the regulation of international flows of money a daunting task. In Russia, for example, it is estimated that by the mid-1990s organized crime groups had set up 700 legal and financial institutions to launder their money.

Faced with this growing tide, governments have stepped up their efforts to clamp down on rogue international banking, tax havens, and money laundering. The imminent, large-scale introduction of e-money—cards with microchips that can store large amounts of money and thus can be easily transported outside regular channels or simply exchanged among individuals—will only magnify this challenge.

WHY GOVERNMENTS CAN'T WIN

The fundamental changes that have given the five wars new intensity over the last decade are likely to persist. Technology will continue to spread widely; criminal networks will be able to exploit these technologies more quickly than governments that must cope with tight budgets, bureaucracies, media scrutiny, and electorates. International trade will continue to grow, providing more cover for the expansion of illicit trade. International migration will likewise grow, with much the same effect, offering ethnically based gangs an ever growing supply of recruits and victims. The spread of democracy may also help criminal cartels, which can manipulate weak public institutions by corrupting police officers or tempting politicians with offers of cash for their increasingly expensive election campaigns. And ironically, even the spread of international law—with its growing web of embargoes, sanctions, and conventions—will offer criminals new opportunities for providing forbidden goods to those on the wrong side of the international community.

Even the spread of international law will offer criminals new opportunities for providing forbidden goods to those on the wrong side of the international community.

These changes may affect each of the five wars in different ways, but these conflicts will continue to share four common characteristics:

They are not bound by geography.

Some forms of crime have always had an international component: The Mafia was born in Sicily and exported to the United States, and smuggling has always been by definition international. But the five wars are truly global. Where is the theater or front line of the war on drugs? Is it Colombia or Miami? Myanmar (Burma) or Milan? Where are the battles against money launderers being fought? In Nauru or in London? Is China the main theater in the war against the infringement of intellectual property, or are the trenches of that war on the Internet?

They defy traditional notions of sovereignty.

Al Qaeda's members have passports and nationalities—and often more than one—but they are truly stateless. Their allegiance is to their cause, not to any nation. The same is also true of the criminal networks engaged in the five wars. The same, however, is patently *not* true of government employees—police officers, customs agents, and judges—who fight them. This asymmetry is a crippling disadvantage for governments waging these wars. Highly paid, hypermotivated, and resource-rich combatants on one side of the wars (the criminal gangs) can seek refuge in and take advantage of national borders, but combatants of the other side (the governments) have fewer resources and are hampered by traditional notions of sovereignty. A former senior CIA official reported that international criminal gangs are able to move people, money, and weapons globally faster than he can move resources inside his own agency, let alone worldwide. Coordination and information sharing among government agencies in different countries has certainly improved, especially after September 11. Yet these tactics fall short of what is needed to combat agile organizations that can exploit every nook and cranny of an evolving but imperfect body of international law and multilateral treaties.

They pit governments against market forces.

In each of the five wars, one or more government bureaucracies fight to contain the disparate, uncoordinated actions of thousands of independent, stateless organizations. These groups are motivated by large profits obtained by exploiting international price differentials, an unsatisfied demand, or the cost advantages produced by theft. Hourly wages for a Chinese cook are far higher in Manhattan than in Fujian. A gram of cocaine in Kansas City is 17,000 percent more expensive than in Bogotá. Fake Italian valves are 40 percent cheaper because counterfeiters don't have to cover the costs of developing the product. A well-funded guerrilla group will pay anything to get the weapons it needs. In each of these five wars, the incentives to successfully overcome government-imposed limits to trade are simply enormous.

They pit bureaucracies against networks.

The same network that smuggles East European women to Berlin may be involved in distributing opium there. The proceeds of the latter fund the purchase of counterfeit Bulgari watches made in China and often sold on the streets of Manhattan by illegal African immigrants. Colombian drug cartels make deals with Ukrainian arms traffickers, while Wall Street brokers controlled by the U.S.-based Mafia have been known to front for Russian money launderers. These highly decentralized groups and individuals are bound by strong ties of loyalty and common purpose and organized around semiautonomous clusters or "nodes" capable of operating swiftly and flexibly. John Arquilla and David Ronfeldt, two of the best known experts on these types of organizations, observe that networks often lack central leadership, command, or headquarters, thus "no precise heart or head that can be targeted. The network as a whole (but not necessarily each node) has little to no hierarchy; there may be multiple leaders.... Thus the [organization's] design may sometimes appear acephalous (headless), and at other times polycephalous (Hydra-headed)." Typically, governments respond to these challenges by forming interagency task forces or creating new bureaucracies. Consider the creation of the new Department of Homeland Security in the United States, which encompasses 22 former federal agencies and their 170,000 employees and is responsible for, among other things, fighting the war on drugs.

RETHINKING THE PROBLEM

Governments may never be able to completely eradicate the kind of international trade involved in the five wars. But they can and should do better. There are at least four areas where efforts can yield better ideas on how to tackle the problems posed by these wars:

Develop more flexible notions of sovereignty.

Governments need to recognize that restricting the scope of multilateral action for the sake of protecting their sovereignty is often a moot point. Their sovereignty is compromised daily, not by nation-states but by stateless networks that break laws and cross borders in pursuit of trade. In May 1999, for example, the Venezuelan government denied U.S. planes authorization to fly over Venezuelan territory to monitor air routes commonly used by narcotraffickers. Venezuelan authorities placed more importance on the symbolic value of asserting sovereignty over air space than on the fact that drug traffickers' planes regularly violate Venezuelan territory. Without new forms of codifying and "managing" sovereignty, governments will continue to face a large disadvantage while fighting the five wars.

Strengthen existing multilateral institutions.

The global nature of these wars means no government, regardless of its economic, political, or military power, will make much progress acting alone. If this seems obvious, then why does Interpol, the multilateral agency in charge of fighting international crime, have a staff of 384, only 112 of whom are police officers, and an annual budget of $28 million, less than the price of some boats or planes used by drug traffickers? Similarly, Europol, Europe's Interpol equivalent, has a staff of 240 and a budget of $51 million.

One reason Interpol is poorly funded and staffed is because its 181 member governments don't trust each other. Many assume, and perhaps rightly so, that the criminal networks they are fighting have penetrated the police departments of other countries and that sharing information with such compromised officials would not be prudent. Others fear today's allies will become tomorrow's enemies. Still others face legal impediments to sharing intelligence with fellow nation-states or have intelligence services and law enforcement agencies with organizational cultures that make effective collaboration almost impossible. Progress will only be made if the world's governments unite behind stronger, more effective multilateral organizations.

Devise new mechanisms and institutions.

These five wars stretch and even render obsolete many of the existing institutions, legal frameworks, military doctrines, weapons systems, and law enforcement techniques on which governments have relied for years. Analysts need to rethink the concept of war "fronts" defined by geography and the definition of "combatants" according to the Geneva Convention. The functions of intelligence agents, soldiers, police officers, customs agents, or immigration officers need rethinking and adap-

tation to the new realities. Policymakers also need to reconsider the notion that ownership is essentially a physical reality and not a "virtual" one or that only sovereign nations can issue money when thinking about ways to fight the five wars.

Move from repression to regulation.

Beating market forces is next to impossible. In some cases, this reality may force governments to move from repressing the market to regulating it. In others, creating market incentives may be better than using bureaucracies to curb the excesses of these markets. Technology can often accomplish more than government policies can. For example, powerful encryption techniques can better protect software or CDs from being copied in Ukraine than would making the country enforce patents and copyrights and trademarks.

In all of the five wars, government agencies fight against networks motivated by the enormous profit opportunities created by other government agencies. In all cases, these profits can be traced to some form of government intervention that creates a major imbalance between demand and supply and makes prices and profit margins skyrocket. In some cases, these government interventions are often justified and it would be imprudent to eliminate them—governments can't simply walk away from the fight against trafficking in heroin, human beings, or weapons of mass destruction. But society can better deal with other segments of these kinds of illegal trade through regulation, not prohibition. Policymakers must focus on opportunities where market regulation can ameliorate problems that have defied approaches based on prohibition and armed interdiction of international trade.

Ultimately, governments, politicians, and voters need to realize that the way in which the world is conducting these five wars is doomed to fail—not for lack of effort, resources, or political will but because the collective thinking that guides government strategies in the five wars is rooted in wrong ideas, false assumptions, and obsolete institutions. Recognizing that governments have no chance of winning unless they change the ways they wage these wars is an indispensable first step in the search for solutions.

Moisés Naím is editor of FOREIGN POLICY *magazine.*

From *Foreign Policy*, January/February 2003, pp. 28–37. © 2002 by the Carnegie Endowment for International Peace. Reprinted by permission.

Will Globalization Go Bankrupt?

Global integration is driven not by politics or the Internet or the World Trade Organization or even—believe it or not—McDonald's. No, throughout history, globalization has been driven primarily by monetary expansions. Credit booms spark periods of economic integration, while credit contractions quickly squelch them. Is today's world on the verge of another globalization bust?

By Michael Pettis

Only the young generation which has had a college education is capable of comprehending the exigencies of the times," wrote Alphonse, a third-generation Rothschild, in a letter to a family member in 1865. At the time the world was in the midst of a technological boom that seemed to be changing the globe beyond recognition, and certainly beyond the ability of his elders to understand. As part of that boom, capital flowed into remote corners of the earth, dragging isolated societies into modernity. Progress seemed unstoppable.

Eight years later, however markets around the world collapsed. Suddenly, investors turned away from foreign adventures and new technologies. In the depression that ensued, many of the changes eagerly embraced by the educated young—free markets, deregulated banks, immigration—seemed too painful to continue. The process of globalization, it seems, was neither inevitable nor irreversible.

What today we call economic globalization—a combination of rapid technological progress, large-scale capital flows, and burgeoning international trade—has happened many times before in the last 200 years. During each of these periods (including our own), engineers and entrepreneurs became folk heroes and made vast fortunes while transforming the world around them. They exploited scientific advances, applied a suc-

cession of innovations to older discoveries, and spread the commercial application of these technologies throughout the developed world [see box]. Communications and transportation were usually among the most affected areas, with each technological surge causing the globe to "shrink" further.

But in spite of the enthusiasm for science that accompanied each wave of globalization, as a historical rule it was primarily commerce and finance that drove globalization, not science or technology, and certainly not politics or culture. It is no accident that each of the major periods of technological progress coincided with an era of financial market expansion and vast growth in international commerce. Specifically, a sudden expansion of financial liquidity in the world's leading banking centers—whether an increase in British gold reserves in the 1820s or the massive transformation in the 1980s of illiquid mortgage loans into very liquid mortgage securities, or some other structural change in the financial markets—has been the catalyst behind every period of globalization.

If liquidity expansions historically have pushed global integration forward, subsequent liquidity contractions have brought globalization to an unexpected halt. Easy money had allowed investors to earn fortunes for their willingness to take risks, and the wealth generated by rising asset values and new investments made the liberal ideology behind the rapid market expansion seem unassailable. When conditions changed, however, the outflow of money from the financial centers was reversed. Investors rushed to pull their money out of risky ventures and into safer assets. Banks tightened up their lending requirements and refused to make new loans. Asset values collapsed. The costs of globalization, in the form of social disruption, rising income inequality, and domination by foreign elites, became unacceptable. The political and intellectual underpinnings of globalization, which had once seemed so secure, were exposed as fragile, and the popular counterattack against the logic of globalization grew irresistible.

THE BIG BANG

The process through which monetary expansions lead to economic globalization has remained consistent over the last two centuries. Typically, every few decades, a large shift in income, money supply, saving patterns, or the structure of financial markets results in a major liquidity expansion in the rich-country financial centers. The initial expansion can take a variety of forms. In England, for example, the development of joint-stock banking (limited liability corporations that issued currency) in the 1820s and 1830s—and later during the 1860s and 1870s—produced a rapid expansion of money, deposits, and bank credit, which quickly spilled over into speculative investing and international lending. Other monetary expansions were sparked by large increases in U.S. gold reserves in the early 1920s, or by major capital recyclings, such as the massive French indemnity payment after the Franco-Prussian War of 1870, the petrodollar recycling of the 1970s, or the recycling of Japan's huge trade surplus in the 1980s and 1990s. Monetary expansions also can result from the conversion of assets into more liquid instruments, such as with the explosion in U.S. speculative real-estate lending in the 1830s or the creation of the mortgage securities market in the 1980s.

The expansion initially causes local stock markets to boom and real interest rates to drop. Investors, hungry for high yields, pour money into new, nontraditional investments, including ventures aimed at exploiting emerging technologies. Financing becomes available for risky new projects such as railways, telegraph cables, textile looms, fiber optics, or personal computers, and the strong business climate that usually accompanies the liquidity expansion quickly makes these investments profitable. In turn, these new technologies enhance productivity and slash transportation costs, thus speeding up economic growth and boosting business profits. The cycle is self-reinforcing: Success breeds success, and soon the impact of rapidly expanding transportation and communication technology begins to cause a noticeable impact on social behavior, which adapts to these new technologies.

But it is not just new technology ventures that attract risk capital. Financing also begins flowing to the "peripheral" economies around the world, which, because of their small size, are quick to respond. These countries then begin to experience currency strength and real economic growth, which only reinforce the initial investment decision. As more money flows in, local markets begin to grow. As a consequence of the sudden growth in both asset values and gross domestic product, political leaders in developing countries often move to reform government policies in these countries—whether reform consists of expelling a backward Spanish monarch in the 1820s, expanding railroad transportation across the Andes in the 1860s, transforming the professionalism of the Mexican bureaucracy in the 1890s, deregulating markets in the 1920s, or privatizing bloated state-owned firms in the 1990s. By providing the government with the resources needed to overcome the resistance of local elites, capital inflows enable economic-policy reforms.

This relationship between capital and reform is frequently misunderstood: Capital inflows do not simply respond to successful economic reforms, as is commonly thought; rather, they create the conditions for reforms to take place. They permit easy financing of fiscal deficits, provide industrialists who might oppose free trade with low-cost capital, build new infrastructure, and generate so much asset-based wealth as to mollify most members of the economic and political elite who might ordinarily oppose the reforms. Policymakers tend to design such reforms to appeal to foreign investors, since policies that encourage foreign investment seem to be quickly and richly rewarded during periods of liquidity. In reality, however, capital is just as likely to flow into countries that have failed to introduce reforms. It is not a coincidence that the most famous "money doctors"—Western-trained thinkers like French economist Jean-Gustave Courcelle-Seneuil in the 1860s, financial historian Charles Conant in the 1890s, and Princeton University economist Edwin Kemmerer in the 1920s, under whose influence many developing countries undertook major liberal reforms—all exerted their maximum influence during these periods. During the 1990s, their modern counterparts advised Argentina on its currency board, brought "shock therapy" to Russia, convinced China of the benefits of membership in the World Trade Organization, and everywhere spread the ideology of free trade.

Globalization takes place largely because sudden monetary expansions encourage investors to embrace new risks.

The pattern is clear: Globalization is primarily a monetary phenomenon in which expanding liquidity induces investors to take more risks. This greater risk appetite translates into the financing of new technologies and investment in less developed markets. The combination of the two causes a "shrinking" of the globe as communications and transportation technologies improve and investment capital flows to every part of the globe. Foreign trade, made easier by the technological advances, expands to accommodate these flows. Globalization takes place, in other words, largely because investors are suddenly eager to embrace risk.

THE BIG CRUNCH

As is often forgotten during credit and investment booms, however, monetary conditions contract as well as expand. In fact, the contraction is usually the inevitable outcome of the very conditions that prompted the expansion. In times of growth, financial institutions often overextend themselves, creating distortions in financial markets and leaving themselves vulnerable to external shocks that can force a sudden retrenchment in credit and investment. In a period of rising asset prices, for example, it is often easy for even weak borrowers to obtain collateral-based loans, which of course increases the risk to the banking system of a fall in the value of the collateral. For example, property loans in the 1980s dominated and ultimately brought down the Japanese banking system. As was evident in Japan, if the financial structure has become sufficiently fragile, a retrenchment can lead to a collapse that quickly spreads throughout the economy.

Since globalization is mainly a monetary phenomenon, and since monetary conditions eventually must contract, then the process of globalization can stop and even reverse itself. Historically, such reversals have proved extraordinarily disruptive. In each of the globalization periods before the 1990s, monetary contractions usually occurred when bankers and financial authorities began to pull back from market excesses. If liquidity contracts—in the context of a perilously overextended financial system—the likelihood of bank defaults and stock market instability is high. In 1837, for example, the U.S. and British banking systems, overdependent on real estate and commodity loans, collapsed in a series of crashes that left Europe's financial sector in tatters and the United States in the midst of bank failures and state government defaults.

The same process occurred a few decades later. Alphonse Rothschild's globalizing cycle of the 1860s ended with the stock market crashes that began in Vienna in May 1873 and spread around the world during the next four months, leading, among other things, to the closing of the New York Stock Exchange (NYSE) that September amid the near-collapse of American railway securities. Conditions were so bad that the rest of the decade after 1873 was popularly referred to in the United States as the Great Depression. Nearly 60 years later, that name was reassigned to a similar episode—the one that ended the Roaring Twenties and began with the near-breakdown of the U.S. banking system in 1930–31. The expansion of the 1960s was somewhat different in that it began to unravel during the early and mid-1970s when, thanks partly to the OPEC oil price hikes and subsequent petrodollar recycling, a second liquidity boom occurred, and lending to sovereign borrowers in the developing world continued through the end of the decade. However, the cycle finally broke down altogether when rising interest rates and contracting money engineered by then Federal Reserve Chairman Paul Volcker helped precipitate the Third World debt crisis of the 1980s. Indeed, with the exception of the globalization period of the early 1900s, which ended with the advent of World War I, each of these eras of international integration concluded with sharp monetary contractions that led to a banking system collapse or retrenchment, declining asset values, and a sharp reduction in both investor risk appetite and international lending.

Following most such market crashes, the public comes to see prevalent financial market practices as more sinister, and criticism of the excesses of bankers becomes a popular sport among politicians and the press in the advanced economies. Once capital stops flowing into the less developed, capital-hungry countries, the domestic consensus in favor of economic reform and international integration begins to disintegrate. When capital inflows no longer suffice to cover the short-term costs to the local elites and middle classes of increased international integration-including psychic costs such as feelings of wounded national pride-support for globalization quickly wanes. Populist movements, never completely dormant, become reinvigorated. Countries turn inward. Arguments in favor of protectionism suddenly start to sound appealing. Investment flows quickly become capital flight.

Following market crashes, the public comes to see financial markets as more sinister, and criticism of bankers becomes a popular sport among politicians and the press.

This pattern emerged in the aftermath of the 1830s crash, when confidence in free markets nose-dived and the subsequent populist and nationalist backlash endured until the failure of the muchdreaded European liberal uprisings of 1848, which saw the earliest stirrings of communism and the publication of the *Communist Manifesto*. Later, in the 1870s, the economic depression that followed the mass bank closings in Europe, the United States, and Latin America was accompanied by an upsurge of political radicalism and populist outrage, along with bouts of protectionism throughout Europe and the United States by the end of the decade. Similarly, the Great Depression of the 1930s also fostered political instability and a popular revulsion toward the excesses of financial capitalism, culminating in burgeoning left-wing movements, the passage of anti-bank legislation, and even the jailing of the president of the NYSE.

PROFITS OF DOOM

Will these patterns manifest themselves again? Indeed, a new global monetary contraction already may be under way. In each of the previous contractions, stock markets fell, led by the collapse of the once-high-flying technology sector; lending to emerging markets dried up, bringing with it a series of sovereign defaults; and investors clamored for safety and security. Consider the crash of 1873, a typical case: Then, the equivalent of today's high-tech sector was the market for railway stocks and bonds, and the previous decade had seen a rush of new stock and bond offerings that reached near-manic proportions in the early 1870s. The period also saw rapid growth of lending to

Latin America, southern and eastern Europe, and the Middle East. Wall Street veterans had expressed nervousness about market excesses for years leading up to the crash, but the exuberance of investors who believed in the infinite promise of the railroads, at home and abroad, coupled with the rising prominence of bull-market speculators like Jay Gould and Diamond Jim Brady, swept them aside. When the market collapsed in 1873, railway securities were the worst hit, with many companies going bankrupt and closing their doors. Major borrowers from the developing world were unable to find new financing, and a series of defaults spread from the Middle East to Latin America in a matter of months. In the United States, the Congress and press became furious with the actions of stock market speculators and pursued financial scandals all the way to President Ulysses S. Grant's cabinet. Even Grant's brother-in-law was accused of being in cahoots with a notorious group that attempted a brutal gold squeeze.

Today, we see many of the same things. The technology sector is in shambles, and popular sentiment has turned strongly against many of the Wall Street heroes who profited most from the boom. Lending to emerging markets has all but dried up. As of this writing, the most sophisticated analysts predict that a debt default in Argentina is almost certain—and would unleash a series of other sovereign defaults in Latin America and around the world. The yield differences between risky assets and the safest and most liquid assets are at historical highs. In short, investors seem far more reluctant to take on risk than they were just a few years ago.

This lower risk tolerance does not bode well for poor nations. Historically, many developing countries only seem to experience economic growth during periods of heavy capital inflow, which in turn tend to last only as long as the liquidity-inspired asset booms in rich-country financial markets. Will the international consensus that supports globalization last when capital stops flowing? The outlook is not very positive. While there is still broad support in many circles for free trade, economic liberalization, technological advances, and free capital flows—even when the social and psychic costs are acknowledged—we already are witnessing a strong political reaction against globalization. This backlash is evident in the return of populist movements in Latin America; street clashes in Seattle, Prague, and Quebec; and the growing disenchantment in some quarters with the disruptions and uncertainties that follow in the wake of globalization.

The leaders now gathered in opposition to globalization—from President Hugo Chávez in Venezuela to Malaysian Prime Minister Mahathir bin Mohamad to anti-trade activist Lori Wallach in the United States—should not be dismissed too easily, no matter how dubious or fragile some of their arguments may seem. The logic of their arguments may not win the day, but rather a global monetary contraction may reverse the political consensus that was necessary to support the broad and sometimes disruptive social changes that accompany globalization. When that occurs, policy debates will be influenced by the less emotional and more thoughtful attacks on globalization by the likes of Robert Wade, a professor of political economy at the London School of Economics, who argues forcefully that glo-

balization has actually resulted in greater global income inequality and worse conditions for the poor.

Investing in the Future

In past periods of monetary expansion and globalization, Western societies have experienced the rapid development and commercial application of new technologies.

1822–37: expansion of canal building, first railway boom, application of steam power to the manufacturing process, advances in machine tool design, invention of McCormick's reaper, first gas-lighting enterprises, and development of the telegraph

1851–73: advances in mining, second railway boom, developments in shipping, and rapid growth in the number of corporations in continental Europe

1881–1914: explosive productivity growth in Europe and the United States, improvement in steel production and heavy chemical manufacturing, first power station, spread of electricity, development of the internal combustion engine, another railway boom, innovation in newspaper practice and technology, and developments in canning and refrigeration

1922–30: commercialization of automobiles and aircraft, new forms of mass media, rising popularity of cinema and radio broadcasting, spread of artificial fibers and plastics, widespread use of electricity in U.S. factories, the creation and sale of a variety of new electric appliances, and expanded telephone ownership

1960–73: development and application of transistor technology, advances in commercial flying and shipping, and the spread of telecommunications and software

1985–present: ubiquity of information processing, explosion in computer memory, advances in biotechnology and medical technologies, and commercial application of the Internet

---

If a global liquidity contraction is under way, antiglobalization arguments will resonate more strongly as many of the warnings about the greed of Wall Street and the dangers of liberal reform will seem to come true. Supposedly irreversible trends will suddenly reverse themselves. Further attempts to deepen economic reform, spread free trade, and increase capital and labor mobility may face political opposition that will be very difficult to overcome, particularly since bankers, the most committed supporters of globalization, may lose much of their prestige and become the target of populist attacks following a serious stock market decline. Because bankers are so identified with globalization, any criticism of Wall Street will also implicitly be a criticism of globalizing markets.

Financiers, after all, were not the popular heroes in the 1930s that they were during the 1920s, and current events seem to mirror past backlashes. Already the U.S. Securities and Exchange Commission, which was created during the Great Depression of the 1930s, is investigating the role of bankers and analysts in misleading the public on the market excesses of the 1990s. In June 2001, the industry's lobby group, the Securities Industry Association, proposed a voluntary code, euphemized as "a compilation of best practices... to ensure the ongoing integrity of securities research and analysis," largely to head off an expansion of external regulation. Increasingly, experts bewail the conflicts of interest inherent among the mega-banks that dominate U.S. and global finance.

Globalization itself always will wax and wane with global liquidity. For those committed to further international integration within a liberal economic framework, the successes of the recent past should not breed complacency since the conditions will change and the mandate for liberal expansion will wither. For those who seek to reverse the socioeconomic changes that globalization has wrought, the future may bring far more progress than they hoped. If global liquidity contracts and if markets around the world pull back, our imaginations will once again turn to the increasingly visible costs of globalization and away from the potential for all peoples to prosper. The reaction against globalization will suddenly seem unstoppable.

Michael Pettis is an investment banker and professor of finance at Columbia University. He is author of The Volatility Machine: Emerging Economies and the Threat of Financial Collapse *(New York: Oxford University Press, 2001).*

Want to Know More?

Charles P. Kindleberger's *A Financial History of Western Europe*, 2nd ed. (New York: Oxford University Press, 1993) is probably the best single volume for anyone interested in understanding the financial history of globalization. For a discussion of how structural changes in financial systems can lead to overextension and banking crises, consult Hyman P. Minsky's *Can "It" Happen Again? Essays on Instability and Finance* (Armonk: M.E. Sharpe, 1982). Also see Paul W. Drake's, ed., *Money Doctors, Foreign Debts, and Economic Reforms in Latin America from the 1890s to the Present* (Wilmington: SR Books, 1994) and Christian Suter's *Debt Cycles in the World Economy: Foreign Loans, Financial Crises, and Debt Settlements, 1820–1990* (Boulder: Westview Press, 1992). Michael Pettis's *The Volatility Machine: Emerging Economies and the Threat of Financial Collapse* (New York: Oxford University Press, 2001) identifies the specific events that set off liquidity booms in prior periods of global integration.

Frank Griffith Dawson's *The First Latin American Debt Crisis: The City of London and the 1822–25 Loan Bubble* (New Haven: Yale University Press, 1990) offers a wonderful account of an early era of globalization, although his description of events 180 years ago is too familiar for comfort. Matthew Josephson's famous *The Robber Barons: The Great American Capitalists, 1861–1901* (New York: Harcourt, Brace & World, 1962) is as good a place as any to start reading up on the railway booms of the late 19th century. Harold James's *The End of Globalization: Lessons from the Great Depression* (Cambridge: Harvard University Press, 2001) is one of the best recent books on economic history and discusses the globalizing period of the 1920s and the subsequent backlash.

Kevin H. O'Rourke and Jeffrey G. Williamson recently completed a major work on globalization at the end of the 19th century and the backlash against it in *Globalization and History: The Evolution of a Nineteenth-Century Atlantic Economy* (Cambridge: MIT Press, 1999). Finally, Niall Ferguson's *The House of Rothschild: Money's Prophets, 1798–1848* (New York: Viking Press, 1998) and *The World's Banker: The History of the House of Rothschild* (London: Weidenfeld & Nicolson, 1998) tell the story of a family intimately involved with every aspect of globalization.

• For links to relevant Web sites, as well as a comprehensive index of related FOREIGN POLICY articles, access **www.foreignpolicy.com**.

Croesus and Caesar

The Essential Transatlantic Symbiosis

Richard Rosecrance

A VENGEFUL United States is snubbing its German allies for opposing the war against Iraq. Prickly France has become a porcupine sitting on America's lap; Moscow and Beijing have followed Paris' lead since the United States and Britain subdued Ba'athi Iraq. Some opine, and others worry, that a belligerent United States or a reactive European Union will throw NATO into the dust bin of fast-moving history. The United States may think it does not need Europe, and Europe may reject association with the American hyperpower. Some Continentals now call more openly for European balancing against a "rogue United States."

This, however, will not happen. The EU will not seek to become another military superpower to contend with or threaten America. The defeat of Iraq will not lead to the end of the American era, with Europe as America's residuary legatee. Nor is the altercation between Europe and America properly cast as a spat between different visions of the social order—with Brussels and the European Commission representing the "paradise" of social democracy and Washington the "power" of the preponderant state. Europe, in fact, does not play Venus to America's Mars.

Amid much bluster on both sides, the paramount truth is that the two major centers are both powerful, but act in different spheres—and they desperately need each other. Neither can, or at any rate should, talk cavalierly of going it alone. Most specifically, America needs Europe's financial power, and Europe requires America's military protection.

America Needs Europe

T HE MOST appropriate metaphor for comparing Europe and America, for those who insist on them, has nothing to do with planets. It rather compares Europe's Croesus to America's Caesar, the paradigmatic example of the power of Rome. Croesus, King of Lydia in the 6th century B.C.E., accumulated vast wealth, and so does Europe today. America is a new Rome that bestrides the world politically and militarily, but rests on shaky financial foundations. Power requires wealth, and wealth cannot protect itself without power. In Washington, Americans berate Europeans for failing to share the burdens of military spending and for free-riding on American largesse. The Bush Administration vilifies most of its NATO allies for lacking military might and having to rely on American air and sea power to get their slender forces to any theater of operations. The so-called European "rapid reaction force" is not rapid, nor will it ever be much of a force. American pundits declare that only European rearmament will remedy this deficiency and restore equality to the link with the United States.

This view is as mistaken as it is common. Military clout is not the appropriate way to measure the European contribution to NATO or to America. With little fanfare, the Europeans (and their Japanese cohorts) have shored up American power against the force of financial tides, enhancing Washington's strength and resiliency. Without the help of Europe and Japan, the United States could not have undertaken or sustained its frequent international military operations. Lacking this financial shield, U.S. foreign and security policy would have been checked and doomed to failure. So it has been for decades.

European strength has always rested on monetary foundations. In the past, Holland and then Great Britain commanded the world economy and plucked imperial spoils as a willful child eats chocolate. These nations could finance operations overseas, run commercial economies at home and still expand their territorial orbits. As Niall Ferguson points out, Britain won contest after contest because its money market was deeper and more resilient than those of its rivals.[1]

The United States inherited this role after World War I and enlarged it after World War II, when the reigning American trade surplus emptied European coffers and American credit refloated capitalist growth. Brimming with wealth, the United States could finance both its military role and expand American private capitalism worldwide while contributing to Europe's rehabilitation. But the American financial reservoir ran dry in the 1960s: since then, the United States has not been able to bankroll its military and economic sectors at the same time.

So Europe (and Japan) has. Since the late 1960s, the Europeans have been called on repeatedly to bail America out economically, and they have done so. In 1971, Europe permitted the United States to go off the gold standard, devalue its currency and levy import surcharges against its trading partners without retaliation. Europe did not punish America for its breach of faith: instead it helpfully maintained the high value of the franc and the deutschemark. In 1978-79, American inflation rates rose above the interest rate, negating any incentive to save in the United States. This led, predictably, to a run on the dollar. Spurred by German Chancellor Helmut Schmidt, Federal Reserve Chief Paul Volcker moved to collar inflation with a high interest rate policy, and the dollar regained its footing. After the dollar had climbed to prodigious heights under Ronald Reagan (fostering a huge trade deficit in the United States in the bargain), Europe helped to engineer a soft landing in both 1985 and 1987 as the dollar returned to earth. This prudent division of Western labor represents a still underappreciated reason for the West's victory in the Cold War. The Soviets, in contradistinction, had nothing comparable to it.

Nor did Europe ever take advantage of American weakness when the United States ran into financial trouble. In 1987, the Europeans stood ready to help the Federal Reserve halt the collapse of the New York stock market on October 19. On that date the Dow-Jones Average suffered its largest one-day decline in history, plunging 22 percent in a single session. When the new Fed Chief Alan Greenspan lowered interest rates and ploughed $100 billion of liquidity into the financial system—floating struggling firms and the exchange itself—the Bundesbank did not devalue the deutschemark in response. Europe and Japan continued to hold funds in New York and did not raise interest rates to counter the inflationary effect of American credit. Germany asked the United States to cut its fiscal deficit but remained ready to support America even when forced in effect to import U.S. inflation. Japan also resumed buying dollars.

Such past services from Europe were, of course, self-interested in the larger sense, but that takes nothing away from their importance. Nonetheless, these services pale beside the tasks the European Croesus will have to undertake in the future. The United States is now running unparalleled trade and fiscal deficits. The trade deficit alone stands at more than $500 billion a year, almost 5 percent of GDP. America is now borrowing 5 percent of world savings, sucking $1.5 billion a day into the United States. The budget deficit, fanned by President Bush's tax cuts, may rise to $500 billion a year, reaching $2 trillion in the next several years. If foreign capital does not enter the United States in huge amounts, the dollar could go into free fall, resulting in renewed inflation. The Fed would then have to raise interest rates, and the American economy would likely plunge into deep recession.

This crisis could occur well before 2004, and President Bush's re-election would then be up for grabs. The political outcome in the United States could thus depend upon what politicians and politically influential bankers in Brussels, Paris, Frankfurt, London and Tokyo decide to do to help—or not help, as the case may be.

It is true that investing in America is still in Europe's own economic interest. Investments yield more in New York than they do in Paris. In the short term, though, Europeans and Japanese have lost money in the United States. The New York stock market has not gone up, and European returns on U.S. investments are cancelled by currency losses. As Europe enlarges, the expansion of the European market will draw money back to Brussels. In the longer term, however, U.S. growth will return, and Europe will want to share in its benefits. As the dollar falls, Europe will substitute new foreign direct investment for trade with the United States. If one takes even a casual glance at the size of such investments in the United States, one sees immediately that their value dwarfs that of commerce. This is not going to change radically or soon. Europe will still place large funds in the American economy. But enough funds, and on what terms?

Europe Needs America

As a "peaceful power", Europe does not want, or at any rate should not want, to separate from the United States and rebuild its own major military capabilities. It needs American defense protection while integrating with a still unstable region stretching beyond east-central Europe to Central Asia. Robert Mundell, the father of the euro, believes that as many as fifty countries will eventually join the EU—even some from North Africa—and adopt the euro as their national currency. America, not Europe, will defend these expanded borders—and Europeans recognize that they must play their financial part to receive help in protecting their growing perimeter.

To be sure, the euro will probably rise to become a competitive world currency with the dollar, but this prospect ought not be feared. It will give America an opportunity to export to European markets in a major way. Washington will also then have a partner in providing liquidity to other nations through deficits in the balance of trade. Europe's economic power will probably grow, despite its demographic doldrums, equaling or even surpassing that of the United States. But this does not mean that Europe will shoulder the world's, or even its own, military burden—and the richer it gets the more it will need U.S. military protection.

Yet Europe as a pacific economic giant will not be inward-looking. It will seek to influence others to follow its peaceful mission. Other nations may well emulate its strikingly successful institutions and policies. By expanding, the new Europe will bring peace and prosperity to eastern Europe and perhaps even to Russia, but it will not police the world. The NATO link with the United States will still be essential in all of this. That is why it is both erroneous and self-defeating to claim that America does not need Europe or Europe the United States. America and Europe are essential and reciprocal counterparts to one another in the structure of world power.

Then there is the continuing threat of terrorism. This threat will not disappear now that the war with Iraq is over; terrorism may even spread, at least in the short run. Neither the technical revolution in military affairs that has occurred in the United

States nor the growing imperial reach of America's conventional forces can vanquish terrorism. For that, the United States will need cooperation and the shared intelligence of many other countries, particularly those of an enlarged European Union. There are no superpowers in the war against terrorism—every nation can be a military theater where battles may be won, and Pakistan's cooperation may be as important as Russia's. International cooperation between the United States and Europe is essential to win the long bout against terrorism.

EUROPEANS and Americans alike have been predicting the demise of NATO and the end of U.S. and European cooperation since Suez in 1956. The alliance, however, has not collapsed—and it won't. In the last thirty years, Europe has carried the primary financial burden, allowing the United States to maintain an essentially unbalanced economy while acting as the world's gendarme. Europe has not built up its military strength, but instead has done something much more important: it has created the financial conditions that have allowed the United States to act.

Americans may complain about the recent actions of Monsieur Chirac and Herr Schröder, but if they insist on remaining oblivious to Europe's critical financial role, they only contribute to an atmosphere in which that role is weakened. And Europe, which will not act militarily on its own in any serious way, should stop squawking about the use of American force as though, cases aside, there is something wrong about it in principle. The Atlantic economic and military partnership serves both Americans and Europeans, and it will become firmer as the true basis of the relationship dawns on both sides of the Atlantic. One hopes this happens sooner rather than later, for Caesar needs Croesus, and Croesus needs Caesar, now.

Notes

1. See, at the least, "True Cost of Hegemony: Huge Debt", *New York Times*, April 20, 2003.

Richard Rosecrance is professor of political science at UCLA and project director of the UCLA-Carnegie study on the effect of globalization on national self-determination..

Reprinted with permission of *The National Interest*, Summer 2003, pp. 31-34. © 2003 by The National Interest, Washington, DC.

Render Unto Caesar: Putin and the Oligarchs

MARSHALL I. GOLDMAN

"Because of the flawed manner in which the privatization process was carried out, the new owners will always lack the legitimacy necessary for a stable political climate and sustained economic investment and growth. Having constructed a faulty foundation, the builders must live with the possibility that their edifice of privatization will periodically shift and crack and may even collapse."

In 1991, a small group of Russians emerged from the collapse of the Soviet Union to claim ownership of some of the world's most valuable oil, natural gas, and metal deposits. This resulted in one of the greatest transfers of wealth ever seen. By 1997, five of these individuals, who in the 1980s had only negligible net worth, were listed by *Forbes* as among the world's richest billionaires. This year seventeen of these men were on the *Forbes* list, including two who merged their companies to create what promises to be an oil conglomerate worth $35 billion. Almost all of these self-styled oligarchs have been accused of guile, intimidation, and even murder in reaping their rewards. Not surprisingly, their rise—and for some, their fall—mirrors the travails of the Russian economy.

SHOCKING THE SYSTEM

The oligarchs gained their wealth by taking advantage of economic policies forged in the early 1990s by President Boris Yeltsin and the reformers he brought into Russia's first post-Soviet government. Why did Yeltsin and his acting prime minister, Yegor Gaidar, whom he put in charge of the Russian economy, choose the policies that they did?

The reformers were determined to transform Russia from what had been the largest republic in the centrally planned communist Soviet Union into an independent, democratic, market-based country. But, given Russia's near-impoverished condition, Gaidar realized he had to act quickly to prevent not only a complete economic collapse but also what his fellow reformer, Anatoly Chubais, feared might be an effort to reconstitute a communist, centrally planned system. It was imperative, they concluded, to resuscitate the economy so that it could pro-

vide jobs and supply the needs of the public. At the same time they wanted to create a sense of ownership and personal wealth among Russians generally so that these newly empowered stakeholders could resist any effort to reintroduce state control and ownership of factories and stores.

To do all this, Gaidar and his staff opted for a set of reforms often referred to as shock therapy. This involved embracing market mechanisms and ending price, labor, and production controls and privatizing ownership of state enterprises.

It was not a painless process. But shock therapy had proved reasonably successful in a few other countries, such as Bolivia and Poland. While higher prices meant that there would be more consumers who could no longer afford to purchase various goods, they would also mean higher profits and thus more output. Fewer consumers and more producers would soon bring an end to queues and bring about what economists call "equilibrium," which would obviate the need for rationing.

As for privatization, it was generally accepted among Western economists as well as Russia's new reformers that privately owned businesses would be run more efficiently and with higher productivity than state-owned enterprises. Private entrepreneurs were more likely to make decisions based on market imperatives, instead of the political and patronage considerations that had almost always characterized state-owned enterprises. Arranging for Russia's factory directors and their staffs to acquire up to 50 percent of the stock of the newly privatized entities, the reformers reasoned that these new "owners" were less likely to strip the enterprise of its assets or allow others to steal from "their" company.

To inculcate the same possessive attitude among the general public, the Yeltsin government arranged to issue vouchers to every citizen. These vouchers could be converted into shares of stock of one of the newly privatized companies. Approximately 49 percent of a company's stock would be set aside for outsiders. Each citizen would also become a stakeholder, a form of people's capitalism.

At first glance, this seemed a reasonable way to dismantle the communist system. However, the plan had several serious flaws. For example, the reformers assumed that once industrial ministries and the planning organizations were eliminated and state enterprises were privatized, the formerly government-controlled companies would quickly begin to think and operate like private businesses in a competitive environment. Unfortunately, nothing of the sort occurred. It was simply too much to expect that a market system would suddenly materialize out of nothing. After all, for 70 years the state and the Communist Party had done everything they could to stamp out any remnant of market behavior, including discouraging the start-up of new businesses and services and eliminating all market-oriented business codes.

Moreover, without a competitive infrastructure, the state enterprises, which for the most part were monopolies, continued to operate as monopolies—only now they were private, not state, monopolies. Thus their behavior differed little from what it had been before privatization. As with most monopolies, whenever they could, they cut back production and raised prices. The combined effect of relatively few start-ups, the dominance of state monopolies, and the sudden downsizing of the military-industrial complex as a consequence of the end of the cold war devastated the economy. GDP fell between 40 and 50 percent from 1990 to 1998 while prices soared; it took almost 1,170,000 rubles in 1998 to buy what 100 rubles did in 1990.

THE NEW OWNERS EMERGE

In the near anarchy created by the reforms, it was inevitable that a certain amount of institutional chaos would follow, especially in sorting out who would end up in control of the state-owned enterprises. The new owners can be divided into three categories: the former factory directors, the nomenklatura, and the non-nomenklatura, or upstarts.

The former factory directors formed the largest group of new owners. It was hoped that they would share control with the rest of the workers, and in some cases they did. But because the workers were usually poorly organized and because they had been told since birth that stock certificates were a fraud, they had no understanding of what stock ownership might mean. Most workers were only too happy to swap their stock for a bottle of vodka. When a vodka-for-stock swap did not work, some especially determined factory directors resorted to threats and violence to obtain the stock they wanted.

The newly constituted factory owners with their newly acquired net wealth quickly became known as the "new Russians" because of their nouveau riche ways and exhibitionism. One or two became millionaires and almost none fell below the poverty line—unlike one-third of the rest of the population. But it was the other two categories of enterprise owners that are the most interesting. Both have come to be known as oligarchs.

Initially it was generally thought that almost all the very rich oligarchs had emerged from what had been the nomenklatura. This was the official Soviet-era list of names (nomenclature) of all the senior-level people in the country. This included top echelons from the Communist

Party, the KGB, the army, the ministries, and the arts. They were the big brothers. Nearly all the members of this group did not have much in the way of personal wealth but they had no need for it since the country was essentially theirs to use as they pleased. They had access to special stores as well as special restaurants, hospitals, and housing denied to the public. Because in the initial phases of privatization many of the early and blatant oligarchs had been members of the nomenklatura, it was assumed that most of the oligarchs possessed a similar background. After all, they were the officials who had been in control of the country's assets, and it seemed like a simple matter for them to take over and privatize the industries they had controlled.

Three of the most prominent nomenklatura oligarchs are Vagit Alekperov, Viktor Chernomyrdin, and Rem Vyakirev. Alekperov had been minister of the petroleum ministry, so it was relatively simple for him to set aside three producing fields—Langepaz, Urengoi, and Kogalym—and combine them into Lukoil, making it the largest of the privatized petroleum companies. Even more brazen, Chernomyrdin and Vyakirev, the minister and deputy minister, respectively, of the gas industry, transformed not selected parts, but the entire ministry into a private company they called Gazprom. While the state continued to hold on to 38 percent of the stock, Vyakirev, after Chernomyrdin became prime minister, ended up controlling the largest number of Gazprom's shares and ran the company as if it were his personal fiefdom.

A fourth member of the nomenklatura, Vladimir Potanin, was an official of the Foreign Trade Organization, which traded nonferrous metals in the Ministry of Foreign Trade. Potanin in effect privatized the FTO into a trading company and built it into a commercial bank and financial holding company. Eventually his holding company acquired control and ownership of Norilsk Nickel, Russia's primary producer of nickel and precious metals such as platinum and palladium. All four men, Alekperov, Chernomyrdin, Potanin and Vyakirev, made the *Forbes* list in 1997.

It is the third category of oligarchs, the non-nomenklatura, that is most intriguing. Unlike those who were members of the nomenklatura, these upstarts were anything but part of the Communist establishment. Many were regarded as living on the margins of society, at best on the fringe of it and in some cases beyond. Members of minority groups for the most part, they were considered untrustworthy and therefore excluded from any responsible official positions in the Soviet era, including the Communist Party, the military, the KGB, and the ministerial bureaucracies.

If Russia is to attract investment from both inside as well as outside the country, some effort has to be made to curb the arbitrary use of state power and the intimidation of enemies and opponents.

Excluded from positions within the power elite, many of those with extra energy or initiative chose instead to go into private business. This meant dealing in the black market for jeans, exchanging rubles, or engaging in construction work and other services on a private contractual basis. There are many examples. Alexander Smolensky was caught selling Bibles that he produced after-hours on a state printing press. Vladimir Gusinsky used his private car as a cab and Mikhail Fridman washed windows. All of them provided construction services. They became adroit at finding and providing goods and services in short supply. Since the economic planning system seldom worked effectively, the opportunities were considerable.

Unfortunately, the state treated all private business activities as economic crimes. Since economic crimes were often punished by imprisonment (reportedly Smolensky was arrested for his activities) and even death, there was a bit of a downside. Offsetting that, the possibility of punishment discouraged competition, which in turn made it all the more lucrative for those daring and determined enough to take this risk.

To those dealers and speculators who persevered, the decision by Soviet leader Mikhail Gorbachev in 1987 to authorize the establishment of cooperative and private businesses meant that what they had been doing had suddenly become legitimate overnight. Not only could they go about their businesses openly without fear of arrest, they actually had an advantage over members of the nomenklatura. In contrast to the upstarts, the nomenklatura had heretofore played by the rules and obtained and supplied goods according to state plans through state agencies. They had virtually no experience in securing goods in short supply or dealing in markets.

This was a serious handicap. In January 1992, with the end of central planning and the shift to the market, the supply system all but disintegrated. Store shelves were empty, and shortages abundant. This was ideal for the upstarts. They knew how to obtain goods in short supply; those in the nomenklatura did not. Suddenly the men who had been social outcasts and on the fringe of economic life were now at its very center.

With their ability to scout out scarce supplies, the former upstarts were able to respond to the hunger that Russian consumers had for goods long denied to them, especially foreign products. In no time they became extremely wealthy. It helped that there was little or no competition from the former state sector and little or no regulation by the now disorganized and disoriented state.

After they began to amass all this new wealth, most of it in cash, these businessmen began to establish their own banks. This too became a possibility after 1987 and for a time all it took was as little as $75,000 in capital. These banks also proved to be an important tool in providing the financing needed for the vouchers and shares of stock that many of the would-be oligarchs began to accumulate. Before long, Boris Berezovsky was able to boast that seven of the oligarchs, six of them "upstarts," had come to control 50 percent of the economy. That was an exaggeration, but if Alekperov and Vyakirev were included, it would not be too far from the truth.

THE BACKLASH

The accumulation of so much wealth by so few at a time of hyperinflation and economic depression inevitably gave rise to political resentment and social dysfunction. Criminal gangs began to appear throughout Russia. These mafia groups, as the Russians called them, were said to control as much as 70 percent of the Russian economy. Equally troublesome, the breakdown in discipline emboldened government officials seeking to exploit their positions. Many bureaucrats justified the breakdown in morality and the growing ubiquity of bribes with the rationalization that their government salaries had fallen far behind the rate of hyperinflation. Given the political and economic tumult, most anyone with money sent a substantial portion of it outside the country. The result was massive capital flight, estimated to exceed $1 billion per month. This more than exceeded the funds provided by international financial organizations in an effort to help finance Russia's economic and investment recovery.

Various forms of capital flight occurred. The most common technique was for an oil or raw materials company to keep in the West the payments for its exports. In other cases, enterprise managers would strip assets from their companies (usually where the state remained a part owner of the company), then transfer the proceeds to another company located offshore and wholly owned by the enterprise manager or his relatives. ITERA, for a time Russia's second-largest producer of natural gas, is a prime example. Located in Jacksonville, Florida, ITERA has received enormous gifts from Gazprom, including control of natural gas fields and access to natural gas supplies. Worth billions of dollars, these assets were often provided free of charge; if there was a payment, it was usually far below the market value.

As members of the public became aware of the magnitude of the theft and the misuse of what once belonged to the state, they began to blame the government and Boris Yeltsin in particular. Harnessing that anger, in December 1995, the Communist Party scored a major victory in the parliamentary elections for the Duma, the lower house of parliament. A month later, in January 1996, Yeltsin's approval rating in opinion polls had dropped to less than 10 percent. This was far below that of his chief opponent in the June 1996 presidential race, Gennady Zyuganov, the head of the Communist Party. Chubais, one of the original architects of the economic reforms, was right: it did look as if the Communist Party might return to power. Not because of the failure to privatize, but because the privatization process was seen to have been implemented unfairly.

The Oligarchs

Roman Abramovich
Major stockholder in Sibneft Oil as well as Millhouse Capital and Russian Aluminum (RUSAL, which controls 70 percent of Russian aluminum production). Elected to the Duma in December 1999 and governor of the Chukotka region in 2000.

Vagit Alekperov
President of Lukoil.

Petr Aven
President of Alfa Bank (see Mikhail Fridman).

Boris Berezovsky
Held effective operating control of ORT (Russia's largest television network), Sibneft Oil, Logovaz, a bank, a large auto dealership, and the newspaper *Nezavisimaia Gazeta*. Dismissed by Yeltsin from leadership of the Commonwealth of Independent States. Elected to Duma but resigned and went into exile. Wanted for questioning. Sold interest in ORT.

Vladimir Bogdanov
President of Surgutneftgas.

Viktor Chernomyrdin
Minister of gas industry and chairman of Gazprom. Former prime minister.

Anatoly Chubais
President of United Energy System.

Oleg Deripaska
General Director of RUSAL and Base Element. Major holdings in Sibneft and RusProm Auto, which in turn controls the GAZ automobile works as well as bus and truck and diesel engine manufacturing plants.

Mikhail Fridman
Chairman, Alfa Group (controls Alfa Bank and Tyumen Oil).

Vladimir Gusinsky
Founder, MOST-Bank and Media-MOST (controls NTV, second-largest TV network) and *Segodnya* newspaper. MOST-Bank collapsed; switched interest to Media-MOST. Arrested and jailed for three days. Fled to Spain. Borrowed from Gazprom, which then took over ownership of Media-MOST.

Mikhail Khodorkovsky
Founder, Menatep Bank and Rosprom (controls Yukos Oil). Being sued for sleight-of-hand tricks at Yukos Oil and subsidiaries. Now says he wants to be honest. State raided offices in 2003.

Vitaly Malkin
President, Russian Credit Bank. Russian Credit Bank's assets shifted to IMPEX Bank.

Vladimir Potanin
President, Oneximbank and Interros Trading Company. Oneximbank assets shifted to Rosbank. Owner of Norelsk Nickel.

Alexander Smolensky
Chairman of SBS AGRO Bank. Assets shifted to Soyuz and First Mutual Trust Society with subsidy from central bank. In hiding in Vienna.

Rem Vyakhirev
President of Gazprom. Stripped assets and was fired, May 2001.

M. I. G.

Concerned that a Zyuganov victory might mean the loss not only of their acquisitions but of their lives, seven of the oligarchs agreed to put aside whatever feuds they may have had with each other and unite to work for Yeltsin's victory. (According to one report, Berezovsky had earlier ordered a contract killing of Gusinsky.) Since these seven oligarchs controlled the major television networks and newspapers, they thought they could engineer Yeltsin's election, but it was not easy. Yeltsin was often drunk and had serious health problems; he had suffered several heart attacks. To erase these images, the oligarchs resorted to selective editing. Whenever Yeltsin appeared in the media, he was shown in a favorable light. In contrast Zyuganov was always depicted negatively. One of the more notable scenes from television at the time was a sequence of Yeltsin doing the twist with two young women in miniskirts. Shortly thereafter, Yeltsin suffered a heart seizure that somehow was omitted from any current news reports.

Ultimately, the oligarchs managed to shift public opinion and Yeltsin won. In reward for their help, Yeltsin agreed to repay his backers. Gusinsky was permitted to acquire more television outlets, allowing him to create a nationwide network. Potanin and Berezovsky were appointed to senior government positions, and Yeltsin agreed to expand a "loans for shares" program that Potanin had proposed in 1995. This proved to be the biggest asset giveaway of all.

Ostensibly, loans for shares was a generous offer by several banks to assist the government with its ever-growing budget deficit. To help the government pay its bills, several of the oligarchs agreed that their banks would lend the government the money it needed. The banks asked the government to use as collateral for the loans government-owned stock it still held in several companies. The government refrained from completely selling off several of the energy and raw materials businesses (which eventually turned out to be the most valuable portions of its portfolio). According to the plan, once the government repaid its loans, the collateral would be returned. In the event the government found itself unable to repay its debt, the banks would organize a series of auctions, the stock would be sold at a good price, and the government would collect the proceeds.

Of course, everyone understood that the government would not be able to repay the loans: revenues would remain inadequate because tax evasion was widespread. What was not anticipated, however, was that the bankers would rig the bids so that in almost every instance, the bank conducting the auction would end up as the only bidder (disguised but in fact a "straw" for the bank). As a result, the price was rarely much above the minimum price that had been determined in advance.

No one was mugged in the holdup, but the state found its pockets picked of hundreds of millions, even billions, of dollars in the process. As an example, Mikhail Khodorkovsky paid not much more than $300 million for Yukos Oil, which by 2003 was worth between $20 billion and $25 billion. Similarly, Berezovsky paid only slightly more than $100 million for Sibneft Oil, which in 2003, when it agreed to merge with Yukos, was valued at over $10 billion.

RENDER UNTO CAESAR, OR ELSE

As good as the early years of the Yeltsin presidency had been for the oligarchs, the meltdown of the economy on August 17, 1998, brought about the weakening, if not downfall, of most of them, especially those who had neglected to include energy and raw materials companies in their portfolios. Almost all their banks except for one closed down. After many years, there was some restitution, but the public was outraged when it was discovered that several of the banks, including Khodorkovsky's, had managed to funnel most of their remaining good assets into newly created entities. As a result, these funds were put beyond the reach of the banks' depositors and those who had lent money to or invested in these banks.

The economic climate began to improve in 1999 and Russia began to show the first signs of a prolonged economic revival. Some attributed this growth to Vladimir Putin, who had been appointed prime minister in August that year. A few of the oligarchs, especially Berezovsky, were among those who had urged Yeltsin to appoint Putin as prime minister. Initially they welcomed what they saw as Putin's firm hold over the economy.

Putin's stable presence, both as prime minister and, after the March 2000 elections, as president, stood in sharp contrast to Yeltsin's unstable ways. But the turnaround in the economy had little to do with anything Putin did. Putin was appointed as prime minister in August 1999; the turnaround in industrial production began several months earlier, in March 1999.

Two changes made the difference. The first was a threefold jump in world oil prices from about $11 a barrel to almost $30 a barrel. The second was the extreme devaluation of the ruble that followed in the wake of the August 17, 1998, financial collapse. On August 1, only slightly more than 6 rubles would buy one dollar; a few months later, more than 24 rubles were required. Since Russia was the world's second-largest exporter of oil, the increase in prices provided a windfall not only for the oil oligarchs, but for the government, which collected an export tax on oil. Of course, the price increase was a consequence of world economic forces and not anything that Putin or Russia had done. The cheaper ruble was also a windfall for Russian manufacturers, who moved into markets heretofore dominated by foreign imports. Almost immediately, imports shrank by 50 percent.

The combined effect of all this in 2000 was to produce a 10 percent increase in GDP. Since then economic growth has fallen to a range of 4 to 6 percent annually. That is still robust compared to the rest of the world, but is generally below the 7 percent rate that Putin needs if the Russian GDP is to double in 10 years—his proclaimed goal.

For Russia to grow that fast, it will take more than exhortation. Part of the problem is Russia's traditional over-reliance on raw materials such as oil, gas, and ferrous and nonferrous metals. The dollar revenues that the oil exports, for example, generate are converted into rubles. This increased demand for rubles pushes the price of the ruble up relative to the dollar, which gradually erodes the advantage that Russian manufacturers enjoyed after the August 1998 devaluation. As a result, domestic manufacturers, once again, find themselves facing stiff foreign competition.

To generate the growth he wants, Putin must restructure the country's economy and institutions to make investing in Russia more attractive to foreign investors and Russians, including the oligarchs. To his credit, once installed as president, Putin did try to facilitate economic growth by initiating some needed structural changes.

Following the advice of his economic advisers, German Gref and Andrei Illarionov, Putin introduced legislation eliminating high marginal personal income tax rates and instead created a flat tax of 13 percent. Many economists in the West oppose a flat tax because it imposes a disproportionate burden on the poor. In the Russian context, however, it may not be such a bad idea because so few of the Russian rich actually paid the 35 percent rate they were supposed to pay under the old tax system. For Russia, a flat tax is better than no tax.

Russian chinovniks (bureaucrats) have survived the czar and the Soviets; they seem determined to outlive current reform attempts.

Putin also restructured business taxes, arranged for the private ownership of land, rationalized the pension system, and tried to shrink or otherwise curb the power of the bureaucracy. These were all steps in the right direction. It remains to be seen, however, if adopting the laws will lead to their implementation, especially where the laws call for a reduction in the power of the bureaucracy.

As for his treatment of the oligarchs, Putin initially tried to reassure the public as well as the oligarchs who survived the 1998 crash that he would attempt to maintain an evenhanded policy. Admittedly it was a bit awkward that he owed his appointment in part to Berezovsky, but he insisted that he would treat the oligarchs no differently from anyone else. As he put it, "It is asked what then should be the relationship with the so-called oligarchs? The same as with anybody else. The same as with the owner of a small bakery or a shoe repair shop...." This implied a tough policy aiming to ensure equal competitive opportunities for everyone, as well as a lack of distinctive privileges for specific businessmen. Although it sounds very nice, there have been no reports of Putin meeting with bakers or cobblers; he has, however, met regularly with oligarchs.

It is true that on occasion some of the oligarchs have found these meetings rather ominous. And not all the oligarchs have liked what they have heard from Putin. Moreover, despite his stated disavowal, Putin has played favorites. He has allowed some former KGB officers to expand their private business empires and build communications monopolies. Nor has Putin been too concerned about the type of person he deals with. For example, he allowed Roman Abramovitch and Oleg Deripaska to establish RUSAL, a company that now controls 70 percent of Russia's aluminum production. The United States refuses to allow Deripaska into the country because of reports that when he decides to "eliminate" competition, he really does.

Putin did eventually move against some of the oligarchs. His decisions to fire Rem Vyakirev and Viktor Gerashchenko were the least controversial. Vyakirev had been looting Gazprom and diverting billions of dollars worth of assets to entities such as ITERA, from which he personally profited. Gerashchenko, the chairman of the Russian Central Bank, had been accused of lying to the International Monetary Fund. He also seemed more interested in using the bank to dole out favors than in conducting a measured monetary policy.

The arrest of Gusinsky in June 2000 and the exile of Berezovsky, as well as this summer's hounding and interrogation of the billionaire executives running Yukos Oil, are more suspect. Whatever the official explanation, the real reason for their harassment was that they had strayed beyond the boundaries of their businesses into the realm of politics.

In 2000, after he took over as president, Putin signaled to the oligarchs that if they stayed out of politics, he would not go back and reexamine their credentials and how they came to gain control of what once was state property. When Gusinsky and Berezovsky allowed or encouraged the television networks they controlled to criticize Putin for his handling of the Chechen conflict and the sinking of the submarine *Kursk*, it was considered a lapse

into politics. Similarly, as Putin saw it, Khodorkovsky's open acknowledgement that he planned to spend as much as $100 million to support political parties (reportedly even the Communists) that were opposed to Putin constituted a blatant challenge that had to be addressed.

To the outsider, it is hard to know whom to support. Even though several of the oligarchs, including Gusinsky, Berezovsky, and Khodorkovsky, have tried in various ways to atone for whatever past sins they may have committed, none are self-made men, and even more important, all have violated numerous codes of ethics, not to mention laws. Almost all of their wealth derives from taking over ownership of what had been state resources. (Gusinsky to some extent is a partial exception because he created his own media network.) None of this wealth is the result of new inventions or the creation of new manufacturing processes.

FAULTY FOUNDATIONS

The episodic harassment of one oligarch or another by the present and probably future presidents of Russia hangs over the Russian economy and potential investors, both domestic and foreign. In a sense it also represents a rebuke to the architects of reform such as Yeltsin's finance minister, Anatoly Chubais, and his Western advisers: Jeffrey Sachs, Andre Shleifer, and Anders Aslund. They felt strongly that the most important step was to transform state enterprise into private enterprise as quickly as possible. Any shortcomings would be resolved with time.

But that has not been the case. Because of the flawed manner in which the privatization process was carried out, the new owners will always lack the legitimacy necessary for a stable political climate and sustained economic investment and growth. Having constructed a faulty foundation, the builders must live with the possibility that their edifice of privatization will periodically shift and crack and may even collapse. Even with a statute of limitations that limits judicial redress, the owners of privatized Russian businesses live with constant uncertainty, knowing that sooner or later they may be charged with having used illegal methods to gain control of their enterprises.

If Russia is to attract investment from both inside as well as outside the country, some effort has to be made to curb the arbitrary use of state power and the intimidation of enemies and opponents and move instead to adhere to the rule of law. Resort to state harassment by the police and tax authorities, experienced by Gusinsky, Berezovsky, and Khodorkovsky, is a throwback to the pervasive use of intimidation and fear that was the hallmark of the Soviet era.

It is hardly reassuring that the KGB agents associated with Putin seem to have come to terms with the collapse of communism and now want to share in the spoils of the market system. Several of those in Putin's circle appear to be using state power not only to seize assets not fully privatized but also to take over assets already claimed by private owners. This behavior, as well as Putin's difficulty in shrinking the bureaucracy, underscores how deeply embedded is the presumption among senior Russian officials that those in authority can act without restraint. It also suggests that at best change will come very slowly. Those surrounding Putin as well as Putin himself seem instinctively to embrace the traditional bureaucratic modes of behavior satirized by Gogol's short story "The Inspector General." Russian *chinovniks* (bureaucrats) have survived the czar and the Soviets; they seem determined to outlive current reform attempts as well.

MARSHALL I. GOLDMAN, a Current History *contributing editor, is associate director of the Davis Center for Russian and Eurasian Studies at Harvard University and author of* The Piratization of Russia: Russian Reform Goes Awry (*Routledge, 2003*).

The Fall of the House of Saud

Americans have long considered Saudi Arabia the one constant in the Arab Middle East—a source of cheap oil, political stability, and lucrative business relationships. But the country is run by an increasingly dysfunctional royal family that has been funding militant Islamic movements abroad in an attempt to protect itself from them at home. A former CIA operative argues, in an article drawn from his new book, Sleeping With the Devil, *that today's Saudi Arabia can't last much longer—and the social and economic fallout of its demise could be calamitous.*

Robert Baer

In the decades after World War II the United States and the rest of the industrialized world developed a deep and irrevocable dependence on oil from Saudi Arabia, the world's largest and most important producer. But by the mid-1980s—with the Iran-Iraq war raging, and the OPEC oil embargo a recent and traumatic memory—the supply, which had until that embargo been taken for granted, suddenly seemed at risk. Disaster planners in and out of government began to ask uncomfortable questions. What points of the Saudi oil infrastructure were most vulnerable to terrorist attack? And by what means? What sorts of disruption to the flow of oil, short-term and long-term, could be expected? These were critical concerns. Underlying them all was the fear that a major attack on the Saudi system could cause the global economy to collapse.

The Saudi system seemed—and still seems—frighteningly vulnerable to attack. Although Saudi Arabia has more than eighty active oil and natural-gas fields, and more than a thousand working wells, half its proven oil reserves are contained in only eight fields—including Ghawar, the world's largest onshore oil field, and Safaniya, the world's largest offshore oil field. Various confidential scenarios have suggested that if terrorists were simultaneously to hit only a few sensitive points "downstream" in the oil system from these eight fields—points that control more than 10,000 miles of pipe, both onshore and offshore, in which oil moves from wells to refineries and from refineries to ports, within the kingdom and without—they could effectively put the Saudis out of the oil business for about two years. And it just would not be that hard to do.

The most vulnerable point and the most spectacular target in the Saudi oil system is the Abqaiq complex—the world's largest oil-processing facility, which sits about twenty-four miles inland from the northern end of the Gulf of Bahrain. All petroleum originating in the south is pumped to Abqaiq for pro-

cessing. For the first two months after a moderate to severe attack on Abqaiq, production there would slow from an average of 6.8 million barrels a day to one million barrels, a loss equivalent to one third of America's daily consumption of crude oil. For seven months following the attack, daily production would remain as much as four million barrels below normal—a reduction roughly equal to what *all* of the OPEC partners were able to effect during their 1973 embargo.

Oil is pumped from Abqaiq to loading terminals at Ras Tanura and Ju'aymah, both on Saudi Arabia's east coast. Ras Tanura moves only slightly more oil than Ju'aymah does (4.5 million barrels per day as opposed to 4.3 million barrels), but it offers a greater variety of targets and more avenues of attack. Nearly all of Ras Tanura's export oil is handled by an offshore facility known as The Sea Island, and the facility's Platform No. 4 handles half of that. A commando attack on Platform 4 by surface boat or even by a Kilo-class submarine—available in the global arms bazaar—would be devastating. Such an attack would also be easy, as was made abundantly clear in 2000 by the attack on the USS *Cole*, carded out with lethal effectiveness by suicide bombers piloting nothing more than a Zodiac loaded with plastic explosives.

Another point of vulnerability is Pump Station No. 1, the station closest to Abqaiq, which sends oil uphill, into the Aramah Mountains, so that it can begin its long journey across the peninsula to the Red Sea port of Yanbu. If Pump No. 1 were taken out, the 900,000 barrels of Arabian light and superlight crude that are pumped daily to Yanbu would suddenly stop arriving, and Yanbu would be out of business.

Even the short pipe run from Abqaiq to the Gulf terminals at Ju'aymah and Ras Tanura is not without opportunity. If heavy damage were inflicted on the Qatif Junction manifold complex, which directs the flow of oil for all of eastern Saudi Arabia, the flow would be stopped for months. The pipes that connect the terminals and processing facilities can be replaced off the shelf, but those at Qatif require custom fabrication.

Promoters of Alaskan, Mexican Gulf, Caspian, and Siberian oil all like to point out that the United States has been weaning itself from Saudi Arabian oil, for protection against the effects of just such an attack on the Saudi oil system. Saudi Arabia may sit on 25 percent of the world's known oil reserves, they argue, but it provides somewhere around 18 percent of the crude oil consumed by the United States—and that is down from 28 percent in only a decade. What these people fail to mention is that Saudi Arabia has the world's only important surplus production capacity—two million barrels a day. This keeps the world market liquid. Not only that, but because the Saudis more or less determine the price of oil globally by deciding how much oil to produce, even countries that don't buy Saudi oil would be vulnerable if the flow of that oil were disrupted.

The Saudis have repeatedly used their surplus production capacity to stabilize the international oil market. They used it to break the OPEC embargo (but not before they had enriched themselves by tens of billions of dollars), in 1974. They used it again during the protracted Iran-Iraq war, to keep oil flowing to the industrialized West. They used it during the Gulf War, in 1990-1991; with help from a couple of other Gulf states, they produced an extra five million barrels a day, making up for the loss of Iraqi and Kuwait oil.

And they used it again on September 12, 2001. Less than twenty-four hours after the attacks on the World Trade Center and the Pentagon, the Saudis decided to send nine million barrels of oil to the United States over the next two weeks. The result was that the United States experienced only a slight inflation spike in the wake of the most devastating terrorist attack in history. Had that same surplus capacity been taken out of play with twenty pounds of Semtex, all bets would have been off. The U.S. Strategic Petroleum Reserve can support the domestic market for only about seventy days. And if Saudi Arabia's contribution to the world's oil supply were cut off, crude petroleum could quite realistically rise from around $40 a barrel today to as much as $150 a barrel. It wouldn't take long for other economic and social calamities to follow.

Americans have long considered Saudi Arabia the one constant in the Arab Middle East. The Saudis banked our oil under their sand, and losing Saudi Arabia would be like losing the Federal Reserve. Even if the Saudi rulers one day did turn anti-American, the argument went, they would never stop pumping oil, because that would mean cutting their own throats. This, at any rate, is the way we looked at the matter before fifteen Saudis and four other terrorists launched their suicide attacks on September 11; before Osama bin Laden suddenly became for the Arab world the most popular Saudi in history; before *USA Today* reported last summer that nearly four out of five hits on a clandestine al Qaeda Web site came from inside Saudi Arabia; and before a recent report commissioned by the UN Security Council indicated that Saudi Arabia has transferred $500 million to al Qaeda over the past decade.

Five extended families in the Middle East own about 60 percent of the world's oil. The Saud family, which rules Saudi Arabia, controls more than a third of that amount. This is the fulcrum on which the global economy teeters, and the House of Saud knows what the West is only beginning to learn: that it presides over a kingdom dangerously at war with itself. In the air in Riyadh and Jidda is the conviction that oil money has corrupted the ruling family beyond redemption, even as the general population has grown and gotten poorer; that the country's leaders have failed to protect fellow Muslims in Palestine and elsewhere; and that the House of Saud has let Islam be humiliated—that, in short, the country needs a radical "purification."

We can try to wish this away all we want. But the reality is getting harder and harder to ignore. Per capita income in Saudi Arabia fell from $28,600 in 1981 to $6,800 in 2001. The country's birth rate has soared, becoming one of the highest in the world. Its police force is corrupt, and the rule of law is a sham. Saudi Arabia almost certainly leads the world in public beheadings, the venue for which is often a Riyadh plaza popularly known as Chop-Chop Square. Illegal arms routinely flow into and out of the country. Taking into account its murky "off-budget" defense spending, Saudi Arabia may spend more per capita on defense than any other country in the world (some estimates put the figure at 50 percent of its total revenues), and the House of Saud believes this is necessary for its personal protection. The regime is threatened by increasingly hostile neighbors—and by determined enemies within the country's borders. Popular preachers all over Saudi Arabia call openly for a *jihad* against the West—a designation that clearly includes the royal family itself—in terms as vitriolic as anything heard in Iran at the height of the Islamic revolution there. The kingdom's mosque schools have become a breeding ground for militant Islam. Recent attacks in Bali, Bosnia, Chechnya, Kenya, and the United States, not to mention those against U.S. military personnel within Saudi Arabia, all point back to these schools—and to the House of Saud itself, which, terrified at the prospect of a militant uprising against it, shovels protection money at the fundamentalists and tries to divert their attention abroad.

Recent examples of Saudi support for the fundamentalists abound. In 1997 a high-ranking member of the royal family coordinated a $100 million aid package for the Taliban. In Los Angeles two of the 9/11 hijackers met with a Saudi working for a company contracted to the Ministry of Defense. A raid on the Hamburg apartment of a suspected accomplice of the hijackers turned up the business card of a Saudi diplomat attached to the religious-affairs section of the embassy in Berlin. Most of the more than 650 al Qaeda prisoners being held at the Guantanamo Bay Naval Base in Cuba—"the worst of the worst," according to Secretary of Defense Donald Rumsfeld—are rumored to be Saudis.

I served for twenty-one years with the CIA's Directorate of Operations in the Middle East, and during all my years there I accepted on faith my government's easy assumption that the money the House of Saud was dumping into weaponry and national security meant that the family's armed forces and bodyguards could keep its members—and their oil—safe. "The royal family is like the fingers of a hand," my colleagues at the State Department liked to say. "Threaten it, and they become a fist." I no longer believe this. Saudi Arabia is more and more a breathtakingly irrational state. For a surprising number of Saudis, including some members of the royal family, taking the kingdom's oil off the world market—even for years, and at the risk of destroying their own economy—is an acceptable alternative to the status quo.

Saudi Arabia has existed as a formal nation only since 1932, when the tribal leader Abdul Aziz ibn Saud gained control of much of the Arabian Peninsula, named the territory after his clan, and proclaimed himself king. But the House of Saud had been powerful in the region ever since the eighteenth century, when the radical cleric Muhammad ibn Abdul Wahhab, the founder of the puritanical Wahhabi movement, wandered into Dar'iya, near present-day Riyadh, and made a bargain with its ruler, Muhammad ibn Saud. The Saud family would provide the generals, and the Wahhabis would provide the foot soldiers. Until recently it was a marriage made in heaven.

An attack on Saudi Arabia's largest oil-processing facility could create a reduction in the flow of oil roughly equal to what *all* of OPEC effected in its 1973 embargo.

If I had to pick a single moment when the House of Saud truly began to fall apart, it would be when Abdul Aziz ibn Saud's son Fahd, who has been king since 1982, suffered a near fatal stroke, in 1995. As soon as the royal family heard about Fahd's stroke, it went on high alert. From all over Riyadh came the *thump-thump* of helicopters and the sirens of convoys converging on the hospital where Fahd had been taken.

Among the first to arrive were Jawhara al-Ibrahim, Fahd's fourth and favorite wife, and their spoiled, megalomaniac twenty-nine-year-old son Abdul Aziz—or "Azouzi" ("Dearie"), as Fahd called him. Anyone who knew how Fahd's court ran knew the extent to which Fahd had come to depend on Jawhara, who helped him with everything from remembering his medicine to handling intricate problems of foreign policy. If a prince wanted a matter immediately brought to Fahd's attention, he called Jawhara. As for Abdul Aziz, he was the youngest of Fahd's children and the apple of his father's eye. Fahd indulged him in everything. Stories circulated widely about Abdul Aziz's riding a Harley-Davidson inside his father's palace, chasing servants and smashing furniture. Most of the royal family found the king's indulgence strange. Abdul Aziz was pimply, craven, a bit slow. Apparently, though, he was regarded as the king's good-luck charm. Fahd's favorite soothsayer had once told him that as long as Abdul Aziz was by his side, the king would have a long, fulfilling life. So Fahd did not complain when Abdul Aziz spent $4.6 billion on a sprawling palace and theme park outside Riyadh, because Abdul Aziz was "interested" in history. The property includes a scale model of old Mecca, with actors attending mosque and chanting prayers twenty-four hours a day, and also replicas of the Alhambra, Medina, and half a dozen other Islamic landmarks.

Next to arrive at the hospital, in a great show of solidarity, were Fahd's full brothers—Sultan, the Defense Minister; Nayef, the Interior Minister; and Salman, the governor of Riyadh province. To outsiders, they were a tight bunch. Their mother, from the Sudayri clan, had taught them from an early age to stick together or risk being elbowed out by the forty-odd other children of their father.

Other princes—the children and grandchildren of Ibn Saud's children—hurried to the hospital too, from all over the kingdom and the rest of the world. Private executive jets were lined up wing to wing at Riyadh's airport. These princes couldn't get anywhere near Fahd, but by being close at hand they could pick up more-reliable news and, just as important, demonstrate their fealty. Most of them lived off his largesse—royal stipends, which ran from $800 to $270,000 a month. The princes knew they were breaking the treasury—all told, their brethren numbered 10,000 to 12,000. Would Crown Prince Abdullah—Fahd's half brother, a seventy-one-year-old reformer who was next in line for the throne—cut back on their stipends, or even eliminate them if Fahd died? They had to stick around to find out.

A recent report commissioned by the UN Security Council indicated that Saudi Arabia has transferred $500 million to al Qaeda over the past decade.

At this point Fahd's brothers were calling doctors in the United States and Europe. They wanted to know not whether Fahd would ever recover his mental capacities, or what kind of life he would be able to live, but what it would take to keep his heart beating and his body warm. Money, of course, wasn't a problem. They told the doctors they were prepared to lease as many Boeing 747 cargo jets as needed to bring in mobile hospitals and medical teams. The doctors couldn't understand the reasoning behind the questions—but only because they didn't understand the politics of the kingdom. What the family knew and the doctors didn't was that Crown Prince Abdullah had long been eager to take power. The only way to keep him at bay was to keep Fahd alive—God willing, until Abdullah died.

Abdullah had always been the odd prince out. To begin with, his mother was from the Rashid tribe, traditional enemies of the Saud. Ibn Saud had married her to cement a trace with the Rashid, and although the Rashid were now loyal subjects, Abdullah was still mistrusted by Fahd's full brothers. Almost alone among the top members of the royal family, Abdullah had chosen the way of the desert, turning his back on the luxuries of Riyadh, Jidda, and Ta'if. He never vacationed lavishly in Europe, unlike King Fahd and his entourage, who typically spent $5 million a day during visits to the palace at Marbella, on the Spanish Riviera. Abdullah preferred to spend his time in a tent, drinking camel's milk and eating dates. He interspersed his conversation with Bedouin aphorisms and turns of phrase. All his children were raised according to the customs of the desert. It is Abdullah who has recently called publicly for democratic reforms, the reining in of the conservative clergy, and military disengagement from the United States.

The royal family hated being reminded that they had abandoned their Bedouin roots, but they hated still more that Abdullah was trying to cut back royal corruption and entitlements. Aping the senior members of the family, the lesser princes had fantastic financial expectations, and their stipends didn't suffice. The third-generation princes were getting only about $19,000 a month—a fraction of what they needed for the lifestyles they sought. To keep even a modest yacht on the French Riviera requires a million dollars a year. What were they supposed to do? In order to make ends meet they had been getting into nastier and nastier business, taking bribes from construction firms (mostly the bin Laden family's) seeking government contracts, getting involved in arms deals, expropriating property from commoners, and selling Saudi visas to guest workers. Another trick they'd discovered was borrowing money from private banks and simply refusing to pay it back. It wasn't as if the larger family could somehow discipline or shame them. There were so many princes that they didn't even all know one another.

Abdullah had made no secret of his intention to put an end to the thievery when he became king—and for a while it looked as if he might get his way even before becoming king. In the mid-1990s, as Saudi Arabia was facing increasingly dire financial difficulties, he persuaded King Fahd to appoint a handful of reformist ministers. Abdullah first had them zero in on expropriations. The practice had become so widespread among the lesser princes that it was completely alienating Saudi Arabia's traditional merchant class and fledgling middle class. A prince might walk into a restaurant, see that it was doing well, and write out a check to buy the place, usually well below market price. There was nothing the owner could do. He knew that if he resisted, he'd end up in jail on trumped-up charges.

The senior princes used their government positions to do the same thing, but on a much grander scale. One of them would pick out a valuable piece of property—maybe a particularly good location for a shopping mall or a new road—and then order a court to condemn it in the name of the state, which would clear the way for the king to award it to him. The money to be earned was staggering, and senior princes had started to rely on the practice to maintain their ever more bloated personal budgets. In Abdullah's view, however, crooked property deals and the like were only a small part of the problem. The off-budget deals were a much bigger part. In off-budget spending, revenue from oil sales goes directly to special accounts, bypassing the Saudi treasury altogether. The money is then used to pay for pet projects, from defense procurement to construction, with no government audits or accountability of any sort. Commissions and bribes are enormous.

As a reformer, Abdullah was kept out of the tight circle that gathered around Fahd after his stroke. Bitterness against Abdullah within the family was so deep that he was in fact blamed for the stroke. One version had it that Fahd and Abdullah had been on the telephone, arguing about who would attend a meeting of the Gulf Cooperation Council in Oman. It was a fundamentally unimportant decision, but relations between the two men had become so toxic, it was said, that Fahd's anger brought on the event. Another rumor in circulation held that Fahd and

Abdullah had been arguing about what they always argued about—looming financial collapse. There were even whispers that Abdullah had intentionally provoked Fahd, knowing his health wouldn't withstand a shouting match.

It eventually became clear that Fahd would live, but the extent of his impairment also became clear—embarrassingly so when, during a therapy session not long after the stroke, Fahd defecated in his pool in front of his family. His mind was affected too. Those close to him knew that he would never truly rule again, though he is still led out for ceremonial appearances.

A year and a half after Fahd's stroke Sultan had come to so despise Abdullah that he stopped attending cabinet meetings chaired by him. For Abdullah, the feeling was mutual. In July of 1997 he simply bypassed the Council of Ministers, which was heavily stacked in favor of the Sudayri, and tried to get Fahd to sign off on decrees and laws he thought needed passing. Jawhara and Abdul Aziz teamed up to thwart him.

Mind you, it is not as if the rest of the Fahd clan is united. Sultan, Salman, and Nayef may have arrived at the hospital together in a show of solidarity, but they got a rude shock once they pushed through the front doors. Jawhara and Abdul Aziz blocked them from seeing their brother. The two had set up camp outside Fahd's hospital room and were deriding who and what would or would not get in. That included ministers, senior princes, and doctors, along with petitions, decrees, and everything else.

Saudi succession doesn't operate according to primogeniture. By tradition, senior princes come to a consensus on succession, usually choosing one from their ranks who is thought to have the necessary experience and wisdom. So far the system had served the royal family well, even though Abdullah had become a gadfly, but now Fahd's brothers were afraid that Abdul Aziz was trying to circumvent custom and place himself higher in the line of succession. For one thing, he had started getting more and more involved in national security, from foreign affairs to intelligence. Even the Americans noticed it. When the commander of U.S. forces in the Middle East, General J. H. Binford Peay, came to Riyadh to meet with Fahd, in July of 1997, he was surprised to find Abdul Aziz at Fahd's side, whispering in his father's ear. Where was Abdullah? What had become of Sultan? Peay had to meet with Abdullah separately, and even then Abdullah didn't talk about the issues at hand.

What really worried some members of his family was that Abdul Aziz was funding radical Wahhabi causes and was gaining strength and popularity as a result. They had little doubt that money was going to clerics and causes that were associated with Osama bin Laden. Abdul Aziz hadn't rediscovered his faith, of course: he was courting favor with the Wahhabis because he knew he would need their support to become king. In September of 1997 he helped to coordinate that $100 million aid package for the Taliban, even though the Taliban were protecting bin Laden—a man who not only had vowed to overthrow the House of Saud but also seemed increasingly capable of doing so. Abdul Aziz was buying support wherever he could find it. In December of 1993 Abdul Aziz authorized $100,000 for a Kansas City mosque. On September 15, 1995, he opened the King Fahd Academy, in Bonn, and two days later he dedicated a new mosque there. Nine days after that he invited the head of the Islamic Society of Spain, Mansur Abdul Salam, to Riyadh. In May of 1996 he and Jawhara arranged for King Fahd to release Muhammad al Fasi from prison. Al Fasi had been imprisoned for opposing the Gulf War and the presence of U.S. troops in Saudi Arabia; in other words, he shared some of bin Laden's chief grievances. In December of 1999 the press finally caught wind of Abdul Aziz's penchant for backing radical Islamic causes. One regional account made available by U.S. translation services noted that he was believed to have been funding an associate of bin Laden's, Sa'd al Burayk, who in turn was giving the money to Islamic groups dedicated to killing Russian soldiers and civilians in Chechnya. Nayef promised to put a stop to Abdul Aziz and bring his charity back under control—but he appears to have done nothing.

All the while, throughout the 1990s, the royal family kept growing and growing. A prince might sire forty to seventy children during a lifetime of healthy copulation; however, the resources to support the growing population of the entitled were shrinking, not just in relative terms but in absolute ones. Young royals were pushing up from below, chafing at leaders who were slipping into their late seventies and eighties. The incapacitated King Fahd will turn eighty this year; Crown Prince Abdullah will turn seventy-nine. Many of the most active court intriguers are also in their seventies.

The Saudi royals number 30,000 and may grow to 60,000 in a generation. What would oil have to cost to support even the most basic privileges they now enjoy?

The House of Saud currently has some 30,000 members. The number will be 60,000 in a generation, maybe much higher. According to reliable sources, anecdotal evidence, and the Saudi gossip machine, the royal family is obsessed with gambling, alcohol, prostitution, and parties. And the commissions and other outlays to fund their vices are constant. What would the price of oil have to be in 2025 to support even the most basic privileges—for example, free air travel anywhere in the world on Saudia, the Saudi national airline—that the Saudi royals have come to enjoy? Once the family numbers 60,000, or 100,000, will there even be a spare seat for a mere commoner who wants to fly out of Riyadh or Jidda? Reformers among the royal family talk about cutting back the perks, but that's a hard package to sell.

Saudi Arabia operates the world's most advanced welfare state, a kind of anti-Marxian non-workers paradise. Saudis get free health care and interest-free home and business loans. College education is free within the kingdom, and heavily subsidized for those who study abroad. In one of the world's driest spots water is almost free. Electricity, domestic air travel, gasoline, and telephone service are available at far below cost. Many of the kingdom's best and brightest—the most well-educated

and in theory, the best prepared for the work world—have little motivation to do any work at all.

About a quarter of Saudi Arabia's population, and more than a third of all residents aged fifteen to sixty-four, are foreign nationals, allowed into the kingdom to do the dirty work in the oil fields and to provide domestic help, but also to program the computers and manage the refineries. Seventy percent of all jobs in Saudi Arabia—and close to 90 percent of all private-sector jobs—are filled by foreigners.

Among men, at least, the Saudis have an admirably high literacy rate, especially for a place that only three generations back was inhabited mostly by nomadic tribesmen. About 85 percent of Saudi men aged fifteen and older can read and write, as opposed to less than 70 percent of Saudi women of the same age. But because in recent years the Saudi education system has been largely entrusted to Wahhabi fundamentalists, as a form of appeasement that many in the royal family hope will direct the fundamentalists' animus at foreign targets, its products are generally ill prepared to compete in a technological age or a global economy. Today two out of every three Ph.D.s earned in Saudi Arabia are in Islamic studies. Doctorates are only very rarely granted in computer sciences, engineering, and other worldly vocations. Younger Saudis are being educated to take part in a world that will exist only if the Wahhabi *jihad*ists succeed in turning back the clock not just a few decades but a few centuries.

Then there's the demographic problem. Saudi Arabia has one of the highest birth rates in the world outside Africa—37.25 births for every 1,000 citizens last year, compared with 14.5 per 1,000 in the United States. Ninety-seven percent of all Saudis are sixty-four or younger, and half the population is under eighteen. The simple presence of so many people of working age, and especially so many just now ready to enter the work force, places enormous pressure on an economy—particularly one designed less to accommodate those who want to work than to provide sustenance for those who would rather contemplate original intent in the Koran. A middle class stabilizes society. Saudi Arabia's middle class is imploding.

The functioning of the world's most advanced welfare state is influenced overwhelmingly by fluctuations in the price of oil. In 1981, when the entire kingdom was in effect being put on the dole, oil was selling at nearly $40 a barrel, and the annual per capita income was $28,600. A decade later, just before Iraq invaded Kuwait, refiners were able to buy oil for about $15 a barrel. The Gulf War sent prices back up to about $36 a barrel before they quickly fell. Today a barrel of oil once again fetches around $40, but twenty years' worth of inflation, combined with a population explosion, has brought per capita income down to below $7,000. Because roughly 85 percent of Saudi Arabia's total revenues are oil-based, every dollar increase in the price of a barrel of oil means a gain of about $3 billion to the Saudi treasury. In the early 1980s the kingdom boasted cash reserves on the order of $120 billion; today the figure is estimated to be $21 billion.

Given all these threatening forces, one might think that every map in official Washington would have a red flag sticking out of Riyadh, as a reminder that Saudi Arabia is on life support. The truth is quite the opposite. Before 9/11 the United States never issued an advisory indicating the obvious security problems for Americans traveling to Saudi Arabia. Dependents of U.S. citizens residing there were never advised to leave. According to official Washington, even today the country is stable: its government is in undisputed control of its borders; its police force and army are efficient and loyal; its people are well clothed, well fed, and well educated.

Consider the way the State Department has handled visas for Saudi nationals. Until 9/11, Saudis were not even required to appear at the U.S. embassy in Riyadh or the consulate in Jidda for a visa interview. Under a system called Visa Express a Saudi had only to send his passport, an application, and the application fee to a travel agent. The Saudi travel agent, in other words, stood in for the U.S. government. Just about any Saudi who had the money could book a flight to New York after a mere twenty-hour wait. Until recently Saudis were exempt from the new anti-terrorism entry regulations that apply to citizens of other Middle Eastern countries, despite the fact that most of the 9/11 terrorists were Saudis.

"The Saudi Arabian Government, at all levels, continued to reaffirm its commitment to combating terrorism," the State Department's 1999 report *Patterns of Global Terrorism* soberly asserted. The report went on to state, "The Government of Saudi Arabia continued to investigate the bombing in June 1996 of the Khobar Towers." This was false; Prince Nayef, Saudi Arabia's grim Interior Minister, had been stalling the investigation for years. Nayef told the kingdom's other senior princes that he was reluctant to help the United States with the Khobar investigation. In one heated meeting Nayef ignored Defense Minister Sultan when Sultan warned that stonewalling the FBI would end up causing a rift with the United States. To make his point Nayef went out of his way to avoid meeting the FBI's director, Louis Freeh, when Freeh showed up in Saudi Arabia to see what he could do to get the Khobar investigation going. Nayef put himself out of reach—on his yacht, anchored off the coast near Jidda, in the Red Sea—and turned the chore over to two low-ranking officials in the internal-security service, neither of whom knew anything about the Khobar investigation.

Even after the 1998 attacks on the U.S. embassies in Kenya and Tanzania, which were organized by Osama bin Laden from his bases in Afghanistan, the Saudi royals continued to aid the Taliban and its main supporter in the region, Pakistan. This was hardly a secret: in July of 2000 *Petroleum Intelligence Weekly*, which calls itself the "bible" of the international petroleum industry, reported that Saudi Arabia was sending as many as 150,000 barrels of oil a day to Afghanistan and Pakistan in off-budget foreign aid that had a value of something like $2 million a day. Furthermore, the United States had known since 1994 that the Saudis were supporting Pakistan's nuclear development program, ultimately contributing upwards of a billion dollars. More recently, because Saudi law does not allow foreign agencies to directly question Saudi citizens, the FBI has not been allowed to interview Saudi suspects, including the families of the fifteen Saudi hijackers, about the 9/11 attacks. For more than a year after September 11 Saudi Arabia refused to provide advance manifests for flights coming into the United States—which could have led to a basic and potentially fatal breach of

security. Although there are plenty of possible al Qaeda members awaiting trial, as of this writing there hasn't been a single Saudi arrest related to 9/11—not even of a material witness.

As for the CIA, the Agency let the State Department take the lead and decided simply to ignore Saudi Arabia. The CIA recruited no Saudi diplomats to tell us, for instance, what the religious-affairs sections of Saudi embassies were up to. The CIA's Directorate of Intelligence avoided writing national intelligence estimates—appraisals, drawn from various U.S. intelligence services, about areas of potential crisis—on Saudi Arabia, knowing that such estimates, especially when negative, have a tendency to find their way onto the front pages of U.S. newspapers, where they might have an undesired effect on public opinion. The CIA's line became the same as State's: There's no need to worry about Saudi Arabia and its oil reserves.

No need to worry, of course, means business as usual—and for decades now that's meant that almost every Washington figure worth mentioning has been involved with companies doing major deals with Saudi Arabia. Spending a lot of money was a tacit part of the U.S.-Saudi relationship practically from the very beginning: the Americans would buy Saudi Arabia's oil and would provide the Saudis with protection and security; the Saudis would buy American weapons, construction services, communications systems, and drilling rigs. In the global-economics game this is known as "recycling," and in this case it worked well: two-way trade between Saudi Arabia and the United States grew from $56.2 million in 1950 to $19.3 billion in 2000.

Consider the case of the Carlyle Group—a private investment company, founded in 1987, that almost since its inception has been conducting immensely profitable business with Saudi Arabia. From 1993 to 2002 the chairman of Carlyle was Frank Carlucci, who served first as Ronald Reagan's National Security Adviser and then as his Secretary of Defense. Carlyle's senior counselor is James Baker, who served as Secretary of State under George H.W. Bush—who in his post-presidency also happens to be a Carlyle adviser. Others who hang their hats at Carlyle include Arthur Levitt, the head of the Securities and Exchange Commission under Bill Clinton, and now Carlyle's senior adviser; John Major, a former Prime Minister of Great Britain and the current chairman of Carlyle Europe; William Kennard, who chaired the Federal Communications Commission during the second Clinton Administration; Afsaneh Mashayekhi Beschloss, a former treasurer and chief investment officer of the World Bank; and Richard Darman, who ran the Office of Management and Budget under the first President Bush and also served as deputy secretary of the treasury under Reagan.

Carlyle isn't the only company in this business. Halliburton, run by Dick Cheney between his stints as Secretary of Defense under the first George Bush and Vice President under the second, has been a frequent beneficiary of Saudi money. In late 2001 Halliburton landed a $140 million contract to develop a new Saudi oil field. For many years Condoleezza Rice, now President Bush's National Security Adviser, served on the board of Chevron, which merged in 2001 with Texaco. The new corporation, ChevronTexaco, is a partner with Saudi Aramco in several ventures and has recently joined forces with Nimir Petroleum to develop oil fields in Kazakhstan. Currently on the board of ChevronTexaco are Carla Hills, who served as the Secretary of Housing and Urban Development under Gerald Ford and as a U.S. trade representative under George H.W. Bush; the former Louisiana senator J. Bennett Johnston, who made a specialty of energy issues while in Congress; and the former Georgia senator Sam Nunn, who served most notably as head of the Senate Armed Services Committee.

Elsewhere, Nicholas Brady, the Secretary of the Treasury under the first President Bush, and Edith Holiday, a former assistant to the first President Bush, serve on the board of Amerada Hess, which has teamed with some of Saudi Arabia's most powerful royal-family members to exploit the rich oil resources of Azerbaijan. In 1998 Amerada Hess formed a joint venture, Delta Hess, with the Saudi-owned Delta Oil to explore and exploit petroleum resources in Azerbaijan. The Houston-based Frontera Resources Corporation joined the Azerbaijan hunt the same year, teaming with the newly created Delta Hess. Among the members of Frontera's board of advisers: the former Texas senator former Secretary of the Treasury, and 1988 Democratic vice-presidential candidate Lloyd Bentsen; and John Deutch, a former CIA director.

Just to make sure that no one upsets the workings of this system, perhaps by meddling in internal Saudi affairs, Saudi Arabia now keeps possibly as much as a trillion dollars on deposit in U.S. banks—an agreement worked out in the early eighties by the Reagan Administration, in an effort to get the Saudis to offset U.S. government budget deficits. The Saudis hold another trillion dollars or so in the U.S. stock market. This gives them a remarkable degree of leverage in Washington. If they were suddenly to withdraw all their holdings in this country, the effect, though perhaps not as catastrophic as having a major source of oil shut down, would still be devastating.

The U.S.-Saudi relationship would not be as cozy as it is without there being someone well connected on both sides who can move comfortably between them. That someone is the fifty-four-year-old Prince Bandar. Although he ranks low on the royal bloodline (his father is King Fahd's brother Sultan, the Saudi Defense Minister, but his mother was a house servant), Prince Bandar has been the Saudi ambassador to the United States since 1983. He is the only foreign ambassador to have a security detail assigned to him by the State Department. A daredevil fighter pilot in his younger years, a Muslim with a taste for single-malt Scotch, and an envoy with a perpetually open wallet, Bandar has proved adept at working both the public and the private sides of diplomacy. As the Saudi military attache to the United States, he scored a stunning coup in 1981 by persuading Congress to approve the sale of AWACS air-defense technology to his country, over the objections of AIPAC, the pro-Israeli Washington lobby. Later, as ambassador, Bandar conveyed the kingdom's thanks by secretly placing $10 million in a Vatican City bank, as reported last year in *The Washington Post*; the money, deposited at the request of William Casey, then the director of the CIA, was to be used by Italy's Christian Democratic Party in a campaign against Italian Communists. Later still, in June of 1984, Bandar started paying out $30 million from the royal family so that Lieutenant Colonel Oliver North could buy arms for the Nicaraguan contras.

Leading U.S. corporations hire and rehire known Saudi crooks and known financiers of terrorism, so that they can land deals back in Saudi Arabia.

It is on the personal front, however, where Bandar shines. A visit in the early nineties to the summer home of George H.W. Bush, in Kennebunkport, Maine, earned the prince the affectionate family sobriquet "Bandar Bush." Bandar reciprocated by inviting Bush to hunt pheasant on his estate in England. For good measure he also contributed a million dollars to the construction of the Bush Presidential Library, in College Station, Texas. King Fahd sent another million to Barbara Bush's campaign against illiteracy. (He had donated a million dollars to Nancy Reagan's "Just Say No" campaign against drugs four years earlier.) Bandar was once Colin Powell's racquetball partner.

Press accounts portrayed Bandar as largely on the outside during the Clinton years, passing melancholy weeks at his mountain compound in Aspen, Colorado (more than 50,000 square feet, thirty-two rooms, sixteen bathrooms). If Bandar was less physically present, however, he was his usual useful self. In 1992 he persuaded King Fahd to donate $20 million to the University of Arkansas's new Center for Middle Eastern Studies, a gesture of respect for the Arkansas governor who had just been elected President. He is said to have played a role in persuading the Libyans, in 1999, to turn over two intelligence operatives suspected in the 1988 bombing of Pan Am Flight 103, over Lockerbie, Scotland. As he reportedly does at the end of every administration, whether he is perceived as friend or foe, Bandar also invited each of the Clinton Cabinet members out to dinner, at a restaurant of their choice, in a private room or a public one, depending on their willingness to be seen with him.

Prince Bandar once told associates that he is very careful to look after U.S. government officials when they return to private life. "If the reputation then builds that the Saudis take care of friends when they leave office," Bandar has observed, according to a source cited in *The Washington Post*, "you'd be surprised how much better friends you have who are just coming into office." Practically every deal with the Saudis eventually becomes hard to trace, lost in some desert sandstorm back near the wellheads where the money sprang from in the first place. Many of Washington's lobbyists, PR firms, and lawyers live off Saudi money. Just about every Washington think tank has taken it. So have the John F. Kennedy Center for the Performing Arts, the Children's National Medical Center, and every presidential library built in the past thirty years.

Bandar hurried back to prominence after the election of George W. Bush, occupying a spot somewhere between ambassador and permanently enthroned visiting head of state. But after 9/11 he began to experience some difficulty in maintaining a positive Saudi image. In March of last year agents of the Treasury Department raided the northern-Virginia offices of four Saudi-based charities: the SAAR Foundation, the Safa Trust, the International Institute for Islamic Thought (IIIT), and the International Islamic Relief Organization (IIRO). Also raided was the local headquarters for the Muslim World League, an umbrella group funded by the Saudi government. All five organizations are located only a few miles from Bandar's mansion overlooking the Potomac River. The organizations can point to a long list of genuinely humanitarian causes they have aided and supported; but they also have a long list of alarming associations. Testifying before Congress in August of 2002, Matthew Levitt, a senior fellow with the Washington Institute for Near East Policy, noted that Tarik Hamdi, an IIIT employee, had personally provided Osama bin Laden with batteries for his satellite phone—a critical link in the stateless world that bin Laden inhabits. IIIT and the SAAR Foundation are suspected of helping to finance Hamas and the Palestinian Islamic Jihad, the sponsors of some of the most lethal suicide bombers in the Middle East. From 1986 to 1994 Muhammad Jamal Khalifa, a brother-in-law to Osama bin Laden, ran the IIRO's Philippine office, from which he channeled funds to al Qaeda. Only excellent work by the Indian police prevented another IIRO employee, Sayed Abu Nasir, from bombing the U.S. consulates in Calcutta and Madras.

In mid-2002 word leaked to the press that the semi-official Defense Policy Board, chaired by the notorious cold warrior Richard Perle, had sponsored a report declaring Saudi Arabia to be part of the problem of international terrorism rather than part of the solution. Saudi Arabia, the report stated, was "central to the self-destruction of the Arab world and the chief vector of the Arab crisis and its outwardly-directed aggression." It went on to say, "The Saudis are active at every level of the terror chain, from planners to financiers, from cadre to foot-soldier, from ideologist to cheerleader." Within hours Colin Powell was on the phone to the Saudi Foreign Minister, assuring him—and through him, the royal family—that such apostasy was not and never would be the official stance of the Bush Administration. To reinforce the message, President Bush invited Bandar down to the family ranch at Crawford, Texas.

And yet the image problems have continued. In October of 2001, NATO forces raided the offices of the Saudi High Commission for Aid to Bosnia, founded by Prince Salman, and discovered, among other items, photos of the U.S. embassies in Kenya and Tanzania, before and after they were bombed; photos of the World Trade Center and the USS Cole; information on the use of crop-duster planes; and materials for forging U.S. State Department badges. His job wasn't made any easier when, in the fall of last year, Bandar found himself having to explain away the fact that about $130,000 in charitable contributions from his wife, Princess Haifa, might have ended up with two of the 9/11 hijackers.

In the wake of these revelations a U.S. delegation headed by Alan Larson, President Bush's undersecretary of state for economic affairs, traveled to Riyadh last November, ostensibly to prod the Saudis toward increasing the surveillance of their charities and financial networks. But U.S. and Saudi sources say that

one of the main reasons for Larson's trip was to ensure that if the United States invaded Iraq, the Saudis would guarantee the flow of extra oil into the World market. The U.S. embrace of the House of Saud was as tight as ever.

Washington's answer for Saudi Arabia—apart from repeating that nothing is wrong—is to suggest that a little democracy will cure everything. Talk the royal family into ceding at least part of its authority; support the reform-minded princes; set up a model parliament; co-opt the firebrands with a cabinet position or two, a minor political party, and some outright bribery; send Jimmy Carter in to monitor the first election; and in a few generations Riyadh will be Ankara, maybe even London. The governmental mechanism may be faulty, the Washington view maintains, but the people who administer the government are for the most part committed to rooting out corruption, rounding up terrorists, and recognizing the right of the people to self-government.

It's utter nonsense, of course. If an election were held in Saudi Arabia today, if anyone who wanted to could run for the office of president, and if people could vote their hearts without fear of having their heads cut off afterward in Chop-Chop Square, Osama bin Laden would be elected in a landslide—not because the Saudi people want to wash their hands in the blood of the dead of September 11, but simply because bin Laden has dared to do what even the mighty United States of America won't do: stand up to the thieves who rule the country.

Saudi Arabia today is a mess, and it is our mess. We made it the private storage tank for our oil reserves. We reaped the benefits of a steady petroleum supply at a discounted price, and we grabbed at every available Saudi petrodollar. We taught the Saudis exactly what was expected of them. We cannot walk away morally from the consequences of this behavior—and we *really* can't walk away economically. So we crow about democracy and talk about someday weaning ourselves from our dependence on foreign oil, despite the fact that as long as America has been dependent on foreign oil there has never been an honest, sustained effort at the senior governmental level to reduce long-term U.S. petroleum consumption.

Not all the wishing in the world will change the basic reality of the situation.

- Saudi Arabia controls the largest share of the world's oil and serves as the market regulator for the global petroleum industry.
- No country consumes more oil, and is more dependent on Saudi oil, than the United States.
- The United States and the rest of the industrialized world are therefore absolutely dependent on Saudi Arabia's oil reserves, and will be for decades to come.
- If the Saudi oil spigot is shut off, by terrorism or by political revolution, the effect on the global economy, and particularly on the economy of the United States, will be devastating.
- Saudi oil is controlled by an increasingly bankrupt, criminal, dysfunctional, and out-of-touch royal family that is hated by the people it rules and by the nations that surround its kingdom.

Signs of impending disaster are everywhere, but the House of Saud has chosen to pray that the moment of reckoning will not come soon—and the United States has chosen to look away. So nothing changes: the royal family continues to exhaust the Saudi treasury, buying more and more arms and funneling more and more "charity" money to the *jihad*ists, all in a desperate and self-destructive effort to protect itself.

The fact is that the West, especially the United States, has left the Saudis little choice. Leading U.S. corporations hire and re-hire known Saudi crooks and known financiers of terrorism to represent their interests, so that they can land the deals that will pay the commissions back in Saudi Arabia—commissions that will further erode the budget and thus further divide the ruling class from everyone else. Former CIA directors serve on boards whose members have to hold their noses to cut deals with Saudi companies—because that's business, that's the price of entry, that's the way it's done. Ex-Presidents, former prime ministers, onetime senators and congressmen, and Cabinet members walk around with their hands out, acting as if they're doing something else but rarely slowing down, because most of them know it's an endgame too. But sometime soon, one way or another, the House of Saud is coming down.

Robert Baer served for twenty-one years with the CIA, primarily as a field officer in the Middle East. He resigned from the agency in 1997 and was awarded its Career Intelligence Medal in 1998. This article is adapted from the forthcoming book Sleeping With the Devil *(Crown Publishers), to be published in June.*

Thirty Years of Petro-Politics:

Daniel Yergin

The 1973 oil embargo was the unsheathing of "the Arab Oil weapon." Coming the day after a decision to double prices, it was meant to be an act of retribution—punishing the United States for resupplying arms to Israel, which at that moment was still reeling from the surprise attack on Yom Kippur, several days earlier. Beyond that, it was also meant to pressure the entire Western world, whose oil consumption had been rising rapidly, into supporting the Arab side in the Arab-Israeli confrontation.

Though hardly anyone anticipated such cuts, there had been warnings. Indeed, during the 1967 Arab-Israeli war, some of the Arab exporters had tried to impose an embargo. It failed because there was plenty of extra production capacity that could be called into service. But by 1973 the world market had changed: Every well in the world, it seemed, was producing flat-out, at full capacity. The reason was the torrid growth in demand. The United States, its domestic production having flattened, had turned to the world market, and in a few short years went from minor importer to the world's largest importer. And this time there was no place to go for extra oil.

Are we less vulnerable today than in the 1970s? Oil is still the preeminent strategic commodity. And in a tight market, even without the political motives that brought on the 1973 embargo, the oil market is vulnerable to shocks and disruptions. True, markets are flexible and can adjust more quickly, but extensive turmoil in the Middle East—or other major oil-producing regions—could send new shocks throughout the world. After the last OPEC meeting, oil prices are again hovering around $30 a barrel. All this has had a significant impact on the world economy.

Thirty years ago today, a group of oil ministers gathered in Kuwait City and made a decision that shook world politics and ignited the energy crisis of the 1970s. It continues to shape the way we see energy issues today, when tight oil and gas markets and high prices are a key factor in our economy—and when the strategic significance of oil is once again evident.

The embargo created a massive global panic as buyers competed furiously with each other to get what they could, pushing the price up further. In the United States, it hit home for most consumers in infuriating gas lines—long waits for limited amounts of gasoline. (The lines were in fact largely self-in-flicted, a result of government controls that prevented flexibility and accentuated shortages in the marketplace.)

The whole international order seemed to have been transformed. Now politics was also about economics. On the day the embargo was announced, President Nixon told his advisers, "No one is more keenly aware of the stakes: oil and our strategic position." The vast flood of "petro-dollars" to the exporters turned "petro-power" into a central fact of international politics.

Prices went up fourfold in the crisis; then, a few years later, with the Iranian revolution, they doubled again. The oil crisis marked the end of the postwar economic boom and did much to turn the 1970s into the worst decade, in economic terms, since the Great Depression. Moreover, there were fears that the crisis portended a permanent shortage of oil. People speculated that the price of oil might go to $100 a barrel.

That's not how things turned out, of course. Within less than a decade, the "permanent shortage" turned into a glut, triggering a price collapse that, among other things, hastened the end of the Soviet Union, which had been depending on its oil exports as the lifeline to keep its economy alive.

There are many lessons here. Nations that had taken their energy supplies for granted suddenly realized how important reliable, reasonably priced supplies were to their well-being. Oil became high politics, and energy became part of public policy.

One of the less obvious but lasting lessons is that markets work, even in circumstances as dramatic as these were. Supply and demand adjusted. The United States and other industrial countries have since become much more efficient in the use of oil. Today—SUVs notwithstanding—the United States uses only half as much oil per unit of GDP as it did in the 1970s. New, non-OPEC sources of oil, led by Alaska and the North Sea, came on stream quickly. And the world switched from oil to other energy sources.

It is also now clear that the starting point for energy security is diversification of supplies—that is, production coming from many sources. The United States now imports oil from a large number of countries; it has a Strategic Petroleum Reserve as a supply source of last resort and can coordinate with other countries through the International Energy Agency.

The response to the crisis also demonstrates the power of technology. Technological advances have brought both greater efficiency in oil consumption and greater range in production. The oil industry is able to accomplish feats—such as drilling beneath the ultra-deep waters of the Gulf of Mexico—that were simply inconceivable in the 1970s.

Exporters learned their own powerful lesson: Customers matter. When prices skyrocketed, the exporters found that they were losing market share; customers had choices. After the price collapse, most of the exporters set out to rebuild their credibility as reliable suppliers, which meant making oil nonpolitical again. Consumers worry about security of supply, but producers—who often depend overwhelmingly on petroleum earnings for their national budgets—learned they need to worry about security of demand.

In November 1973, a few weeks after the embargo went into effect, President Nixon announced Project Independence to make the United States self-reliant in energy. Thirty years of rhetoric later, we are no closer to that goal—indeed, we're farther away. At the time of the embargo, the United States was importing a third of its oil; today it's almost 60 percent. We will continue to hear much about energy independence, but the real challenge is how to manage our dependence through diversification, efficiency, technological advances and the stability of relations with a wide range of suppliers.

The writer is chairman of Cambridge Energy Research Associates and author of "The Prize: the Epic Quest for Oil, Money, and Power."

India and China

Two systems, one grand rivalry

DELHI

As India's prime minister goes to China, Indians should learn that they have less to fear from their giant neighbour than they think

SOME comparisons are stark enough to generate a national inferiority complex. In 1980, India had about 687m people, 300m fewer than China. Living standards, as measured by purchasing power per head, were roughly the same. Then, as China embraced modernity with a sometimes ugly but burning passion, it left India behind. In the next 21 years, India outperformed its neighbour in almost nothing but population growth.

By 2001, India had 1,033m people against China's 1,272m. But China's national income per head, according to the World Bank, was $890, nearly double India's $450. Adjusted for purchasing power, the Chinese were still 70% wealthier than Indians were. In the ten years from 1992, India's GDP per head grew at 4.3% a year, China's twice as fast. Some 5% of Chinese now live below the national poverty line, compared with 29% of Indians.

Economic strength begets strategic clout. An accident of history made China one of the five permanent, veto-wielding members of the United Nations Security Council, but that seat now seems to belong to it as of right. India, feeling it should have one too, is just one of a number of big countries with a claim, and laments its comparative geopolitical weakness. As Raja Mohan writes in a new book on Indian foreign policy[1], "In all the new comparisons between the two Asian neighbours who entered the modern world at around the same time, India appeared to fall behind in every single indicator of national power and social prosperity."

Many Indian businessmen, politicians and even diplomats now often ask westerners, in an aggrieved tone, why the West is so obsessed with China's economic suc-cess. But the obsession is India's, too. Comparison with China has become a distorting mirror in which Indians see their country's shortcomings grotesquely magnified. Many are looking at—often misunderstood—versions of the "Chinese model", to see what India can learn.

For Indians, the "Chinese threat" comes in at least three forms: the geopolitical panic that rivalry with China may one day lead to another war between them; the economic nightmare of an India of underemployed farm labourers spending their meagre earnings on imported Chinese goods; and the ideological doubt that maybe India's heroic experiment with democracy has exacted an even higher price than has China's erratic dictatorship.

Bye-bye, Bhai Bhai

This was not the way Jawaharlal Nehru planned it. India's relations with China are still scarred by the bitterness that ended its first prime minister's dream of *Hindi-Chini Bhai Bhai*, Indo-Chinese brotherhood, sealed in a treaty in 1954. Sibling tension soon surfaced, and sharpened when India gave sanctuary in 1959 to the Dalai Lama and 100,000 of his followers as they fled China's suppression of an uprising in Tibet. It ended, in humiliating betrayal for Mr Nehru and India, in the war of 1962.

In the late 1950s the Chinese army was found to have built a road across what India regarded as part of Ladakh, in its state of Jammu and Kashmir. It met Indian protests with a territorial claim of its own, far to the east, to what is now the state of Arunachal Pradesh. A brief war fought on three fronts resulted in a comprehensive Chinese victory. China invaded in the east, mostly for effect, before withdrawing a month later to the "line of actual control".

It took a quarter of a century for relations to return to something like normal. In 1988 the two prime ministers, Rajiv Gandhi and Li Peng, agreed to set the border dispute to one side. Since then there have been 14 meetings of a joint working group set up to tackle it. Although its energies have been directed mainly to deciding the extent of the dispute—ie, where the line of actual control should run—rather than solving it, the process has allowed what look like cordial relations to develop. Last year Zhu Rongji, then Chinese prime minister, came to India, and his Indian counterpart, Atal Behari Vajpayee, will repay the visit this weekend. On many international issues—such as the war in Iraq—the two countries agree. Bilateral trade has grown from a paltry $338m in 1992 to nearly $5 billion in 2002.

At times, the mask of mutual respect slips. When India exploded a nuclear bomb in 1998, its defence minister, George Fernandes, let it be known that the arsenal was needed not so much because of Pakistan's nuclear ambitions as because of the long-term threat from China. It was certainly true that India's nuclear programme, started in the mid-1960s, was a response to its defeat in the 1962 war and to China's acquisition of the bomb two years later. Even this year, the present foreign minister, Yashwant Sinha, felt he needed to debunk theories of "the inevitability of conflict" between the two countries.

China's rapprochement with India has been complicated by its relations with Pakistan. Mr Vajpayee has now embarked on his "third and final" attempt to build a last-

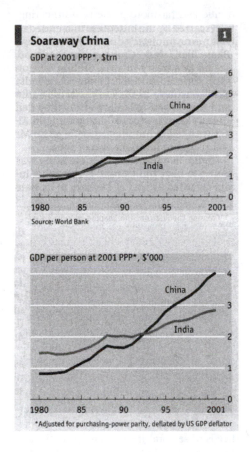

Soaraway China

GDP at 2001 PPP*, $trn

China

India

1980 85 90 95 2001

Source: World Bank

GDP per person at 2001 PPP*, $'000

China

India

1980 85 90 95 2001

*Adjusted for purchasing-power parity, deflated by US GDP deflator

ing peace with Pakistan, despite the two countries' dispute over Kashmir. He is hoping that China will be a moderating influence on Pakistan's generals. But with India still providing hospitality to the Dalai Lama, and China refusing to recognise India's incorporation of Sikkim in 1975 as a state of the Indian union, there is plenty besides the border wrangle to keep the two countries wary of each other.

Mr Mohan argues that, despite these remaining hang-ups, India is now ready for "comprehensive engagement" with China. For some Indian pundits, however, the clouds are still thick. In March, in India's *Outlook* magazine, Rajinder Puri analysed the peace process aimed at ending the Maoist rebellion in Nepal. His conclusion: "The Chinese plan to encircle India with its proxies is near completion." The Maoists, he said, had a "neat plan to make South Asia China's vassal." This is an extreme expression of what Uday Bhaskar, of the Institute of Defence Studies and Analyses in Delhi, says is undeniable: that "Chinese influence on the Indian periphery"—Pakistan, Bangladesh, Myanmar, Nepal and even Sri Lanka—has increased.

A widely held belief colours Indian analysis of China's economic policies as well as its diplomacy: that Beijing has a

grand and cunning plan, which survives all its political turmoil. India, in contrast, is an amiable, bumbling and inconsistent improviser. Mr Puri wound up his column with a little rhyme: "China moves events; India moves files. China never relents; India only smiles."

Tell me it isn't true

Many Indian businessmen and policymakers react to evidence of China's superior economic performance first with denial, and then with anger. Denial takes the form of an assault on the factual basis of the comparison: China, it is well known, stir-fries its books, and India, suffering in comparison, is the victim of geopolitical statistical fraud. The two Chinese numbers most subject to suspicion are, first, its spectacular GDP growth rates and, second, the gargantuan amounts of foreign direct investment (FDI) that China claims to attract.

It is true that both sets of figures are highly dubious. For many years, for example, China's growth figures simply have not added up: aggregated provincial data produce a higher annual growth rate than the national one (which, hardly inviting confidence, is usually published before the end of the year in question). Even some national data—on energy consumption, say, or freight volumes—are inconsistent with such rapid growth.

China's and India's relative success in attracting FDI represents the sharpest contrast of all between the two countries. According to the official data, China received $52.7 billion last year; India got just 4% of that amount, $2.3 billion. Again, some Indian economists have cried foul. India's figures are understated, they say, because they exclude foreigners' reinvested profits, the proceeds of foreign stockmarket listings, intra-company loans, trade credits, financial leases and so on. China's, on the other hand, are inflated by "round-tripping" of domestic investment through Hong Kong.

On both counts, there is little doubt that the statistics flatter China. Many economists argue that its growth figures overstate the truth by one or two percentage points. As for FDI, Sadhana Srivastava, in an article in India's *Economic and Political Weekly*, has recalculated both India's and China's figures for the year 2000 to make a fairer comparison. He found that China's FDI fell by half, while India's more than tripled. This meant that, as a

percentage of GDP, there was not much difference: 1.7% for India against 2% for China.

However, even on this basis, India was still attracting just 40% of the amount of foreign investment that went to China. Much of the gap is attributable to the activities of overseas Chinese—in Taiwan, Hong Kong, South-East Asia and America. They have ploughed far more of their money back into the motherland than have non-resident Indians, despite Indians' economic success in many countries.

Making things better

Nor, statistical quibbling aside, can there be much argument about the relative pace of growth fuelled by such investment. Some parts of India—Bangalore and Hyderabad, for example, or the industrial outskirts of Delhi—display the same feverish building that is under way in China. But in comparison with the massive construction site that is China's eastern seaboard, they seem like isolated pockets. China's growth may be patchy, localised and exaggerated. But all the evidence of the senses suggests that it is far faster than India's.

That is especially true of industrial growth, and above all of manufacturing, which in 2002 made up just 15% of India's GDP, compared with 35% of China's. Indian manufacturers scratch their heads in bafflement at China's ability to undercut them. Rahul Gupta, managing director of Phoenix Lamps, a lighting-products firm employing 1,300 people near Delhi, says that whenever he comes up against Chinese competition, he finds that the Chinese price is no more than the cost of Phoenix's raw materials.

He blames this, in part, on hidden support in the form of subsidised raw materi-

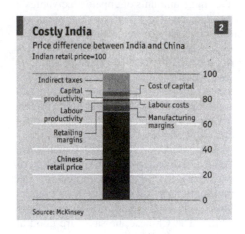

Costly India

Price difference between India and China
Indian retail price=100

Indirect taxes
Capital productivity
Labour productivity
Retailing margins
Chinese retail price

Cost of capital
Labour costs
Manufacturing margins

Source: McKinsey

als; "huge amounts of unpaid receivables"; the absence of any concept of earning a return on capital invested; and export subsidies, which, in his business, he calculates to be worth 17% of the price. Phoenix, he says, simply cannot compete, and has retreated altogether from some products and some markets in Europe and South America. But customers, he thinks, will soon realise how shoddy Chinese products are.

In a similar vein, another Indian firm, Paper Products, a Pune-based maker of packaging for big Indian companies and multinationals, sees, in the words of its boss, Suresh Gupta, "zero threat" from China. The Chinese, he says, are good at selling standardised commodities—which his company buys, causing grief for Indian suppliers—but are poor at "bespoke" products, such as unique packaging.

Both firms' experiences are commonplace. A report commissioned by the Confederation of Indian Industry, a national lobby organisation, sought to explode what it said were five widespread Indian myths about China: that its growth stems entirely from investment, not improvements in productivity; that manufacturing is driven primarily by exports; that low Chinese prices are the result of flawed accounting; that exports are priced more or less at cost; and that Chinese products are shoddy[2]. Indian industry, it seems, suffers from a curious mixture of panic and complacency about China.

The report found that neither is justified. It explains lower Chinese prices largely in terms of a tedious accumulation of minor cost disadvantages borne by Indian industry (see chart). The biggest, accounting for as much as half the difference, are sales and excise taxes, followed by the cost of capital. India's much higher import duties—a trade-weighted average of around 24% compared with China's 13%—also push up the cost of inputs. The price of copper, for example, is 25% higher in India than in China.

If the report is right that the roots of Chinese manufacturing superiority lie in the fiscal and legal framework, that might worsen Indian industry's sense of inadequacy. But it also means that policy changes could do much to help India catch up: cutting import duties; simplifying and cutting indirect taxes by introducing a nationwide value-added tax (VAT); reducing the list of industries "reserved" for small

companies; reforming the power industry by breaking down the state electricity boards into separate generation, transmission and distribution companies; easing labour laws to make hiring and firing and the use of contract workers easier. Indeed some of these reforms are already, slowly, under way, or at least under consideration.

But almost all of them are politically difficult. Bringing in a national VAT, for example, has been postponed three times, most recently this month, after state governments failed to lay the groundwork. In a number of places, the proposed tax was resisted by merchants, who shut shops in protest. With state and general elections looming, the government was loth to antagonise them. Not only do their votes matter; they can also be a vital source of party funds and, with their strike threats, of irritation to other voters. Typical, say reformists, of the political obstacles India faces.

China as a model?

"We must import either Chinese products", according to one gloomy Indian manufacturer, "or the Chinese model." Many Indians believe that a large part of the blame for their country's inferior economic performance must be borne by the political system. China, the argument goes, is a dictatorship where the government and the businesses it favours can do what they want—change laws, build infrastructure, secure licences, fiddle their books—all without brooking any opposition. In India, however, not only does every step require dealing with an inept, corrupt and intrusive bureaucracy, but the democratic system itself also imposes extra costs and delays. For every important and helpful reform, such as VAT, there is a powerful lobby that will oppose it.

Some make this argument in a spirit of pride: slower economic growth is a price worth paying for democratic freedom. For others it is an analysis of despair: what poor, backward India needs is less democracy and more discipline.

Such a political comparison, however, contains many misperceptions. First, as those who have done business in China know, decision-making there is far less monolithic, far more ponderous, far more

erratic and far more prone to profiteering by rent-seeking officials than it appears to some envious Indians.

Second, much that holds India's economy and businesses back has little to do with democracy as such: corruption, fiscal mismanagement, a lack of international ambition and a history of over-protection at home. Where India overcomes these obstacles, and has a clear competitive advantage—as in software and other information-technology services—it can be a huge success. Indian firms such as Tata Consultancy Services and Wipro leave their Chinese rivals far behind, and are preparing to compete with the best multinationals.

The way India's democracy has evolved has undoubtedly complicated policymaking. Coalition governments give disproportionate power to regional and caste-based minority parties. Worse, the electoral timetable, with many important states holding polls between the national votes, tends to put governments into cautious, electioneering mode for much of their time in office.

Yet these handicaps should not be exaggerated. The political system does offer the hope that a party might aim over the heads of sectoral-interest groups to the aspirations of all Indians for a better life. Already, Indians have the enormous benefit of being able to turf out a government that performs appallingly. India may not have had a Communist Party-led boom, but nor has it endured a Cultural Revolution or a Great Leap Forward. Indians may also have the chance, one day, of being able to elect a government that does the right things. Some of them would involve, in Mr Mohan's phrase, moving from "envy to emulation" of China. In terms of the political system itself, however, the envy and emulation should go the other way.

Notes

1. "Crossing the Rubicon: The Shaping of India's New Foreign Policy". Viking/Penguin Books India, 2003

2. "Learning from China to unlock India's manufacturing potential". Confederation of Indian Industry and McKinsey, October 2002

Overcoming Japan's China Syndrome

By Chi Hung Kwan

Introduction

More and more people in Japan have come to perceive the rise of China as a threat, prompted by the sharp contrast between the growth performances of the two countries in recent years. It should, however, be noted that the economic relations between Japan and China can be characterized as complementary rather than competitive, reflecting the prevailing gap in the level of development. Both sides can benefit by promoting a division of labor according to comparative advantage, with China specializing in labor-intensive products and Japan specializing in high-tech products.

Although Japan has lagged behind the United States and Europe in penetrating the Chinese market through foreign direct investment (FDI), it has benefited from importing cheap products from China. With China's import tariffs coming down after World Trade Organization (WTO) entry, Japanese companies now also have better access to the fast-growing Chinese market not only through FDI but also through exporting from their headquarters.

Do not Confuse "Made in China" with "Made by China"

Recently, Japanese imports of manufactured goods from China have surged and the reputation of Chinese products has improved substantially, giving rise to concern that China will soon replace Japan as the "factory of the world." An objective evaluation of China's industrial strength, however, suggests that there is still a long way to go before it will become a truly advanced industrial country on a par with Japan.

First of all, the high proportion of labor-intensive products in China's exports means that its trade structure is typical of a Newly Industrializing Economy (NIE). This is different from that of developed countries, where the major export items, such as machinery, are technology-intensive. Although China is increasing its share of the global market for manufactured goods, including some information technology products that are classified as high-tech, Chinese exports are still highly concentrated in lower-end products. In the case of televisions, for instance, Japan specializes in high-definition and other higher-end models, while China produces standard models whose unit values are much lower.

> With China's import tariffs coming down after WTO entry, Japanese companies have better access to the fast-growing Chinese market through FDI and exporting from their headquarters

Reflecting China's emphasis on processing trade, goods "made in China" contain large numbers of foreign components, some of which are made in Japan. According to official Chinese statistics, increasing exports by $1 million requires importing intermediate goods and components worth over half a million dollars, which do not form part of China's gross domestic product (GDP). Moreover, the proportion of this imported content is higher for high-tech than for low-tech products. A computer labeled "made in China" is likely to contain a large portion of imported contents including an Intel CPU, Microsoft Windows operating system, and a liquid crystal display made in Japan or South Korea.

In addition, approximately half of China's exports are produced by subsidiaries of foreign companies, to which dividends, interest charges, royalties and other fees must be paid. Even among Chinese companies with no capital relations with foreign companies, the majority of their exports are processed

Figure 1 Competition between China and Japan

under original equipment manufacturing (OEM) contracts and sold with foreign brand names. Thus only a very small percentage of the added value of products labeled "made in China" is actually "made by China." The latter corresponds to the concept of China's gross national product (GNP), and excludes import charges on intermediate goods and investment income paid to foreign countries.

China is so heavily dependent on foreign partners that it has yet to develop its own cutting-edge technology and internationally recognized brand names. On the top of this, Chinese companies are inferior to their foreign counterparts in virtually every aspect, be it capital, human resources, or business management. As a result, China has no option but to look to cheap labor for its export competitiveness. Indeed, the majority of China's contribution to the added value of its exports lies with the cost of labor, and the very low wages in China averaging about $100 a month imply that this contribution must be very small.

As such, the common assumption that Chinese goods are competitive because the country's wage levels are low holds true only for labor-intensive products, and does not necessarily apply to industry as a whole. Instead, China's low wages should be interpreted as a reflection of the fact that its labor productivity is poor. It is when China's wage levels approach those of Japan, reflecting a rise in productivity, that China will really become a formidable competitor for Japan.

Complementarity between Japan and China

The recent economic relations between Japan and China can be explained in terms of Figure 1. The horizontal axis represents the level of sophistication of export items, and the vertical axis represents the amount of exports corresponding to export items at different levels of sophistication. A country's exports can then be represented by a distribution among products at different levels of sophistication ranging from low-tech products to high-tech products. Based on the assumption that high-value-added products are likely to be exported from high-income countries, while low-value-added products are likely to be exported from low-income countries, the product sophistication index for each product can be calculated as the weighted average of the per capita GDP of its exporters. The distribution for Japan's exports is expected to be larger than that of China, reflecting its larger volume. It should also be located more to the right, reflecting the fact that high-tech products make up a larger portion of Japan's total exports. The size of the part of the two distributions that overlap one another (C in Figure 1), as a proportion of each country's total exports (A for China and B for Japan), serves as an indicator of the degree of competition between the two countries. The greater the area of overlap between the two distributions as a percentage of Japanese exports (that is, C/B), the more China is a competitor of Japan. Conversely, the smaller the overlap, the more likely that China has an export structure complementary to that of Japan. For China, the degree of competition with Japan is given by C/A. (Figure 1)

There is no question that the size of exports from Japan is bigger than that from China, and that Japan's export structure is more advanced than that of China. However, there has been rising concern in Japan that the distribution representing China is expanding rapidly and moving fast to the right. In contrast, the Japanese distribution has been static and the prospect for restarting the engine of growth has remained dim. Against this background, many people in Japan have come to believe that China has already become a strong competitor for Japan, and

Table 1 Asian Countries' Competition with China in the U.S. Market

	1990	1995	2000
Japan	3.0%	8.3%	16.3%
South Korea	24.0%	27.1%	37.5%
Taiwan	26.7%	38.7%	48.5%
Hong Kong	42.5%	50.5%	55.9%
Singapore	14.8%	19.2%	35.8%
Indonesia	85.3%	85.5%	82.8%
Malaysia	37.1%	38.9%	48.7%
Philppines	46.3%	47.8%	46.1%
Thailand	42.2%	56.3%	65.4%

Source: Calculated by the author based on U.S. Department of Commerce, *U.S. Import History*

that in the near future Japan will be eclipsed by China. Japan's China syndrome is merely an expression of this fear.

Although the total amount of exports from China has been increasing, labor-intensive products still feature largely in the export structure and the level of competition with Japan today is not necessarily high. It is clear that the export structures of China and Japan are complementary to, rather than competing with, each other—just as the big difference in their levels of economic development as well as their export volumes would lead one to expect. Based on the framework laid down in Figure 1, we use U.S. imports from individual countries with detailed breakdown by product (covering 10,000 manufactured goods according to the Harmonized System Commodity Classification) as proxies for their global exports to confirm this point. Our estimates show that China and Japan competed for only about 16.3% of their exports to the United States in value terms in 2000, although the percentage has been growing over time (from 3.0% in 1990 and 8.3% in 1995.)

These results show merely the extent to which products exported from Japan and China overlap, and two additional factors have to be considered to evaluate more accurately the degree of competition between the two countries. First of all, even though certain products are classified in the same category, in many cases Japan specializes in products for an upscale market and China specializes in low-priced products. TVs are a typical case in point, and the price tags for high-definition TVs exported by Japan are many times higher than those for the standard TVs made in China. Also, as noted above, Chinese exports include many more imported parts and components than Japanese exports.

Thus, the degree of actual competition between Japan and China is likely to be even lower than what the result of the calculations would indicate. In addition, the competition between Japan and China exists only in relatively low value-added products, in which Japan no longer enjoys any comparative advantage.

For the sake of comparison, we also calculate the level of competition with China for major Asian economies. Our estimates show that the Association of South-East Asian Nations (ASEAN) countries, whose income levels are still low, tend to compete more with China than Japan and the Asian NIEs, which are at a more advanced stage of economic development. (Table 1)

The Rise of China as a Business Opportunity for Japan

The potential complementarity between China and Japan, however, has not been fully exploited. Many Japanese companies view the expansion of China's production capacity as a threat, while at the same time finding little attractiveness in the Chinese market. As suggested by the GDP identity, income, and thus the size of the market, should grow at the same pace as output. Moreover, due to the high degree of complementarity between the two countries, Japan should enjoy advantages in penetrating the Chinese market. Market information may be biased as companies making money in China prefer to remain mute while those incurring losses are crying out in a loud voice, causing increased pessimism. If in fact market expansion is lagging behind production in China, it may reflect the following three factors.

First, foreign affiliates hold a large share of production as well as corporate earnings in China, so that the country's GNP is far below its GDP. The dividends paid by these foreign affiliates do not become income for the Chinese people, and may actually be transferred overseas. This makes China more attractive as a production base for exports than as a market. Among the foreign companies in China, if U.S. firms and European firms are making money while Japanese firms are not, then the latter should reexamine their corporate strategies.

Second, at the macro level aggregating the household, corporate and government sectors, China has a high rate of savings, so that expenditures are far lower than income. The difference does not translate into demand for goods, but rather is used by the monetary authorities to build up foreign exchange reserves. In this case, Japan should persuade China's monetary authorities to invest a larger proportion of its foreign exchange reserves in yen assets. Unfortunately, the lion's share of this "China money" is flowing to the United States instead of to Japan.

Third, China's terms of trade have been deteriorating as rising exports drive down export prices, so that the same amount of exports can be exchanged for less and less imports. This fall in purchasing power has been reflected in the yuan's sharp depreciation over time, and the slow growth in GDP in dollar terms. Under these circumstances, Japan should benefit by importing cheap products and components from China, which would allow lower prices for consumers while cutting costs at Japanese companies.

Thus even if the Chinese market is not growing as fast as production, there are various ways that Japanese companies can take advantage of China's growing economy. While importing goods produced in China through OEM and other schemes is going relatively well, Japanese investment in China has stayed at a very low level.

On the other hand, the U.S. and European countries view the emergence of China as a business opportunity, rather than as a threat, and success stories of their companies in China are on the rise. Indeed they now lead the list of the top ten foreign companies in China, which does not include a single Japanese company. Among automakers, for example, Germany's Volkswagen has a market share of 50%, while the mobile phone market has been dominated by Motorola of the United States, Nokia of Finland and Ericsson of Sweden. Even in China's electronics sector, Japanese companies are losing share as many Chinese firms emerge. Thus Japan should worry more about being left out of the fast-growing Chinese market, rather than about the hollowing out of its industry as more and more Japanese companies move to China.

Exports Versus Investment as a Means to Access the Chinese Market

One way for Japan to compromise between the concerns of hollowing out and losing the Chinese market is to put more emphasis on exporting to, than on producing in, China. Indeed, the sharp cut in tariffs following China's WTO accession should favor the former over the latter, particularly in areas where Japan enjoys comparative advantage while China does not.

One example is the automobile industry. Until now, the Chinese government has encouraged foreign automakers to produce in China by allowing auto sales in the domestic market on the one hand, while imposing high tariffs on auto imports on the other. Under this "swapping market for technology" strategy, foreign automakers would hit a wall of high tariffs when they tried to export to China, but the very same wall would protect them if their production were inside China. But things are changing. Following its entry to the WTO, China will abolish import quotas and lower import tariffs on finished cars from 80–100% to 25% by mid-2006. As a result, it may become cheaper to export finished cars to China rather than producing them there.

Should Japanese automakers opt for exports as a way to increase their presence in the Chinese market, they will be able to

shift their investment in China away from production facilities and into the reinforcement of sales networks, aftercare services and the establishment of research and development facilities for improving auto designs to better fit local tastes. Meanwhile, expanding production at home to satisfy the rising demand in China would reduce the pace of the hollowing out of Japanese industry.

Moreover, the Chinese auto industry faces so many handicaps that cheap labor in China does not necessarily translate into low production costs. Most automakers there are so small that they are unable to benefit from the economies of scale. Their productivity and research and development (R&D) capability also remain low and the quality of their products is far below global standards. As a consequence, the costs of local auto production are often higher than international prices for cars of equivalent quality. Utilizing their technology and financial resources, Japanese automakers may be able to overcome some of the drawbacks to producing in China. But it is doubtful whether they can compete with imported cars produced by their foreign rivals when China's import tariffs on auto imports come down to 25%.

The majority view among Japanese automakers is that they should produce close to the market so as to improve their brand image and meet the needs of consumers. But consumers rarely do their shopping by visiting car factories, and their needs can be served better by expanding local R&D facilities and aftercare services. Indeed, European automakers that are expanding their share of the Japanese market do not produce in Japan.

Also, for expensive goods like cars, other things being equal, Chinese consumers would certainly prefer a "made-in-Japan" to a "made-in-China" label. Above all, if Japanese automakers try to increase their production in China, they should also be prepared for the risk of overcapacity problems, which they may face in the near future as a result of too many foreign automakers rushing into the Chinese market. Instead of jumping on the bandwagon and building new plants in China, Japanese carmakers should consider the alternative option of exporting from their headquarters.

Chi Hung Kwan is a Senior Fellow at the Research Institute of Economy, Trade and Industry. He specializes in such fields as China's economic reform, regional integration in Asia and the yen bloc.

From the *Journal of Japanese Trade & Industry*, September/October 2002, pp. 26-29. © 2002 by Journal of Japanese Trade & Industry. Reprinted by permission.

What's Wrong With This Picture?

THE RISE OF THE MEDIA CARTEL HAS BEEN A LONG TIME COMING. THE CULTURAL EFFECTS ARE NOT NEW IN KIND, BUT THE PROBLEM HAS BECOME CONSIDERABLY LARGER.

MARK CRISPIN MILLER

For all their economic clout and cultural sway, the ten great multinationals—AOL Time Warner, Disney, General Electric, News Corporation, Viacom, Vivendi, Sony, Bertelsmann, AT&T and Liberty Media—rule the cosmos only at the moment. The media cartel that keeps us fully entertained and permanently half-informed is always growing here and shriveling there, with certain of its members bulking up while others slowly fall apart or get digested whole. But while the players tend to come and go—always with a few exceptions—the overall Leviathan itself keeps getting bigger, louder, brighter, forever taking up more time and space, in every street, in countless homes, in every other head.

The rise of the cartel has been a long time coming (and it still has some way to go). It represents the grand convergence of the previously disparate US culture industries—many of them vertically monopolized already—into one global superindustry providing most of our imaginary "content." The movie business had been largely dominated by the major studios in Hollywood; TV, like radio before it, by the triune axis of the networks headquartered in New York; magazines, primarily by Henry Luce (with many independent others on the scene); and music, from the 1960s, mostly by the major record labels. Now all those separate fields are one, the whole terrain divided up among the giants—which, in league with Barnes & Noble, Borders and the big distributors, also control the book business. (Even with its leading houses, book publishing was once a cottage industry at both the editorial and retail levels.) For all the democratic promise of the Internet, moreover, much of cyberspace has now been occupied, its erstwhile wildernesses swiftly paved and lighted over by the same colossi. The only industry not yet absorbed into this new world order is the newsprint sector of the Fourth Estate—a business that was heavily shadowed to begin with by the likes of Hearst and other, regional grandees, flush with the ill-gotten gains of oil, mining and utilities—and such absorption is, as we shall see, about to happen.

Thus what we have today is not a problem wholly new in kind but rather the disastrous upshot of an evolutionary process whereby that old problem has become considerably larger—and that great quantitative change, with just a few huge players now co-directing all the nation's media, has brought about enormous qualitative changes. For one thing, the cartel's rise has made extremely rare the sort of marvelous exception that has always popped up, unexpectedly, to startle and revivify the culture—the genuine independents among record labels, radio stations, movie theaters, newspapers, book publishers and so on. Those that don't fail nowadays are so remarkable that they inspire not emulation but amazement. Otherwise, the monoculture, endlessly and noisily triumphant, offers, by and large, a lot of nothing, whether packaged as "the news" or "entertainment."

Of all the cartel's dangerous consequences for American society and culture, the worst is its corrosive influence on journalism. Under AOL Time Warner, GE, Viacom et al., the news is, with a few exceptions, yet another version of the entertainment that the cartel also vends nonstop. This is also nothing new—consider the newsreels of yesteryear—but the gigantic scale and thoroughness of the corporate concentration has made a world of difference, and so has made this world a very different place.

Let us start to grasp the situation by comparing this new centerfold with our first outline of the National Entertainment State, published in the spring of 1996. Back then, the national TV news appeared to be a tidy tetrarchy: two network news divisions owned by large appliance makers/weapons manufacturers (CBS by Westinghouse, NBC by General Electric), and the other two bought lately by the nation's top purveyors of Big Fun (ABC by Disney, CNN by Time Warner). Cable was still relatively immature, so that, of its many enterprises, only CNN competed with the broadcast networks' short-staffed newsrooms; and its buccaneering founder, Ted Turner, still seemed to call the shots from his new aerie at Time Warner headquarters.

Today the telejournalistic firmament includes the meteoric Fox News Channel, as well as twenty-six television stations owned outright by Rupert Murdoch's News Corporation (which holds majority ownership in a further seven). Although ultimately thwarted in his bid to buy DirecTV and thereby dominate the US satellite television market, Murdoch wields a pervasive influence on the news—and not just in New York, where he has two TV stations, a major daily (the faltering *New York Post*) *and* the Fox News Channel, whose inexhaustible platoons of shouting heads attracts a fierce plurality of cable-viewers. Meanwhile, Time Warner has now merged with AOL—so as to own the cyberworks through which to market its floodtide of movies, ball games, TV shows, rock videos, cartoons, standup routines and (not least) bits from CNN, CNN Headline News, CNNfn (devised to counter GE's CNBC) and CNN/Sports Illustrated (a would-be rival to Disney's ESPN franchise). While busily cloning CNN, the parent company has also taken quiet steps to make it more like Fox, with Walter Isaacson, the new head honcho, even visiting the Capitol to seek advice from certain rightist pols on how, presumably, to make the network even shallower and more obnoxious. (He also courted Rush Himself.) All this has occurred since the abrupt defenestration of Ted Turner, who now belatedly laments the overconcentration of the cable business: "It's sad we're losing so much diversity of thought," he confesses, sounding vaguely like a writer for this magazine.

Whereas five years ago the clueless Westinghouse owned CBS, today the network is a property of the voracious Viacom—matchless cable occupier (UPN, MTV, MTV2, VH1, Nickelodeon, the Movie Channel, TNN, CMT, BET, 50 percent of Comedy Central, etc.), radio colossus (its Infinity Broadcasting—home to Howard Stern and Don Imus—owns 184 stations), movie titan (Paramount Pictures), copious publisher (Simon & Schuster, Free Press, Scribner), a big deal on the web and one of the largest US outdoor advertising firms. Under Viacom, CBS News has been obliged to help sell Viacom's product—in 2000, for example, devoting epic stretches of *The Early Show* to what lately happened on *Survivor* (CBS). Of course, such synergistic bilge is commonplace, as is the tendency to dummy up on any topic that the parent company (or any of its advertisers) might want stifled. These journalistic sins have been as frequent under "longtime" owners Disney and GE as under Viacom and Fox [see Janine Jaquet, "The Wages of Synergy"]. They may also abound beneath Vivendi, whose recent purchase of the film and TV units of USA Networks and new stake in the satellite TV giant EchoStar—could soon mean lots of oblique self-promotion on *USAM News*, in *L'Express* and *L'Expansion*, and through whatever other news-machines the parent buys.

Such is the telejournalistic landscape at the moment—and soon it will mutate again, if Bush's FCC delivers for its giant clients. On September 13, when the minds of the American people were on something else, the commission's GOP majority voted to "review" the last few rules preventing perfect oligopoly. They thus prepared the ground for allowing a single outfit to own both a daily paper and a TV station in the same market—an advantage that was outlawed in 1975. (Even then, pre-existing cases of such ownership were grandfathered in, and any would-be owner could get that rule waived.) That furtive FCC "review" also portended the elimination of the cap on the percentage of US households that a single owner might reach through its TV stations. Since the passage of the Telecommunications Act of 1996, the limit had been 35 percent. Although that most indulgent bill was

dictated by the media giants themselves, its restrictions are too heavy for this FCC, whose chairman, Michael Powell, has called regulation per se "the oppressor."

And so, unless there's some effective opposition, the several-headed vendor that now sells us nearly all our movies, TV, radio, magazines, books, music and web services will soon be selling us our daily papers, too—for the major dailies have, collectively, been lobbying energetically for that big waiver, which stands to make their owners even richer (an expectation that has no doubt had a sweetening effect on coverage of the Bush Administration). Thus the largest US newspaper conglomerates—the New York Times, the Washington Post, Gannett, Knight-Ridder and the Tribune Co.—will soon be formal partners with, say, GE, Murdoch, Disney and/or AT&T; and then the lesser nationwide chains (and the last few independents) will be ingested, too, going the way of most US radio stations. America's cities could turn into informational "company towns," with one behemoth owning all the local print organs—daily paper(s), alternative weekly, city magazine—as well as the TV and radio stations, the multiplexes and the cable system. (Recently a federal appeals court told the FCC to drop its rule preventing any one company from serving more than 30 percent of US cable subscribers; and in December, the Supreme Court refused to hear the case.) While such a setup may make economic sense, as anticompetitive arrangements tend to do, it has no place in a democracy, where the people have to know more than their masters want to tell them.

That imperative demands reaffirmation at this risky moment, when much of what the media cartel purveys to us is propaganda, commercial or political, while no one in authority makes mention of "the public interest"—except to laugh it off. "I have no idea," Powell cheerily replied at his first press conference as chairman, when asked for his own definition of that crucial concept. "It's an empty vessel in which people pour in whatever their preconceived views or biases are." Such blithe obtuseness has marked all his public musings on the subject. In a speech before the American Bar Association in April 1998, Powell offered an ironic little riff about how thoroughly he doesn't get it: "The night after I was sworn in [as a commissioner], I waited for a visit from the angel of the public interest. I waited all night, but she did not come." On the other hand, Powell has never sounded glib about his sacred obligation to the corporate interest. Of his decision to move forward with the FCC vote just two days after 9/11, Powell spoke as if that sneaky move had been a gesture in the spirit of Patrick Henry: "The flame of the American ideal may flicker, but it will never be extinguished. We will do our small part and press on with our business, solemnly, but resolutely."

Certainly the FCC has never been a democratic force, whichever party has been dominant. Bill Clinton championed the disastrous Telecom Act of 1996 and otherwise did almost nothing to impede the drift toward oligopoly. (As *Newsweek* reported in 2000, Al Gore was Rupert Murdoch's personal choice for President. The mogul apparently sensed that Gore would happily play ball with him, and also thought—correctly—that the Democrat would win.)

What is unique to Michael Powell, however, is the showy superciliousness with which he treats his civic obligation to address the needs of people other than the very rich. That spirit has shone forth many times—as when the chairman genially compared the "digital divide" between the information haves and have-nots to a "Mercedes divide" between the lucky few who can afford great cars and those (like him) who can't. In the intensity of his pro-business bias, Powell recalls Mark Fowler, head of Reagan's FCC, who famously denied his social obligations by asserting that TV is merely "an appliance," "a toaster with pictures." And yet such Reaganite *bons mots*, fraught with the anti-Communist fanaticism of the late cold war, evinced a deadly earnestness that's less apparent in General Powell's son. He is a blithe, postmodern sort of ideologue, attuned to the complacent smirk of Bush the Younger—and, of course, just perfect for the cool and snickering culture of TV.

Although such flippancies are hard to take, they're also easy to refute, for there is no rationale for such an attitude. Take "the public interest"—an ideal that really isn't hard to understand. A media system that enlightens us, that tells us everything we need to know pertaining to our lives and liberty and happiness, would be a system dedicated to the public interest. Such a system would not be controlled by a cartel of giant corporations, because those entities are ultimately hostile to the welfare of the people. Whereas we need to know the truth about such corporations, they often have an interest in suppressing it (as do their advertisers). And while it takes much time and money to find out the truth, the parent companies prefer to cut the necessary costs of journalism, much preferring the sort of lurid fare that can drive endless hours of agitated jabbering. (Prior to 9/11, it was Monica, then *Survivor* and Chandra Levy, whereas, since the fatal day, we have had mostly anthrax, plus much heroic footage from the Pentagon.) The cartel's favored audience, moreover, is that stratum of the population most desirable to advertisers—which has meant the media's complete abandonment of working people and the poor. And while the press must help protect us against those who would abuse the powers of government, the oligopoly is far too cozy with the White House and the Pentagon, whose faults, and crimes, it is unwilling to expose. The media's big bosses want big favors from the state, while the reporters are afraid to risk annoying their best sources. Because of such politeness (and, of course, the current panic in the air), the US coverage of this government is just a bit more edifying than the local newscasts in Riyadh.

Against the daily combination of those corporate tendencies—conflict of interest, endless cutbacks, endless trivial pursuits, class bias, deference to the king and all his men—the public interest doesn't stand a chance. Despite the stubborn fiction of their "liberal" prejudice, the corporate media have helped deliver a stupendous one-two punch to this democracy. (That double whammy followed their uncritical participation in the long, irrelevant *jihad* against those moderate Republicans, the Clintons.) Last year, they helped subvert the presidential race, first by prematurely calling it for Bush, regardless of the vote—a move begun by Fox, then seconded by NBC, at the personal insistence of Jack Welch, CEO of General Electric. Since the coup, the corporate media have hidden or misrepresented the true story of the theft of that election.

And having justified Bush/Cheney's coup, the media continue to betray American democracy. Media devoted to the public interest would investigate the poor performance by the CIA, the FBI, the FAA and the CDC, so that those agencies might be improved for our protection—but the news teams (just like Congress) haven't bothered to look into it. So, too, in the public interest, should the media report on all the current threats to our security—including those far-rightists targeting abortion clinics and, apparently, conducting bioterrorism; but the telejournalists are unconcerned (just like John Ashcroft). So should the media highlight, not play down, this government's attack on civil liberties—the mass detentions, secret evidence, increased surveillance, suspension of attorney-client privilege, the encouragements to spy, the warnings not to disagree, the censored images, sequestered public papers, unexpected visits from the Secret Service and so on. And so should the media not parrot what the Pentagon says about the current war, because such prettified accounts make us complacent and preserve us in our fatal ignorance of what people really think of us—and why—beyond our borders. And there's much more—about the stunning exploitation of the tragedy, especially by the Republicans; about the links between the Bush and the bin Laden families; about the ongoing shenanigans in Florida—that the media would let the people know, if they were not (like Michael Powell) indifferent to the public interest.

In short, the news divisions of the media cartel appear to work *against* the public interest—and *for* their parent companies, their advertisers and the Bush Administration. The situation is completely un-American. It is the purpose of the press to help us run the state, and not the other way around. As citizens of a democracy, we have the right and obligation to be well aware of what is happening, both in "the homeland" and the wider world. Without such knowledge we cannot be both secure and free. We therefore must take steps to liberate the media from oligopoly, so as to make the government our own.

Mark Crispin Miller is a professor of media studies at New York University, where he directs the Project on Media Ownership. He is the author of The Bush Dyslexicon: Observations on a National Disorder *(Norton).*

LETTER FROM BOLIVIA

LEASING THE RAIN

The world is running out of fresh water, and the fight to control it has begun.

BY WILLIAM FINNEGAN

In April of 2000, in the central plaza of the beautiful old Andean city of Cochabamba, Bolivia, the body of Víctor Hugo Daza lay on a makeshift bier. Daza, a seventeen-year-old student, had been shot in the face by the Army during protests sparked by an increase in local water rates. These protests had been growing for months, and unrest had also erupted in other parts of the country. The national government had just declared martial law. In Cochabamba, a city of eight hundred thousand, the third largest in Bolivia, a good part of the population was now in the streets, battling police and soldiers in what people had started calling *la guerra del agua*—the Water War. Peasants from the nearby countryside manned barricades, sealing off all roads to the city. The protesters had captured the central plaza, where thousands milled around a tiled fountain and the catafalque of Víctor Daza. Some of their leaders had been arrested and taken to a remote prison in the Amazon; others were in hiding.

The chief demand of the water warriors, as they were called, was the removal of a private, foreign-led consortium that had taken over Cochabamba's water system. For the Bolivian government, breaking with the consortium—which was dominated by the United States-based Bechtel Corporation—was unthinkable, politically and financially.

Bolivia had signed a lucrative, long-term contract. Renouncing it would be a blow to the confidence of foreign investors in a region where national governments and economies depend on such confidence for their survival. (Argentina's recent bankruptcy was caused in large part by a loss of credibility with international bankers.) The rebellion in Cochabamba was setting off loud alarms, particularly among the major corporations in the global water business. This business has been booming in recent years—Enron was a big player, before its collapse—largely because of the worldwide drive to privatize public utilities.

For opponents of privatization, who believe that access to clean water is a human right, the Cochabamba Water War became an event of surpassing interest. There are many signs that other poor communities, especially in Third World cities, may start refusing to accept deals that put a foreign corporation's hand on the neighborhood pump or the household tap. Indeed, water auctions may turn out to test the limits of the global privatization gold rush. And while the number of populists opposing water privatization seems effectively inexhaustible—the leaders of the Cochabamba rebellion included peasant farmers and an unassuming former shoemaker named Óscar Olivera—the same cannot be said of the world's water supply. There was a great

deal more than local water rates riding on the outcome of this strange, passionate clash in Bolivia.

The world is running out of fresh water. There's water everywhere, of course, but less than three percent of it is fresh, and most of that is locked up in polar ice caps and glaciers, unrecoverable for practical purposes. Lakes, rivers, marshes, aquifers, and atmospheric vapor make up less than one percent of the earth's total water, and people are already using more than half of the accessible runoff. Water demand, on the other hand, has been growing rapidly—it tripled worldwide between 1950 and 1990—and water use in many areas already exceeds nature's ability to recharge supplies. By 2025, the demand for water around the world is expected to exceed supply by fifty-six percent.

Some of the resource depletion is visible from outer space. The Aral Sea, in central Asia, was until recently the world's fourth-largest lake. Then Soviet planners dammed and diverted its source waters for cotton irrigation. The Aral has since lost half its area and three-fourths of its volume. Its once great fisheries have vanished; all twenty-four species native to the lake are believed to be extinct. The local climate has changed, and dust storms now plague the region.

Aquifer depletion, though less visible, is an even more serious problem. There is sixty times as much fresh water stored underground as in lakes and rivers aboveground. And yet parts of northern China, to take one example, are approaching groundwater bankruptcy. Beijing's water table has dropped more than a hundred feet in the past forty years. In the United States, the Ogallala Aquifer, which reaches from Texas to South Dakota and is indispensable to farming on the Great Plains, is being drained eight times faster than it can naturally recharge. In vast areas of India, Mexico, the Middle East, and California's Central Valley the story is the same.

Meanwhile, more than a billion people have no access to clean drinking water, and nearly three billion live without basic sanitation. Five million people die each year from waterborne diseases such as cholera, typhoid, and dysentery. This enormous, slow-motion public-health emergency is, in large measure, a result of rapid, chaotic urbanization in the nations of the Global South. Traditional water sources have been polluted, destroyed, overtaxed, or abandoned.

Annual rainfall is not always a measure of water wealth. Poland, for instance, gets plenty of rain, but its lakes, rivers, and groundwater are so polluted that it has as little usable water as Bahrain. Arid regions with the means to pay (Southern California, the Persian Gulf States) already pipe water in from wetter areas. New technologies are being hurriedly developed: huge fabric bags holding millions of gallons of fresh water are being hauled by barges across the Mediterranean, and there are businessmen in Alaska who believe that the state's earnings from fresh water will eventually dwarf its earnings from oil.

For strategic planners at some of the world's largest corporations, the global freshwater shortage coincides opportunely with privatization. According to Johan Bastin, of the European Bank for Reconstruction and Development, "Water is the last infrastructure frontier for private investors." In the past fifteen years, municipal and regional water systems have been steadily coming onto the international market. Two French corporations, Vivendi Environment and Suez, lead the industry: Vivendi runs eight thousand systems in a hundred countries; Suez has operations in a hundred and thirty countries. The biggest American player, Bechtel, whose directors include former Secretary of State George P. Shultz, has always been notable for its political connections. The United States is itself a field for direct foreign investment in water. Suez is running Atlanta's water system, and Vivendi recently bought U.S. Filter, a national water-services group, for more than six billion dollars.

But the main push is in the Global South, where, over the past twenty years, the World Bank and the International Monetary Fund have effectively taken control of the economies of scores of nations that are heavily in debt. The Bank and the I.M.F. have been requiring these countries to accept "structural adjustment," which includes opening markets to foreign firms and privatizing state enterprises, including utilities. The Bank once had a quite different approach to public works: it was an enthusiastic financier of monumental projects, and would typically lend the money to build large dams. Many of the dams were spectacular failures, delivering few, if any, benefits (except to politicians and construction firms) while displacing millions of people and leaving behind environmental destruction and public debts. The Bank is now getting out of the dam business and into water privatization. It often works closely with the conglomerates, helping them to acquire the water assets of debtor nations.

The idea behind privatization is to bring market discipline and efficiency to bear on a crucial and frequently corrupt sector. Supporters argue that only private capital—which means, in practice, multinational corporations—can afford to expand water and sanitation networks sufficiently to reach the underserved poor. Since corporations are in business to make money, they often increase water rates. But, in theory, higher water rates can also help to promote conservation. Indeed, privatization advocates say, any valuable commodity—and this includes health care and education—that is provided free eventually gets taken for granted and wasted. According to this argument, turning water into a tradable commodity may even be the only practical way to avoid worldwide shortages and environmental disasters. Public subsidies for essential services such as water may sound like humane policy, but in the real world subsidies benefit the powerful, because they have the resources to manipulate them.

In Cochabamba, which has a chronic water shortage, this unintended consequence was grotesquely clear. Most of the poorest neighborhoods were not hooked up to the network, so state subsidies to the water utility went mainly to industries and middle-class neighborhoods; the poor paid far more for water of dubious purity from trucks and handcarts. In the World Bank's view, it was a city that was crying out for water privatization.

I went to Villa San Miguel, a ramshackle settlement on an arid hillside a couple of miles south of Cochabamba, to find out how the global water business looks from the ground. My guide was a student named Fredy Villagomez, who grew up in Villa San Miguel and helped organize an independent water coöperative for the barrio. The coöperative is one of dozens that have been formed in recent years, in part with international aid. We rode out of the old city in a clattering taxi, down a road jammed with trucks, buses, minivans, donkeys, and pedestrians. It was a hot, clear afternoon; Mt. Tunari, a seventeen-thousand-foot peak, glittered in the northwest. Cochabamba sits in a wide, fertile valley at eight thousand feet—only a middling elevation in Bolivia, where the capital, La Paz, is at twelve thousand feet. To the east, beyond the mountains, lies the Amazon rain forest. Cochabamba is more than four hundred years old, but a recent influx of migrants from the countryside has caused its population to quadruple since 1976. Today, Cochabamba is ringed by dozens of *barrios marginales*—dusty, impoverished settlements that have sprung up to house the newcomers. Basic services—electricity, transportation, sanitation, water—are catch-as-catch-can in the *barrios marginales*.

"We started digging our well in 1994," Villagomez told me. Stocky, soft-spoken, with Indian features and thick eyeglasses, he was leading me down a rocky path to a small cinder-block pump house. "The planning took years. It was an expensive project, and a lot of work. All the residents helped, and we finished in 1997." The coöperative received technical assistance from Danish aid workers; there is a dirt road in the barrio—the Avenida Dinamarca—named for them. Villagomez urged me to peer into the dim pump house, which contained a single electric pump. "The well is a hundred and twelve metres deep," he said.

I thought that sounded awfully deep. Villagomez agreed. "Before, the water under this valley was at only twenty metres."

The well made a major difference to Villa San Miguel. Clean water was suddenly plentiful and relatively cheap—households paid the water coöperative between two and five dollars a month to cover the costs of running the pump and maintaining the system. "It gives water to two hundred and ten families," Villagomez said. "We felt a lot of pride in this achievement."

Then, in 1999, the Bolivian government conducted an auction of the Cochabamba water system as part of its privatization program. The auction drew only one bidder: a consortium called Aguas del Tunari. The controlling partner in the consortium was International Water, a British engineering firm that was then wholly owned by Bechtel. (An Italian company later bought a half interest.) But the government, unfazed by its own weak bargaining position, decided to proceed.

The terms of the two-and-a-half-billion-dollar, forty-year deal reflected the lack of competition for the contract. Aguas del Tunari would take over the municipal water network and all the smaller systems—industrial, agricultural, and residential—in the metropolitan area, and would have exclusive rights to all the water in the district, even in the aquifer. The contract guaranteed the company a minimum fifteen-per-cent

annual return on its investment, which would be adjusted annually to the consumer price index in the United States. On coöperative wells such as Villa San Miguel's—which the government hadn't even helped build—the new water company could install meters and begin charging for water. Residents would also be charged for the installation of the meters. These expropriations were legal under a new water law that had been rushed through the Bolivian parliament.

The first Cochabambinos to question the terms of the water privatization were not the small water coöperatives that faced expropriation and crushing bills but local professionals—mainly engineers and environmentalists—and a federation of peasant farmers who rely on irrigation. They began calling public meetings to air their concerns. The government ignored them. At a contract-signing ceremony, the President of Bolivia and the mayor of Cochabamba drank champagne with consortium executives. The news of impending expropriations and rumors of big water-price hikes began to circulate. The list of alarmed groups—neighborhood associations, water coöperatives, the labor unions—grew. There were street protests, and a broad coalition emerged, called the Coördinator for the Defense of Water and Life, or simply La Coordinadora, led by Óscar Olivera.

Olivera, who is forty-six, at first seems an unlikely leader. He wears a black leather-billed cap, which makes him look like a gentle, would-be street hood from the Beat era. He is small and sad-faced, and he often looks downward, inward—as if he were thinking extremely hard about something unpleasant. It's an especially striking manner in Bolivia, where virtually everyone presents a placid, reserved face to the world. Olivera's grandfather worked in Bolivia's huge tin mines; his father went to work as a carpenter when he was a child. Olivera started out as a machinist in a shoe factory, then went on to work for the factory's union. He is now the head of a confederation of factory workers' unions. Oddly, his air of distraction doesn't seem to unsettle anyone around him. Working people volubly engage

him wherever he goes, calling him Oscarito.

"*Compañeros*," Olivera would tell crowds. "It's become a fight between David and Goliath, between poor people and a multinational corporation. They have a lot of money, and they want to take away our water."

The World Bank warmly calls Bolivia an "early adjuster." Other poor, indebted countries have had to be forced to accept structural adjustment, but in Bolivia the World Bank and the I.M.F. have enjoyed a deep understanding with successive governments since 1985. Public enterprises—the railways, the telephone system, the national airlines, the great tin mines of Oruro and Potosí—have been sold, mainly to foreign investors. (This fire sale goes on: a Bolivian government Web site lists dozens of factories, refineries, cement plants, paper mills, and municipal utilities that are still available.)

The tin mines, as it happens, had been nationalized after a popular revolution in 1952, which also destroyed the semi-feudal hacienda system that had been in place in Bolivia for centuries. The United States played an unlikely role in that revolution. The Eisenhower Administration, already busy undermining left-wing governments in Guatemala and Guyana, accepted the new government's plea that it was not Communist (even if some of its allies were) while demanding, and getting, a new investment code that permitted American companies to start operating in Bolivia's eastern oil fields. The United States increased food aid to help the new government survive, and saw to it that "state capitalism" became the official economic model.

An Army coup overthrew the elected government in 1964, leading to the first in a long string of military regimes. (Many of the officers involved had received "counter-insurgency" training in the United States.) They were not, except during a period in the late nineteen-seventies, as violent as those in Chile, Argentina, and Brazil. The unions and the left were repressed, but not so severely as to engender a guerrilla movement. By 1982, however, when civilian rule was finally restored, the Bolivian economy, plundered by the generals, was in ruins, and hyperinflation soon took hold, hitting an annual rate of twenty-five thousand percent in 1985.

Enter "the Boys," also known as "the Chicago Boys," after a group of economists, educated at the University of Chicago, who implemented free-market policies (known in Latin America as neoliberalism) in Pinochet's Chile. In Bolivia they were led by Gonzalo Sánchez de Lozada, a wealthy mine owner who was then Minister of Planning (he was later President). Sánchez worked closely with Jeffrey Sachs, the Harvard economist who became famous for the "shock therapy" he designed for post-Communist Poland.

Bolivia's shock treatment was ferocious. The currency was devalued, all price and wage controls were abolished, government spending was cut, and the state-owned tin mines were effectively closed. The economy went into instant recession; unemployment soared. The inflationary spiral, however, was broken, and good relations between the government and the I.M.F. were restored, initiating a new flow of foreign investment and loans.

And for the past sixteen years Bolivia has dutifully followed the dictates of the World Bank and the I.M.F. Most of its people, however, have nothing to show for it. Poverty was never significantly reduced. This is not unusual in Latin America, where the poverty rate is higher today than it was in 1980—after a full generation of nominal democracy and ever-increasing free trade. But Bolivia, like Argentina, really put on what Thomas L. Friedman, the *Times* columnist, calls "the golden straitjacket" of liberalized economic policies. The predicted foreign investment materialized, but the prosperity did not. Landlocked, with a population of eight million and a wretched infrastructure, Bolivia remains the poorest country in South America.

Hugo Banzer, who was Bolivia's military dictator in the nineteen-seventies, became President again after the last election, before stepping down last August in favor of his Vice-President, Jorge Quiroga, because of illness. Quiroga now sits at the head of a "megacoalition." The

political class in Bolivia has always been small, rich, and overwhelmingly white, but rarely have the major parties, all business-aligned, shared a ruling philosophy so peaceably as in recent years. The consensus was that nothing and no one should be exempt from the discipline of the market. Then Bechtel came to Cochabamba and, as the local peasants put it, tried to "lease the rain."

When the first monthly bills from Aguas del Tunari arrived, in January, 2000, stunned business owners and middle-class householders began to join the Coordinadora's protests. Some bills had doubled, and ordinary workers now had water bills that amounted to a quarter of their monthly income.

Aguas del Tunari seemed to have given little thought to how its plans would be received in Cochabamba. The International Water executives who were actually doing the work in the city were engineers, not marketers, and, being newly arrived from abroad, they were not attuned to the problems or passions of the Bolivian public. Geoffrey Thorpe, the company's manager, simply said that if people didn't pay their water bills their water would be turned off.

In truth, the price hikes were not as arbitrary as they seemed. The consortium had agreed, in its contract, to expand the city's water system. This was going to be expensive, as was the large-scale repair job required by the deterioration of the existing system. "We were confident that we could implement this program in a shorter period of time than the one required by the contract," Didier Quint, the managing director of International Water, said. He added, however, "We had to reflect in the tariff increase all the increases that had never been implemented before." The consortium had also agreed to finish a stalled dam project known as Misicuni, which would pipe water through the mountains. This aspect of the deal seemed to make little sense—the World Bank had commissioned studies that pronounced Misicuni uneconomic. But the dam project had less to do with how privatization works in theory than with the reality of how multinational cor-

porations must come to terms with local politics.

Plans for the Misicuni Dam have been around for decades. The mayor of Cochabamba, Manfred Reyes Villa, had campaigned hard to complete it and was undeterred by questions about whether it was worth building. Reyes Villa was a popular mayor. People liked to say that his good looks got him the women's vote—he is better known by the nickname Bombón (Sweetie)—but his political instincts are sharp. Despite the widespread poverty in his city, he had pulled off major vanity projects, including a gargantuan white statue of Christ on a hilltop. (This Cristo de la Concordia is supposedly six feet taller than its rival in Rio de Janeiro.) Reyes Villa had been a real-estate developer before he became mayor, and everyone in Cochabamba was quick to note the proximity of most new roads and parks to Bombón's properties. His political party, the New Republican Force, was a personal vehicle as well as a formidable municipal patronage machine, and people said that N.F.R., its Spanish abbreviation, stood for "Nueva Forma de Robar" ("New Form of Robbery"). Reyes Villa lived in conspicuous splendor, and when I visited, soldiers were guarding his estate. There was a wall-size oil painting of his family. "Poverty is growing here, with wealth being concentrated in very few hands," he told me earnestly while we sat in his mansion. Still, Bombón won elections, his party was in President Banzer's megacoalition, and he had national ambitions. More to the point, some of his main financial backers stood to profit fabulously from the Misicuni Dam's construction. When the central government first tried to lease Cochabamba's water system to foreign bidders, in 1997, and did not include Misicuni in the tender, Bombón stopped it cold. It was only the inclusion of the project in the Aguas del Tunari contract that got the Mayor on board.

The Water War began in earnest in February, 2000. "The people went to the plaza to demonstrate against the contract," Fredy Villagomez told me. "The young men were in the city center, trying to hold the plaza. Others were here, maintaining a barricade across the highway. The women were cooking for those on the barricade. There were many campesinos passing by, walking to Cochabamba to join the rebellion." Crowds of *regantes*—peasant irrigators—arrived in the city, marching under village banners, or *wiphala*. The women, *cholitas*, wore fine straw hats and shimmering pleated velvet dresses, and their black hair was braided in long pigtails woven together in back with bright ribbons.

Most of the troops that Óscar Olivera could personally call out for demonstrations were *jubilados*—retired factory workers, old union men shuffling under battered fedoras, like faded figures on a postcard from some antipodal workers' state of another era. For a great many Bolivians, the labor unions retain an association with a time when workers were organized and proud, and when Bolivia's railroads and airlines and mines belonged to Bolivians. It made symbolic sense, then, that Olivera, the leader of the Cochabamba rebellion, was a union man, even though only a minority of Cochabamba's factory workers are still union members. (In the union confederation's dingy offices on the corner of Cochabamba's central plaza, soccer trophies fill the shelves, but when I asked about them the receptionist explained that the unions could no longer field a team.)

The growing crowds of Coordinadora supporters were drawn instead from the informal sector of pieceworkers, sweatshop employees, and street venders that has expanded enormously since the advent of structural adjustment and the closing of the tin mines. These men and women, most of them young, had more flexible schedules than workers in regular factories. Students from the University of Cochabamba, some of them middle-class anarchists, also joined in, carrying banners denouncing the World Bank, the I.M.F., and neoliberalism. But the front-line troops, particularly after the conflict sharpened and the authorities began to fire live bullets, proved to be the city's street children—an adolescent army of the homeless which has been growing in recent years.

The government's response escalated steadily. After the first protests, a ministerial delegation was sent to Cochabamba. By the end of the month, the water-price hikes had been rolled back, but the protests continued. Then the government sent in troops from Oruro and La Paz. Nearly two hundred protesters were arrested, and seventy civilians and fifty-one policemen were wounded. The Catholic archbishop of Cochabamba tried to mediate. In March, the Coordinadora held an unofficial referendum, counted nearly fifty thousand votes, and announced that ninety-six percent favored the cancellation of the contract with Aguas del Tunari. "There is nothing to negotiate," the government replied.

In April, protesters again occupied Cochabamba's central plaza, and when the Coordinadora's leaders, including Óscar Olivera, arrived at the governor's office for a meeting they were arrested. After they were released the following day, some went into hiding. Then more leaders were arrested, and some were taken to a jungle prison in the Amazon. The house of Óscar Olivera's parents was searched four times.

Protests had begun to break out in other parts of the country—in La Paz, Oruro, and Potosí, and in many rural communities—and national peasant organizations held demonstrations. By this point, most of the country's major highways were blocked.

Bolivia's rulers have always harbored a deep fear that the country's Indian majority might one day rise up and kill them in their beds—or, more realistically, trap them in their cities. In 1781, an Indian rebel army, having killed all the Spaniards in a regional capital, laid siege to La Paz for several months. The rebellion was ultimately defeated by troops brought from Buenos Aires, but white Bolivia's fear of a horizon suddenly filling with angry Indians has never fully dissipated, and on April 8, 2000, the Banzer government declared a national state of siege. This meant martial law, and it allowed for mass arrests. The minister who announced the decree also said—in a remark that Bechtel's spokesman in London quickly picked up—that the uprising in Cochabamba was being financed by *narcotraficantes*.

The state of siege, along with the comments about drug traffickers, backfired in Cochabamba. The small coca-leaf farmers, known as *cocaleros*, from the lowlands east of Cochabamba, had indeed joined the protesters, but ordinary Bolivians draw a sharp distinction between *cocaleros* and the wealthy, Army-bribing, customs-bribing *narcotraficantes*. Many of the *cocaleros* are ex-miners. Water is not their issue—they are more concerned about a coca-eradication program sponsored by the United States— but their natural sympathy was with the protesters. And the *cholitas* in their velvet dresses, and the *jubilados* marching in their rumpled fedoras, and the water warriors in their bandannas did not appreciate the suggestion that they were insincere.

The day the state of siege was declared, the main plaza in Cochabamba was filled with people. The Army fired tear gas into the narrow streets of the old city, where protesters had built barricades. Demonstrators blocked all roads into the city; the government cut off power to local radio and television stations. Middle-class matrons took wounded protesters into their homes and beauty salons to nurse them, and bowls of vinegar mixed with water and baking powder—useful for soaking bandannas for protection from tear gas—appeared outside a thousand respectable doorways.

Then, from behind a line of military police, a sharpshooter in civilian clothes fired a rifle into a crowd of unarmed civilians. He was caught on video by a Bolivian television crew, and was later identified as Robinson Iriarte de la Fuente, a Bolivian Army captain who had been trained in the United States. Víctor Hugo Daza, the seventeen-year-old student, who was on his way home from a part-time job, was, according to eyewitnesses, among the crowd that Iriarte fired into. He was hit in the face and died instantly. Dozens of other people were treated for bullet wounds. By the time Daza had been raised onto his bier and the police and the Army had been repeatedly prevented from seizing his body, there was clearly no future for Aguas del Tunari in Cochabamba.

The company's executives were told that the police could no longer guarantee their safety, and fled Cochabamba for the lowland city of Santa Cruz. They may have noted that, several weeks earlier, Mayor Reyes Villa had left their side. When the people took to the streets en masse, Bombón had assessed their mood and stepped away from Aguas del Tunari so fast that it was as if he had never seen these foreigners before. He was not the only one trying to distance himself; when water privatization collapsed in Cochabamba, the World Bank's representatives insisted that the fiasco had nothing to do with them. The government informed Aguas del Tunari that, because the company had "abandoned" its concession, its contract was revoked. (The company argued that it had not left voluntarily but had been pushed out.) The day after Víctor Hugo Daza's funeral, Óscar Olivera announced the consortium's departure to thousands of exhausted, disbelieving demonstrators from the balcony of his union's offices above the plaza.

The Coordinadora had swept the field so completely that a new national water law was immediately passed—"written from below," as the water-rights campaigners say. Banged together by parliamentarians and water specialists from the Coordinadora who gathered in La Paz, the new law gave legal recognition to usos y costumbres—traditional communal practices—by protecting small independent water systems, guaranteeing public consultation on rates, and giving social needs priority over financial goals. This triumph seemed to the water warriors too good to be true, and it was. Laws in Bolivia are implemented—if, indeed, they are ever implemented— only after bylaws have been attached and approved, and the government soon made it clear that, in the case of the new water law, this process could take years.

After the Bechtel consortium's exit, the management of Cochabamba's water was returned to the old public utility known as SEMAPA, which was thoroughly overhauled. The new board of directors included Coordinadora representatives, who vowed to treat water as a "social good" and not as an ordinary commodity. But so far the fervently envisioned transformation of Cochabamba's water system has been fitful at best. Service is still poor. Even within the existing network, many neighborhoods have service only occasionally, and the valley's aquifer continues to sink. Corruption has reportedly been reduced, but an intolerable situation persists: the poor in Cochabamba, those who are not on the network and who have no well, pay ten times as much for their water as the relatively wealthy residents who are hooked up. The new SEMAPA, having driven away international capitalists, desperately needs new capital. Since simply raising water rates across the board is politically impossible, that means new partners or new loans.

The Bolivian government has little interest in seeing the new SEMAPA succeed. Neither is it likely to get much help from the World Bank. In recent years, the Bank has been widely accused of not fulfilling (or even seriously pursuing) its self-proclaimed mission of fighting poverty, and, in response, has changed its description of its mandate, emphasizing "empowerment" and "pro-poor coalitions" over fiscal discipline. What its officials do when confronted by an actual pro-poor coalition is, of course, another matter. In Bolivia, its representatives have never met with the leaders of the Coordinadora.

The Coordinadora's leadership knows that an old-fashioned state-centered solution to Cochabamba's water crisis would be unsatisfactory. The new SEMAPA has not won the hearts of many in Cochabamba—it is, after all, still the water company. But, its supporters point out, at least it is Bolivian.

"The tragedy is that the solution to Cochabamba's water problem has been pushed off for at least another five years," Michael Curtin, a Washington-based executive who became president of Aguas del Tunari after its removal from Cochabamba, told me. Curtin had been involved in the deal as a consultant for International Water, and represents the interests of Bechtel and its partners in talks with the Bolivian government. In November, 2001, after negotiations had deadlocked, the consortium filed a complaint against the Bolivian government in a World Bank trade court in Washington. The complainant was Aguas del Tunari, but the political weight was supplied by Bechtel. The company's claim is being made under a bilateral investment treaty between Bolivia and, of all places, the Netherlands. It seems that International Water, which was originally registered in the Cayman Islands (which has no comparable investment treaty with Bolivia), moved its registration to Amsterdam soon after the Cochabamba contract was signed. Bechtel and its partners are demanding at least twenty-five million dollars in compensation for the broken contract. While trade-court proceedings are notoriously slow, nobody seems to believe that the Bolivians can afford that kind of money.

Still, the prospect of one more financial burden is the least of the government's worries in connection with what people in Bolivia refer to as the Bechtel case. The truly frightening part is the impact that the Water War has had on the foreign-investment climate. "Right now the situation is 'You can't trust Bolivia,'" Curtin said. The United States Embassy agrees. "It was a pretty significant blow," an Embassy official told me. A subsequent auction to sell the La Paz telephone company drew no bidders at all. The United States has not yet decided, the official added, whether to formally designate the breaking of the contract in Cochabamba an expropriation. That will depend on the settlement, if there is one, that is reached in the case. Having the United States label the episode an expropriation would be a blow to Bolivia's hope of seeing any new foreign investment in the near future.

And yet Jorge Quiroga, Bolivia's new President, is bullish on his country's prospects. Quiroga, who is forty-one, is tall and fair. He graduated in 1981 from the University of Texas, where he studied industrial engineering, and worked for I.B.M. in Austin as a self-described "corporate yuppie" before moving back to Bolivia with his American wife, Virginia. "We will be the vital heart of South America," Quiroga predicted repeatedly when I visited him in his ornate, republican-era office in La Paz. Gas exports will lead the way. A long-awaited transcontinental highway connecting Brazil and Chile will pass through Cochabamba. Fibre-optic cables will be laid. One of the biggest things delaying Bolivia's economic progress? The hypocrisy of the United States and Europe on free trade. "Bolivia is the most open economy in Latin America," Quiroga said. Meanwhile, American and European farm subsidies, along with tariffs on textiles and agricultural products, make it impossible for Bolivia to sell its exports in the Global North. "They tell us to be competitive while tying our arms behind our backs."

When I asked him about the Water War, he looked uncomfortable. "A lot of things certainly could have been different along the way, from a lot of different actors," he said. But, like Michael Curtin, he was certain of one point: "The net effect is that we have a city today with no resolution to the water problem." In the end, he said, it will be "necessary to bring in private investment to develop the water."

Quiroga insists that the World Bank and the I.M.F. are not running Bolivia. If it sometimes looks as if they are, that is because technocrats in the government (he has been one) use what he calls "the I.M.F. blackmail" on politicians—warning them that loans will stop if hard fiscal choices are not made, and thus giving them some cover, when in truth there have been no direct threats from the Bank or the I.M.F.

The World Bank now requires all its client governments to submit a "poverty-reduction strategy," and Quiroga gave me a sketch of Bolivia's. He had recently been to a meeting of the World Economic Forum—he goes almost every year—and his remarks struck me as having an up-to-the-minute cosmopolitan gloss. The Bolivian state has "socialized" its spending away from production toward education and health, he said, because that's the one sure route to real development. Quiroga sighed. "But we'll still be talking about poverty reduction twenty-five years from now, when one of my daughters is sitting here being interviewed."

Some things, certainly, never change. Last month, Captain Robinson Iriarte was acquitted of all responsibility for the death of young Víctor Hugo Daza. The case had been transferred from the civilian criminal-justice system—no judge was willing to hear it—to a military tribunal, which has final jurisdiction over the cases it hears. Upon his acquittal, Iriarte was promoted to the rank of major.

Claudia Vargas, a lawyer in La Paz who worked for Bolivia's fledgling utility-regulation body, travelled often to Cochabamba. The leaders of the new SEMAPA, she told me, were unnecessarily preoccupied with the idea that the central government was going to make another attempt to privatize the city's water. "I tell them 'Don't *worry*. Nobody is going to try that again for a long time.' Really, they have a lot of problems in Cochabamba now, and that is not one of them."

Vargas is young and chic, and she had recently finished a postgraduate course in development administration in thoroughly deregulated Britain. She nonetheless has an unfashionable enthusiasm for regulation. It's still poorly understood in Bolivia, she thinks, and it got a very bad name in Cochabamba when her boss (now retired) appeared to shill both for the government and for Bechtel's consortium, which he was supposed to regulate. Vargas was appalled both by the Aguas del Tunari contract and by the consortium's performance. She thinks that the members of the Coordinadora "are still the good guys in the film," but that some of them are, at best, misguided. "In the name of *usos y costumbres*, a lot of terrible things are done." Water-truck operators, for instance, "drill polluted water and sell it. They waste a lot of water. But the world is changing. Within five years, they will be charged for water—and they will be regulated."

Trying to find a path outside the binary of state or market control, the Coordinadora has no model. Its leaders vow to treat water as a "human right," and yet they know it cannot be provided free. They also know that their water company—and Bolivia generally—cannot survive without foreign investment. Meanwhile, Manfred Reyes Villa has started making his move to regain control of SEMAPA. In elections to be held soon, his party will run candidates for all the seats on the water company's board that are now occupied by the Coordinadora. Their campaigns will be far more extravagant and crowd-pleasing than anything Óscar Olivera and his colleagues are likely to stage. Bombón knows his constituency. He has also announced his candidacy for President.

The juggernaut of water privatization has hardly slowed. Bechtel, through International Water, has closed two major water deals in the past year, winning a thirty-year concession for the port city of Guayaquil, in Ecuador, and a controlling stake in the water company of Tallinn, the capital of Estonia. Many water privatizations seem likely to deliver eventually on their backers' promises of improved service. In Chile, fierce opposition to concessions has been overcome in several cities by innovative price structures, including water vouchers, that assure poor residents of an adequate supply of clean water.

At the same time, water privatizations have been backfiring all over Latin America. In Panama, popular anger about an attempted privatization helped cost the President his bid for reëlection. Vivendi, the French multinational, had its thirty-year water contract with the Argentine province of Tucumán terminated after two years because of alleged poor performance. Major water privatizations in Lima and Rio de Janeiro have had to be cancelled because of popular opposition. Trinidad recently allowed a management contract with a British water giant to expire. Protests against water privatization have also erupted in Indonesia, Pakistan, India, South Africa, Poland, and Hungary.

One large-scale Bolivian water privatization that the World Bank still points to with pride took place a few years ago in La Paz. The concession was awarded to Suez, which honored its commitment to expand the La Paz water network to several poor neighborhoods just outside the city. This area, known as El Alto, is home to nearly three quarters of a million people, virtually all of them Indians recently arrived from the countryside. But a problem emerged. It seemed that the people in El Alto weren't using enough water. Accustomed to Andean peasant life, they were extremely careful with water, never wasting a drop, and they continued to be so even after they had taps installed in their homes. This was good conservation, but it was bad for Suez's bottom line, and the corporation was disappointed in the return on its investment. After it appeared to raise its rates, which were pegged to the dollar, when the local currency was devalued the general happiness with the contract evaporated and residents began to complain about the service. When I was in La Paz, the people of El Alto were marching against Suez. When I asked a World Bank official about the situation, she agreed that there was a basic problem: those Indians needed to learn to use more water.

From *The New Yorker,* April 8, 2002, pp. 43-53. © 2002 by The Condé Nast Publications, Inc.

UNIT 5
Conflict

Unit Selections

Key Points to Consider

- Are violent conflicts and warfare increasing or decreasing today?

- What changes have taken place in recent years in the types of conflicts and who participates?

- How is military doctrine changing to reflect new political realities?

- How is the nature of terrorism different than conventional warfare? What new threats do terrorists pose?

- How is the national security policy of the United States likely to change? What about Russia, India, and China?

 Links: www.dushkin.com/online/
These sites are annotated in the World Wide Web pages.

DefenseLINK
   http://www.defenselink.mil
Federation of American Scientists (FAS)
   http://www.fas.org
ISN International Relations and Security Network
   http://www.isn.ethz.ch
The NATO Integrated Data Service (NIDS)
   http://www.nato.int/structur/nids/nids.htm

Do you lock your doors at night? Do you secure your personal property to avoid theft? These are basic questions that have to do with your sense of personal security. Most individuals take steps to protect what they have, including their lives. The same is true for groups of people, including countries.

In the international arena, governments frequently pursue their national interest by entering into mutually agreeable "deals" with other governments. Social scientists call these types of arrangements "exchanges" (i.e., each side gives up something it values in order to gain something in return that it values even more). On an economic level, it functions like this: "I have the oil that you need and am willing to sell it. In return I want to buy from you the agricultural products that I lack". Whether on the governmental level or the personal level ("If you help me with my homework, then I will drive you home this weekend"), exchanges are the process used by most individuals and groups to obtain and protect what is of value. The exchange process, however, can break down. When threats and punishments replace mutual exchanges, conflict ensues. Neither side benefits and there are costs to both. Further, each may use threats with the expectation that the other will capitulate. But if efforts at intimidation and coercion fail, the conflict may escalate into violent confrontation.

With the end of the cold war, issues of national security and the nature of international conflict have changed. In the late 1980s agreements between the former Soviet Union and the United States led to the elimination of superpower support for participants in low-intensity conflicts in Central America, Africa, and Southeast Asia. Fighting the cold war by proxy is now a thing of the past. In addition, cold war military alliances have either collapsed or have been significantly redefined. Despite these historic changes, there is no shortage of conflicts in the world today.

Many experts initially predicted that the collapse of the Soviet Union would decrease the arms race and diminish the threat of nuclear war. However, some analysts now believe that the threat of nuclear war has in fact increased as control of nuclear weapons has become less centralized and the command structure less reliable. In addition, the proliferation of nuclear weapons into North Korea and South Asia (India and Pakistan) is a growing security issue. Further, there are concerns about both dictatorial governments and terrorist organizations obtaining weapons of mass destruction. What these changing circumstances mean for U.S. policy is a topic of considerable debate.

The unit begins with two perspectives on contemporary U.S. national security issues. First, the fundamental redefinition of U.S. security in the post 9/11 era is described. Then an analysis of the changes in Department of Defense contracting is described along with the policy consequences of war profiteering.

In addition to the pressing issues of terrorism and the conflict in the Middle East, there are additional strategic issues of considerable importance that are examined. The unit concludes with a series of articles that present specific case studies related to security issues in Asia, including the challenges presented by nuclear proliferation.

Like all the other global issues described in this anthology, international conflict is a dynamic problem. It is important to understand that conflicts are not random events, but follow patterns and trends. Forty-five years of cold war established discernable patterns of international conflict as the superpowers deterred each other with vast expenditures of money and technological know-how. The consequence of this stalemate was often a shift to the developing world for conflict by superpower proxy.

The changing circumstances of the post–cold war era generate a series of important new policy questions: Will there be more nuclear proliferation? Is there an increased danger of so-called "rogue" states destabilizing the international arena? Is the threat of terror a temporary or permanent feature of world affairs? Will there be a growing emphasis on low-intensity conflicts related to the interdiction of drugs, or will some other unforeseen issue determine the world's hot spots? Will the United States and its European allies lose interest in security issues that do not directly involve their economic interests and simply look the other way, for example, as age-old ethnic conflicts become brutally violent? Can the international community develop viable institutions to mediate and resolve disputes before they become violent? The answers to these and related questions will determine the patterns of conflict in the twenty-first century.

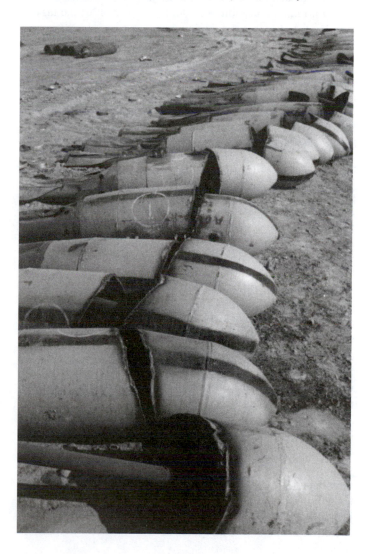

The Transformation of National Security

Five Redefinitions

Philip Zelikow

PERHAPS THE key question of international politics and U.S. national security policy today is whether a genuinely new era has dawned since the end of the Cold War. It has. The attacks of September 11, 2001 did not create the new era, but they were a catalytic moment in our recognition of it. Like previous shocks to the United States in June 1940, December 1941 or June 1950, this shock gave emerging trends a form, brought them into mass consciousness, and forced upon us the task of defining a comprehensive national response.

Such a definition appears in the Bush Administration's recently published *National Security Strategy of the United States*.[1] This essay draws out some of the ideas that appear to undergird the administration's emerging strategy. It focuses on five essential redefinitions of what national security means for the United States in the 21st century—but first a note about the rhetoric of empire that has come to dominate much current discussion.

The Distraction of Empire

ALL NATIONAL security strategies start with a mental image of the world. The image of the new era is properly that of a modern and truly pluralistic international system. In the traditional world, populations, governance, commerce, culture and habits of life evolved slowly and changed gradually. The break point between this traditional world and that of modernity arrives, at different times for different societies, when social and technological change severs the links that had defined the relationships between humanity and the physical resources of the earth; shatters the hitherto ageless ceilings on productivity enforced by the physical limits of humans, animals, wind and water; and transforms our ability to communicate across distances, communities and nations.

These transformations acquired momentum in the 18th century and spread in the 19th until, by the year 1900, the modern world extended to Europe, North America and to their limited veins of settlement and commerce stretched out over the rest of the planet. The 20th century saw the further, nearly global extension of modernity and wrenching, worldwide efforts to adjust to its impact. Indeed, such has been the shock to many societies that the last century has been dominated by great contests over how to conceive, organize and provide moral justification and political order for modern industrial nations.

Those great contests have now subsided. The world is no longer so broken and divided. The militant utopias of class, nation and race have been defeated and discredited. Modernity means that change itself becomes the constant, and the challenge of change has not disappeared. Its pressures have instead been internalized within every society trying to adapt to a quickening pace of change. In other words, today's battle lines are less *international* and more *transnational*.

In dealing with these pressures, some see the United States as a new source of world order—fearing or welcoming that prospect. This accounts in part for the current popularity of "empire" as a reigning metaphor for America's ambitions. The metaphor is seductive, yet vicious.

What is an empire? Once upon a time, "empire" was casually applied as a positive expression, as with Jefferson's "empire of liberty." In recent years "empire" has been used to describe—often with an edge—any circumstance where a powerful country exerts influence over lesser powers, whether direct or indirect, physical, cultural or commercial. This shallow equation of all sorts of economic and cultural influences with "imperialism" was first popularized about a hundred years ago by writers reacting to Britain's war against the Boers in South Africa. John Atkinson Hobson, for example, who had an immense influence on Vladimir Lenin, condemned economic imperialism as a ruling force in the world. Hobson also worked a strong element of his anti-Semitism into his theories, seeing a cabal of Jewish bankers and merchants lurking behind Britain's excesses.

The same distasteful mixture of bad thinking was prominent after World War I. Many writers in Europe, especially in the defeated nations, decried the new "Anglo-Saxon empire" that they accused Britain and America—and the Jews, of course—of having created.[2] The same fetid intellectual waters seep from this old gutter into much anti-American commentary today.

Sadly, even some American conservatives have joined the new anti-imperialists in seeing the United States as the metropole of a world empire. The appeal is understandable. After all, there is a certain gratification in imagining that one is the successor to emperors and proconsuls of past ages, a certain pleasure in opening such a dusty, venerable chest of ideas about how to sort out the affairs of faraway peoples.

But these imperial metaphors, of whatever provenance, do not enrich our understanding; they impoverish it. They use a metaphor of how to rule others when the problem is how to persuade and lead them. Real imperial power is sovereign power. Sovereigns rule, and a ruler is not just the most powerful among diverse interest groups. Sovereignty means a *direct* monopoly control over the organization and use of armed might. It means *direct* control over the administration of justice and the definition thereof. It means control over what is bought and sold, the terms of trade and the permission to trade, to the limit of the ruler's desires and capacities.

In the modern, pluralistic world of the 21st century, the United States does not have anything like such direct authority over other countries, nor does it seek it. Even its informal influence in the political economy of neighboring Mexico, for instance, is far more modest than, say, the influence the British could exert over Argentina a hundred years ago.

The purveyors of imperial metaphors suffer from a lack of imagination, and more, from a lack of appreciation for the new conditions under which we now live. It is easier in many respects to communicate images in a cybernetic world, so that a very powerful United States does exert a range of influences that is quite striking. But this does not negate the proliferating pluralism of global society nor does it suggest a will to imperial power in Washington. The proliferation of loose empire metaphors thus distorts into banal nonsense the only precise meaning of the term imperialism that we have.

The United States is central in world politics today, not omnipotent. Nor is the U.S. Federal government organized in such a fashion that would allow it to wield durable imperial power around the world—it has trouble enough fashioning coherent policies within the fifty United States. Rather than exhibiting a confident will to power, we instinctively tend, as David Brooks has put it, to "enter every conflict with the might of a muscleman and the mentality of a wimp." We must speak of American power and of responsible ways to wield it; let us stop talking of American empire, for there is and there will be no such thing.

Five Redefinitions

THE UNITED States does have unique responsibilities as the greatest power in this pluralistic world, however. Those responsibilities have moved the Bush Administration to rethink the meaning of America's national security and it is a process of thinking that transcends yesterday's partisan differences. Both conservative and liberal orthodoxies are being challenged. To attain lasting influence, these new ideas must pass into the vocabulary and assumptions of many in *both* parties—just as happened with strategies of "containment" in the late 1940s and early 1950s—even as political conflict continues around the edges of this new vision.

- This vision is redefining the geography of national security.
- It is redefining the nexus between principles and power.

- It is redefining the structure of international security.
- It is redefining multilateralism.
- And it is redefining national security threats in the dimension of time.

The New Geography of National Security

IN THE PAST, the geography of national security was defined by foreign frontiers. Dangerous enemies had to possess mass and scale as they first accumulated armies, navies or air forces and then deployed them. Today the frontiers of national security can be everywhere. The point is so obvious in the case of mass-casualty terrorism that it needs no elaboration.

Less obvious is the way the Bush Administration, following on but surpassing the Clinton Administration, has consistently identified poverty, pandemic disease, biological and genetic dangers, and environmental degradation as significant national security threats. All of these dangers have a transnational character, their social origins resting within the turmoil of modern and modernizing societies. In other words, national security threats in the new era are defined more by the fault-lines within societies than by the territorial borders between them. The decisive clashes in this phase of history are not therefore between civilizations but inside them.

The implications of this redefinition touch every major institution of national security in the U.S. government. We are witness to the largest Executive Branch transformation in half a century with the new Department of Homeland Security, the largest reorganization of the FBI in a generation, significant overhauling of the U.S. intelligence community, the creation of an entirely new unified command and other dramatic restructuring in the Department of Defense. Fundamental relationships between Federal, state, regional and local agencies—and the private sector—must change, and are changing, as well. These transformations are just getting under way, but they are already breaking down the core paradigm of the American national security system created in the 1940s and 1950s.

The geographical redefinition of national security profoundly affects the entire U.S. foreign policy agenda. Whether dealing with terrorism or public health, the division of security policy into domestic and foreign compartments is breaking down. U.S. foreign and security policies must delve into societies, into problems from law enforcement to medical care, in novel ways—challenging international institutions and the principles that define them to adapt. More fundamentally, if the United States is to develop national security policies that are aimed at the fault-lines within societies, those policies must transcend physical and material dimensions. They must include positions about fundamental principles.

The New Centrality Of Moral Principles

DURING AND after World War II a conventional image of American interests developed that posed a dichotomy between realism and idealism. Practically every thinker on foreign policy alive today has grown up under the influence of this dichotomy. Realism was usually identified as a cold-eyed focus on calculations of power. Idealism embraced a pre-eminent concern for human rights, global poverty or other facets of human welfare. These two images became a convenient shorthand for labeling political factions or individual leaders. But those stereotypes, overly simple to begin with, no longer even remotely suit our times. They do not capture the nature of the controversies within the present administration, nor do they comprehend the new fusion of power and principle that is now guiding U.S. policy.

The administration's concept of a "balance of power that favors freedom" —to note the marquee concept of the *National Security Strategy*—applies calculations of power to the worldwide capacity to support beneficial principles affecting both relations among states and conditions within them. The administration thus emphasizes both power and a readiness to distinguish good from bad, right from wrong. The administration takes to heart the observation of John Courtney Murray that

> policy is the meeting-place of the world of power and the world of morality, in which there takes place the concrete reconciliation of the duty of success that rests upon the statesman, and the duty of justice that rests upon the civilized nation that he serves.[3]

The administration drives power and principle together around a remarkably straightforward statement of the "nonnegotiable demands of human dignity." Seven of these demands, all originally appearing in the President's January 2002 State of the Union message, are listed in the strategy document: the rule of law; limits on the power of the state; respect for women; private property; free speech; equal justice; and religious tolerance. All seven focus on the relationship of individuals to the state. None deals directly with the form or processes of government to produce these relationships: There is no mention here of democracy. Far from being an assertion of American exceptionalism, therefore, or a call for others to emulate the example of our "city on a hill", both the strategy document and President Bush have stressed that these are universal principles that apply everywhere. This is not an affirmation of the Scottish Enlightenment, but of human civilization itself.

Such a stance is critical in an age that seems to contradict or qualify every universal truth, preferring instead to cultivate irony as the essential human sensibility. The Bush Administration has defined one constant as the essential complement to modernity, what the *National Security Strategy* calls a "single, sustainable model for national success." That model features a linked conception of respect for human dignity and regard for liberating human potential. If modernity implies constant change, then greater personal and economic freedom is the perpetual safety valve, the constant source of adaptation and thus the very source of structural resilience that enables different societies to organize themselves in different successful forms. Influenced by his experiences in the Balkan crises, Tony Blair put the matter well in his October 1, 2002 speech to the Labour Party conference in Blackpool:

> Our values aren't Western values. They're human values and anywhere, anytime people are given the chance, they embrace them. Around these values, we build our global partnership. Europe and America together. Russia treated as a friend and equal. China and India seeking not rivalry but cooperation and for all nations the basis of our partnership—not power alone but a common will based on common values, applied in an even-handed way.

The Bush Administration's similar emphasis on moral reasoning, on uncomfortable (to some) judgments about good and evil, challenges some conservatives and liberals alike. The challenge to conservatives flows from the fact that traditional conservatism is founded on a cultural pessimism, on an abiding skepticism about human improvement. But President Bush (and many of his key advisors) are cultural optimists who, in their own lifetimes, have witnessed great and *positive* upheaval—not least the successful end to the Cold War. Indeed, President Bush's whole education reform agenda ("Leave no child behind!") epitomizes the break with traditional conservatism, and it is an attitude that carries over to world affairs. His administration's efforts on HIV/AIDS, for example, have not bothered with the Clinton Administration's convoluted effort to justify attention to global disease in traditional national security terms. The danger is not dressed up as a threat to "stability" but is portrayed for what it really is: a moral obligation to act when tens of millions face preventable death.

The administration's rhetoric is broader still: "Including all of the world's poor in an expanding circle of development—and opportunity", the *National Security Strategy* states, "is a moral imperative…one of the top priorities of U.S. international policy." Hence the proposal for a 50 percent increase in development assistance, the largest proposed by any administration since that of John F. Kennedy. President Bush is gambling that a notoriously broken foreign aid system can be reformed to make such investments politically saleable, and perhaps thereby legitimize an even greater effort in the future.

The challenge to liberals has been more interesting, if somewhat depressing in its manifestations. The American Left in its various hues once defined itself by its fierce belief in values, a readiness to judge and to act on those judgments. It is now too often defined by dedication to a consensus that disclaims any right to make moral judgments at all—except those that condemn the U.S. government. The Left's "knowing" attitude about America's own failings has evolved into the chronic suspicion of any expression of patriotism, and into the instinctive assumption that anyone who uses words like "evil" must be some Bible-thumping rube pronouncing naive moral judgments. And to think that such a person is President of the United States!

But many liberal Democrats are now wisely pressing for a return to the ancient virtues. What, then, is the net result? One old contrarian of the Left, Christopher Hitchens wrote of it in the October 20, 2002 *Washington Post*:

> As someone who has done a good deal of marching and public speaking about Vietnam, Chile, South Africa, Palestine, and East Timor in his time (and would do it all again), I can only hint at how much I despise a Left that thinks of Osama Bin Laden as a slightly misguided anti-imperialist, or a Left that can think of Milosevic and Saddam as victims. Instead of interna-

tionalism, we find among the Left now a sort of affectless, neutralist, smirking isolationism. In this moral universe, the views of the corrupt and conservative Jacques Chirac—who built Saddam Hussein a nuclear reactor, knowing what he wanted it for—carry more weight than that of persecuted Iraqi democrats.

Of course, the Bush Administration's emphasis on moral judgment is vulnerable to criticisms. One is simply aesthetic. The image of the preachy American abroad, short on talent and long on sanctimony, inspires enough resentment already, and few foreign diplomats want to encourage more such Americans to sally forth. But there are more substantive problems, too. One concerns the so-called slippery slope—and it *is* slippery. Yet it is hard to explain why, because one cannot right wrongs everywhere, one should not try to right them anywhere. If the United States simply deals and trades with all countries as we find them, then how, in an age when the frontiers of national security are increasingly defined by issues inside societies, can the United States take stands on the battle lines that matter most?

Then there is the problem of hypocrisy and double standards. (Saudi Arabia is of late the usual friendly villain hauled into the dock for illustration.) Here the administrations goals and rhetoric do bring with them a real burden. Whether it chooses to be reticent or outspoken, whether its reflections are offered in private or in public, any administration that stresses moral judgment is obliged to display a basic honesty about the character of other governments. And it assumes the burden of a consistent regard for the non-negotiable demands of human dignity, no matter the pull of other interests or the nation s accounts. Sometimes other strategic equities will win pride of place, and the administration should not lie to itself about the tradeoffs—either their difficulty or their durability—that will have to be made.

Confronting a messy world full of frustrating choices, President Bush might draw comfort from the remarks of another wartime president, Franklin Roosevelt:

> I am everlastingly angry only at those who assert vociferously that the four freedoms and the Atlantic Charter are nonsense because they are unattainable. If those people had lived a century and a half ago they would have sneered and said that the Declaration of Independence was utter piffle. If they had lived nearly a thousand years ago they would have laughed uproariously at the ideals of Magna Carta. And if they had lived several thousand years ago they would have derided Moses when he came from the Mountain with the Ten Commandments.
>
> We concede that these great teachings are not perfectly lived up to today, but I would rather be a builder than a wrecker, hoping always that the structure of life is growing—not dying.[4]

The New Structure Of World Politics

ANOTHER reason for thinking hard about the centrality of moral judgment in redefining American national security is because of the way the Bush Administration, following on the work of its predecessors, is trying to integrate universal principles into great power politics. For centuries the structure of world politics has been defined by the rivalry of great powers. It is now possible instead for the United States to form active agendas of cooperation with every major center of global power, founded on imperfect yet extraordinary degrees of agreement about underlying principles in the organization of society.

A circumstance of unprecedented great power harmony is neither an immutable fact nor a deterministic prediction. It is a contingency over which we have limited but not inconsiderable influence. Great power rivalry could return to the foreground, so present circumstances ought to be viewed as an opportunity. It is an opportunity that can be lost, however, if the Bush Administration and its successors allow serious agendas of cooperation to drift into routinized gestures and petty fractiousness. These agendas, moreover, will never be built entirely out of the bilateral relationships between the United States and other great powers. Instead, the United States must challenge its present and future partners to join in common tasks that transcend narrow concerns, offering the networks of American allies in Europe and Asia real opportunities to share the responsibilities of global leadership.[5]

In this regard, the United States and its traditional allies have already given considerable attention to Russia and China as potential great powers in transition. India, set to become not only the most populous country in the world but also its largest and most diverse democracy, deserves comparable notice. If India can balance these attributes with sustained growth and stability, the world must bring that nation, and the distinct and ancient civilization it represents, into every inner circle of global power. That would include at the least permanent representation on the United Nations Security Council and in the Group of Eight.

Varieties Of Multilateralism

THE CARTOON version of America's international policy dilemma poses a choice between unilateralism versus multilateralism, the wild cowboy versus the cooperative diplomat. This depiction is false.

Everything that America does in the world is done multilaterally. That emphatically includes the policies the Bush Administration considers most important, and even those that are the most "military" in character. The global war against terrorism is being conducted through an elaborate, often hidden, network of multilateral cooperation among scores of governments. A large number of players are interacting on intelligence, law enforcement, military action, air transportation, shipping, financial controls and more. Ongoing military operations in Afghanistan involve several countries, and were multilateral even at the height of American military activity, as the United States relied heavily on relationships with Pakistan, Russia, three Central Asian governments and a variety of Afghan factions.

The caricature of the administration's unilateralism usually rests on the recitation of a by now standard list of diplomatic actions that some other governments did not like (Kyoto, the International Criminal Court and so on). Some of these disagreements were handled in a style and manner that seemed insensitive or simply maladroit. Unfortunately, too, the caricature of the administration's unilateralism is willingly fed by some U.S. officials and unofficial advisers who relish the chance to play the role of the truth teller lancing foreign obfuscations. Sometimes they overplay the part, sensing the license they get from working for a plain-spoken president.

President Bush, however, is more sensitive to foreign opinion than some who act in his name. He knows that, to much of the world, "I'm the toxic Texan, right?" He recognizes that

> if you want to hear resentment, just listen to the word unilateralism. I mean, that's resentment. If somebody wants to try to say something ugly about us, 'Bush is a unilateralist, America is unilateral.' You know, which I find amusing.

He finds it amusing because he meets and works with foreign dignitaries almost every day, and sees himself as a "pretty good diplomat" —though he smilingly concedes that "nobody else does." From the time Bush first started making decisions about the Iraq issue, for example, he has worked at every turn through international coalitions, noting how much he had enjoyed building one for the war in Afghanistan.[6]

Europeans, of course, rush to play the part of the cosmopolitan professional who rolls his eyes and offers wittily barbed observations about American innocents abroad. Sometimes they, too, overplay the part. But all this is a very old genre of diplomatic theater; John Adams and the Comte de Vergennes set the archetypes more than two hundred years ago. It has since been performed on many stages with a wonderful variety of scripts and sets. No doubt it will be played often in future, as well.

Beyond problems of tone and execution, the real differences in the multilateral strategies of the Bush Administration and other friendly states are not about "unilateralism" versus "multilateralism." They have to do instead with five contrasting ways of conceiving and operationalizing multilateral action.

First, the Bush Administration prefers an inductive method that draws ideas from many sources and adapts them to specific conditions. Alternative deductive strategies develop abstract principles and develop them into generic, universal solutions. For instance, the painstakingly crafted individual agreement drawn up for handling war crimes in Sierra Leone, blending local and foreign judicial traditions, is preferable to the one-size-fits-all approach taken in designing the International Criminal Court. The former is designed to solve real problems and get real results on a case-by-case basis. The latter, unfortunately, aims at more rarefied ambitions.

Second, the administration prefers international institutions that judge performance and stress accountability rather than those that maintain a detached neutrality in order to preserve a friendly consensus. Too often, for example, the institutions charged with preventing the proliferation of weapons of mass destruction seemed to choose courtesy over candor. The history of IAEA inspections in Iraq—going back to the 1970s—is exemplary

Third, the administration prefers multilateral strategies that rely on the sovereign accountability of states instead of strategies that limit sovereignty in order to link states together in a common enterprise—but which thereby dissipate responsibility. As the European Union attempts to develop a "common foreign and security policy", it can hardly avoid a lowest common denominator approach, for example.

Fourth, the administration takes a view of international law that emphasizes democratic accountability, plainly linking the authority of international officials to constitutional sources of political authority that are essentially national in character. Other nations contemplate and encourage much broader delegation of sovereign powers, where only the initial delegation need occur through a democratic process, so that international officials can have greater freedom of action. Hence the U.S. government believes that the International Criminal Court, as a permanent yet essentially stateless entity, might grow ever more distant from the democratic sources of legitimacy that are an essential source for its claimed right to administer global justice.

Fifth, the administration prefers functional institutions that produce concrete results instead of symbolic measures that might rally more support for an ideal, but at the cost of not doing much to further its attainment. Sometimes, as in the case of the Kyoto Accords, well-meaning but dysfunctional efforts can be worse than useless if they complicate attempts to develop a more effective and sustainable solution. We cannot afford such indulgences in a world that in some ways is more threatening today than it was during the Cold War.

Redefining National Security As A Function Of Time

IMAGINE THREATS as having a cadence—rhythm plus speed. In the past, threats tended to emerge slowly, often visibly, as weapons were forged, armies were conscripted, units were trained and deployed, and enemy forces were massed in position to move. The greatest threats, too, came from large states that could raise and equip the mass armies of the industrial age. Precisely because of their size and elaborate structures, these large states had much to lose in a war. Doctrines of deterrence were developed to confront such states, first in the pre-nuclear and then in the nuclear age.

In today's world, threats can emerge more quickly, without having to accumulate a mass of men and metal. Nor do the greatest threats necessarily come from large states that have much to lose.[7] It is thus hard to quarrel with the essential premise of the Bush Administration's open willingness to consider pre-emption, which is that the strategic military doctrines developed for the Cold War must be adapted to the circumstances of the 21st century.

Note, too, the interaction of the new cadence of threat with the new geography of national security. The fault-lines are now inside societies, where even small political factions might have access to weapons of unprecedented destructiveness. The line between internal and international security becomes blurry in parallel with the acceleration of the cadence of threat.

The basic critique of the Bush strategy of pre-emption is that it is better to wait until threats are so acute and universally apparent that a consensus can form in favor of forceful action against them. The flaw in this argument is that there is today a kind of inverse continuum of threat and vulnerability. As a potential enemy's WMD capability becomes more threatening, it becomes less vulnerable to military disruption. Such programs are most vulnerable when they are immature, but that is when the threats they pose are so ambiguous that it is harder to rally allies to act against them. Yet at the point where such threats are so evident that coalitions readily arise, it may be too late—weapons will have already been used, programs will be difficult or impossible to destroy, or outsiders will be themselves deterred by their fear of retaliation.[8]

There is a long tradition in American history of keeping dangerous threats at bay, if necessary by pre-emptive action. Defending Andrew Jackson's pre-emptive invasion of Spanish Florida, his occupation of Pensacola and his execution of individuals inciting Seminole and Creek raids against the United States, Secretary of State John Quincy Adams informed the Spanish government that

> by all the laws of neutrality and of war, as well as of prudence and humanity, [Jackson] was warranted in anticipating his enemy by the amicable and, that being refused, by the forcible occupation…. There will need no citations from printed treatises on international law to prove the correctness of this principle. It is engraved in adamant on the common sense of mankind. No writer upon the laws of nations ever pretended to contradict it. None, of any reputation or authority, ever omitted to assert it.[9]

More recently the United States was prepared to deal with acute dangers by taking pre-emptive or preventive military action against states with which we were technically at peace—Cuba and the Soviet Union in 1962 and North Korea in 1994—if diplomacy had failed to remove or contain specified dangers. The great debates in the Kennedy Administration were over whether or how to even give diplomacy a chance. But there is abundant evidence of President Kennedy's underlying resolve to remove the Soviet missiles from Cuba, one way or another, to prevent an even more dangerous nuclear crisis from arising the next month over Berlin. In 1993 and 1994 the Clinton Administration readied military options against North Korea with evident purpose, but President Clinton never had to make the final decision on what to do if his diplomatic gambit failed. Actually, Iraq is a less clear case of pre-emptive strategy than either of these historical examples, since the United States and the most concerned allies have been in a state of hostilities against Iraq for years. The United States and others have long been conducting constant, if mostly low-intensity, combat operations ever since Iraq conclusively broke the terms of the 1991 military truce that suspended, but never concluded, the Gulf War.

The Bush Administration's strategy is a more explicit adaptation to the new conditions of international life. That explicitness has drawn fire from some who argue that while it might be proper to hold these views, it is not wise to call attention to them. They do not reject the administration's thinking, and can themselves list many examples of real and potential cases of pre-emption. They rather accuse the administration of tactlessness.

The new strategy is somewhat provocative, but it is deliberately so. It must be provocative if it is to foster the painful worldwide debate that must occur in order to condition the international community to think hard about these new dangers, and about how the cadence of security threats has changed. Although the debate is still raging, the older habits of thought are already changing, subtly, around the world. It remains to be seen how the argument over what is and is not tactful will ultimately be settled.

In any event, the United States has not arrogated to itself some vague right to pursue and pummel anyone it dislikes, as many critics contend. The strategy document lists five criteria that must be met for a state or an organization to cast itself outside of ordinary international protection. Condoleezza Rice observed in her Wriston lecture:

> [T]his approach must be treated with great caution. The number of cases in which it might be justified will always be small. It does not give a green light—to the United States or any other nation—to act first without exhausting other means, including diplomacy. Pre-emptive action does not come at the beginning of a long chain of effort. The threat must be very grave. And the risks of waiting must far outweigh the risks of action.

Moreover, the administration's focus is not just on security threats, but also on opportunities. The terrorists have no vision of the future that can assure Muslim parents that their children will lead a better life; the United States does. There are only a few states that could start a new wave of dangerous WMD proliferation, and deflecting them now may nudge history in the right direction. Failure to do so, however, may condemn millions to needless suffering and the American people to years of living in fear.

Grand Strategy and Small Realities

THE BUSH Administration has helped spur worldwide debate not only about the purposes of American power, but about the objectives of the international system as a whole. The United States is not challenging the necessity of international institutions for common action, but it is pressing other nations to decide what they want, to reconsider how to get it, and to re-evaluate old habits in light of new realities. This is an uncomfortable and necessarily disruptive process, and the U.S. government is bound to add its share of distracted stumbles to its progress. But the international agenda is already changing in positive ways.

Grand strategy usually disappoints when it is carried into action. Measured against this or that phrase, this or any administration will come up short. But it is possible to offer general direction, to set different chains of action and reaction into motion. Critics of the Bush Administration's emerging ideas must either accept the new definitions of national security presented or articulate coherent alternatives, working through the implications of present—not past—realities.

There are always too many problems. But the great powers are working together more than anyone would have considered possible half a century ago. New approaches to economic and humanitarian cooperation offer great promise. The advances in human

Patience, Flexibility, Intelligence

For progress there is no cure. Any attempt to find automatically safe channels for the present explosive variety of progress must lead to frustration. The only safety possible is relative, and it lies in an intelligent exercise of day-to-day judgment.

The problems created by the combination of the presently possible forms of nuclear warfare and the rather unusually unstable international situation are formidable and not to be solved easily. Those of the next decades are likely to be similarly vexing, 'only more so.' The U.S.-USSR tension is bad, but when other nations begin to make felt their full offensive potential weight, things will not become simpler.

Present awful possibilities of nuclear warfare may give way to others even more awful…[and] we should not deceive ourselves: once such possibilities become actual, they will be exploited. It will, therefore, be necessary to develop suitable new political forms and procedures. All experience shows that even smaller technological changes than those now in the cards profoundly transform political and social relationships. Experience also shows that these transformations are not *a priori* predictable and that most contemporary 'first guesses' concerning them are wrong. For all these reasons, one should take neither present difficulties nor presently proposed reforms too seriously.

The one solid fact is that the difficulties are due to an evolution that, while useful and constructive, is also dangerous. Can we produce the required adjustments with the necessary speed? The most hopeful answer is that the human species has been subjected to similar tests before and seems to have a congenital ability to come through, after varying amounts of trouble. To ask in advance for a complete recipe would be unreasonable. We can specify only the human qualities required: patience, flexibility, intelligence.

—John von Neumann
"Can We Survive Technology?"
Fortune (June 1955)

liberty and material well-being just since 1990 have been staggering enough to encourage ambitious thoughts about what might be possible. With new understandings about the real problems the United States and its friends face together, and a flexible, pragmatic approach to achieving them, the prospects have never been better for the project of building a commonwealth of freedom.

Philip Zelikow is the White Burkett Miller Professor of History and Director of the Miller Center of Public Affairs at the University of Virginia. He contributed unofficially to the preparation of the *National Security Strategy of the United States*, but the views expressed in this article are solely his own.

Notes

1. This document represents a more comprehensive articulation of themes that President Bush had already begun to introduce, notably in his speeches to a joint session of Congress (September 20, 2001), on the State of the Union (January 29, 2002) and to graduating cadets at West Point (June 1, 2002). Less well known, but displaying a related set of ideas that antedate the 9/11 attacks, were the President's remarks to the World Bank on July 17, 2001 and, subsequent to those attacks, the address to the Inter-American Development Bank on March 14, 2002 (followed eight days later by the Millennium Challenge Account initiative in Monterrey). Condoleezza Rice also added an important summary of the strategy in the Wriston Lecture she delivered in New York City on October 1, 2002.

2. See Richard Koebner and Helmut Dan Schmidt, *Imperialism: The Story and Significance of a Political Word*, 1840-1960 (Cambridge: Cambridge University Press, 1964).

3. Murray, *Morality and Modern War* (New York: Council on Religion and International Affairs, 1959).

4. From an address in Ottawa delivered after the Quebec strategy conference of 1943. The language was FDR's own, added in his own hand at the end of the draft his speechwriters had prepared. Samuel I. Rosenman, *Working with Roosevelt* (London: Rupert Hart-Davis, 1952), p.356.

5. I have elaborated. these ideas in "American Engagement in Asia", in Robert D. Blackwill & Paul Dibb, eds., *America's Asian Alliances* (Cambridge, MA: MIT Press, 2000); and "The New Concert of Europe", *Survival* (Summer 1992).

6. Quotations are drawn from Bob Woodward, *Bush At War* (New York; Simon & Schuster, 2002), pp. 44, 341-2.

7. An essential argument in Charles Krauthammer, "The Unipolar Moment Revisited", *The National Interest* (Winter 2002/03).

8. The current cases of North Korea and Iraq offer interesting illustrations. For a general consideration of preventive/preemptive strategies, see my essay on "Offensive Military Options", in Robert D. Blackwill & Albert Carnesale, eds., *New Nuclear Nations: Consequences for US. Policy* (New York: Council on Foreign Relations, 1993).

9. Adams to Erving, November 28, 1818, in Worthington Chauncey Ford, ed., *Writings of John Quincy Adams* (New iYork: Macmillan, 1916), p. 483. John Lewis Gaddis called my attention to the significance of Adams' influence in the Goldman Lectures he delivered in New York City in September 2002.

Nuclear Nightmares

Experts on terrorism and proliferation agree on one thing:
Sooner or later, an attack will happen here. When and how is what robs them of sleep.

By Bill Keller

The panic that would result from contaminating the Magic Kingdom with a modest amount of cesium would probably shut the place down for good and constitute a staggering strike at Americans' sense of innocence.

Not If But When Everybody who spends much time thinking about nuclear terrorism can give you a scenario, something diabolical and, theoretically, doable. Michael A. Levi, a researcher at the Federation of American Scientists, imagines a homemade nuclear explosive device detonated inside a truck passing through one of the tunnels into Manhattan. The blast would crater portions of the New York skyline, barbecue thousands of people instantly, condemn thousands more to a horrible death from radiation sickness and—by virtue of being underground—would vaporize many tons of concrete and dirt and river water into an enduring cloud of lethal fall-

out. Vladimir Shikalov, a Russian nuclear physicist who helped clean up after the 1986 Chernobyl accident, envisioned for me an attack involving highly radioactive cesium-137 loaded into some kind of homemade spraying device, and a target that sounded particularly unsettling when proposed across a Moscow kitchen table—Disneyland. In this case, the human toll would be much less ghastly, but the panic that would result from contaminating the Magic Kingdom with a modest amount of cesium—Shikalov held up his teacup to illustrate how much—would probably shut the place down for good and constitute a staggering strike at Americans' sense of innocence. Shikalov, a nuclear enthusiast who thinks most people are ridiculously squeamish about radiation, added that personally he would still be happy to visit Disneyland after the terrorists struck, although he would pack his own food and drink and destroy his clothing afterward.

Another Russian, Dmitry Borisov, a former official of his country's atomic energy ministry, conjured a suicidal pilot. (Suicidal pilots, for obvious reasons, figure frequently in these fantasies.) In Borisov's scenario, the hijacker dive-bombs an Aeroflot jetliner into the Kurchatov

Institute, an atomic research center in a gentrifying neighborhood of Moscow, which I had just visited the day before our conversation. The facility contains 26 nuclear reactors of various sizes and a huge accumulation of radioactive material. The effect would probably be measured more in property values than in body bags, but some people say the same about Chernobyl. Maybe it is a way to tame a fearsome subject by Hollywoodizing it, or maybe it is a way to drive home the dreadful stakes in the arid-sounding business of nonproliferation, but in several weeks of talking to specialists here and in Russia about the threats an amateur evildoer might pose to the homeland, I found an unnerving abundance of such morbid creativity. I heard a physicist wonder whether a suicide bomber with a pacemaker would constitute an effective radiation weapon. (I'm a little ashamed to say I checked that one, and the answer is no, since pacemakers powered by plutonium have not been implanted for the past 20 years.) I have had people theorize about whether hijackers who took over a nuclear research laboratory could improvise an actual nuclear explosion on the spot. (Expert opinions differ, but it's very unlikely.) I've been instructed how to disperse plutonium into the ventilation system of an office building.

The realistic threats settle into two broad categories. The less likely but far more devastating is an actual nuclear explosion, a great hole blown in the heart of New York or Washington, followed by a toxic fog of radiation. This could be produced by a black-market nuclear warhead procured from an existing arsenal. Russia is the favorite hypothetical source, although Pakistan, which has a program built on shady middlemen and covert operations, should not be overlooked. Or the explosive could be a homemade device, lower in yield than a factory nuke but still creating great carnage.

The second category is a radiological attack, contaminating a public place with radioactive material by packing it with conventional explosives in a "dirty bomb" by dispersing it into the air or water or by sabotaging a nuclear facility. By comparison with the task of creating nuclear fission, some of these schemes would be almost childishly simple, although the consequences would be less horrifying: a panicky evacuation, a gradual increase in cancer rates, a staggeringly expensive cleanup, possibly the need to demolish whole neighborhoods. Al Qaeda has claimed to have access to dirty bombs, which is unverified but entirely plausible, given that the makings are easily gettable.

Nothing is really new about these perils. The means to inflict nuclear harm on America have been available to rogues for a long time. Serious studies of the threat of nuclear terror date back to the 1970's. American programs to keep Russian nuclear ingredients from falling into murderous hands—one of the subjects high on the agenda in President Bush's meetings in Moscow this weekend—were hatched soon after the Soviet Union disintegrated a decade ago. When terrorists get around to trying their first nuclear assault, as you can be sure they will, there will be plenty of people entitled to say I told you so.

All Sept. 11 did was turn a theoretical possibility into a felt danger. All it did was supply a credible cast of characters who hate us so much they would thrill to the prospect of actually doing it—and, most important in rethinking the probabilities, would be happy to die in the effort. All it did was give our nightmares legs.

Tom Ridge cupped his hands prayerfully and pressed his fingertips to his lips. "Nuclear," he said simply.

And of the many nightmares animated by the attacks, this is the one with pride of place in our experience and literature—and, we know from his own lips, in Osama bin Laden's aspirations. In February, Tom Ridge, the Bush administration's homeland security chief, visited The Times for a conversation, and at the end someone asked, given all the things he had to worry about—hijacked airliners, anthrax in the mail, smallpox, germs in crop-dusters—what did he worry about most? He cupped his hands prayerfully and pressed his fingertips to his lips. "Nuclear," he said simply.

My assignment here was to stare at that fear and inventory the possibilities. How afraid should we be, and what of, exactly? I'll tell you at the outset, this was not one of those exercises in which weighing the fears and assigning them probabilities laid them to rest. I'm not evacuating Manhattan, but neither am I sleeping quite as soundly. As I was writing this early one Saturday in April, the floor began to rumble and my desk lamp wobbled precariously. Although I grew up on the San Andreas Fault, the fact that New York was experiencing an earthquake was only my second thought.

The best reason for thinking it won't happen is that it hasn't happened yet, and that is terrible logic. The problem is not so much that we are not doing enough to prevent a terrorist from turning our atomic knowledge against us (although we are not). The problem is that there may be no such thing as "enough."

25,000 Warheads, and It Only Takes One My few actual encounters with the Russian nuclear arsenal are all associated with Thomas Cochran. Cochran, a physicist with a Tennessee lilt and a sense of showmanship, is the director of nuclear issues for the Natural Resources Defense Council, which promotes environmental protection and arms control. In 1989, when glasnost was in flower, Cochran persuaded the Soviet Union to open some of its most secret nuclear venues to a roadshow of American scientists and congressmen and invited along a couple of reporters. We visited a Soviet missile cruiser bobbing in the Black Sea and drank vodka with physicists and engineers in the secret city where the Soviets first produced plutonium for weapons.

Not long ago Cochran took me cruising through the Russian nuclear stockpile again, this time digitally. The days of glasnost theatrics are past, and this is now the only way an outsider can get close to the places where Russians store and deploy their nuclear weapons. On his office computer in Washington, Cochran has installed a detailed United States military map of Russia and superimposed upon it high-resolution satellite photographs. We spent part of a morning mouse-clicking from missile-launch site to submarine base, zooming in like voyeurs and contemplating the possibility that a terrorist could figure out how to steal a nuclear warhead from one of these places.

"Here are the bunkers," Cochran said, enlarging an area the size of a football stadium holding a half-dozen elongated igloos. We were hovering over a site called Zhukovka, in western Russia. We were pleased to see it did not look ripe for a hijacking.

"You see the bunkers are fenced, and then the whole thing is fenced again," Cochran said. "Just outside you can see barracks and a rifle range for the guards. These would be troops of the 12th Main Directorate. Somebody's not going to walk off the street and get a Russian weapon out of this particular storage area."

In the popular culture, nuclear terror begins with the theft of a nuclear weapon. Why build one when so many are lying around for the taking? And stealing tends to make better drama than engineering. Thus the stolen nuke has been a staple in the literature at least since 1961, when Ian Fleming published "Thunderball," in which the malevolent Spectre (the Special Executive for Counterintelligence, Terrorism, Revenge and Extortion, a strictly mercenary and more technologically sophisticated precursor to al Qaeda) pilfers a pair of atom bombs from a crashed NATO aircraft. In the movie version of Tom Clancy's thriller "The Sum of All Fears," due in theaters this week, neo-Nazis get their hands on a mislaid Israeli nuke, and viewers will get to see Baltimore blasted to oblivion.

Eight countries are known to have nuclear weapons—the United States, Russia, China, Great Britain, France, India, Pakistan and Israel. David Albright, a nuclear-weapons expert and president of the Institute for Science and International Security, points out that Pakistan's program in particular was built almost entirely through black markets and industrial espionage, aimed at circumventing Western export controls. Defeating the discipline of nuclear nonproliferation is ingrained in the culture. Disaffected individuals in Pakistan (which, remember, was intimate with the Taliban) would have no trouble finding the illicit channels or the rationalization for diverting materials, expertise—even, conceivably, a warhead.

But the mall of horrors is Russia, because it currently maintains something like 15,000 of the world's (very roughly) 25,000 nuclear warheads, ranging in destructive power from about 500 kilotons, which could kill a million people, down to the one-kiloton land mines that would be enough to make much of Manhattan uninhabitable. Russia is a country with sloppy accounting, a disgruntled military, an audacious black market and indigenous terrorists.

It's easier to take the fuel and build an entire weapon from scratch than it is to make one of these things go off.

There is anecdotal reason to worry. Gen. Igor Valynkin, commander of the 12th Main Directorate of the Russian Ministry of Defense, the Russian military sector in charge of all nuclear weapons outside the Navy, said recently that twice in the past year terrorist groups were caught casing Russian weapons-storage facilities. But it's hard to know how seriously to take this. When I made the rounds of nuclear experts in Russia earlier this year, many were skeptical of these near-miss anecdotes, saying the security forces tend to exaggerate such incidents to dramatize their own prowess (the culprits are always caught) and enhance their budgets. On the whole, Russian and American military experts sound not very alarmed about the vulnerability of Russia's nuclear warheads. They say Russia takes these weapons quite seriously, accounts for them rigorously and guards them carefully. There is no confirmed case of a warhead being lost. Strategic warheads, including the 4,000 or so that President Bush and President Vladimir Putin have agreed to retire from service, tend to be stored in hard-to-reach places, fenced and heavily guarded, and their whereabouts are not advertised. The people who guard them are better paid and more closely vetted than most Russian soldiers.

Eugene E. Habiger, the four-star general who was in charge of Amer-

ican strategic weapons until 1998 and then ran nuclear antiterror programs for the Energy Department, visited several Russian weapons facilities in 1996 and 1997. He may be the only American who has actually entered a Russian bunker and inspected a warhead *in situ*. Habiger said he found the overall level of security comparable to American sites, although the Russians depend more on people than on technology to protect their nukes.

The image of armed terrorist commandos storming a nuclear bunker is cinematic, but it's far more plausible to think of an inside job. No observer of the unraveling Russian military has much trouble imagining that a group of military officers, disenchanted by the humiliation of serving a spent superpower, embittered by the wretched conditions in which they spend much of their military lives or merely greedy, might find a way to divert a warhead to a terrorist for the right price. (The Chechen warlord Shamil Basayev, infamous for such ruthless exploits as taking an entire hospital hostage, once hinted that he had an opportunity to buy a nuclear warhead from the stockpile.) The anecdotal evidence of desperation in the military is plentiful and disquieting. Every year the Russian press provides stories like that of the 19-year-old sailor who went on a rampage aboard an Akula-class nuclear submarine, killing eight people and threatening to blow up the boat and its nuclear reactor; or the five soldiers at Russia's nuclear-weapons test site who killed a guard, took a hostage and tried to hijack an aircraft, or the officers who reportedly stole five assault helicopters, with their weapons pods, and tried to sell them to North Korea.

The Clinton administration found the danger of disgruntled nuclear caretakers worrisome enough that it considered building better housing for some officers in the nuclear rocket corps. Congress, noting that the United States does not build housing for its own officers, rejected the idea out of hand.

If a terrorist did get his hands on a nuclear warhead, he would still face the problem of setting it off. American warheads are rigged with multiple PAL's ("permissive action links")—codes and self-disabling devices designed to frustrate an unauthorized person from triggering the explosion. General Habiger says that when he examined Russian strategic weapons he found the level of protection comparable to our own. "You'd have to literally break the weapon apart to get into the gut," he told me. "I would submit that a more likely scenario is that there'd be an attempt to get hold of a warhead and not explode the warhead but extract the plutonium or highly enriched uranium." In other words, it's easier to take the fuel and build an entire weapon from scratch than it is to make one of these things go off.

Then again, Habiger is not an expert in physics or weapons design. Then again, the Russians would seem to have no obvious reason for misleading him about something that important. Then again, how many times have computer hackers hacked their way into encrypted computers we were assured were impregnable? Then again, how many computer hackers does al Qaeda have? This subject drives you in circles.

The most troublesome gap in the generally reassuring assessment of Russian weapons security is those tactical nuclear warheads—smaller, short-range weapons like torpedoes, depth charges, artillery shells, mines. Although their smaller size and greater number makes them ideal candidates for theft, they have gotten far less attention simply because, unlike all of our long-range weapons, they happen not to be the subject of any formal treaty. The first President Bush reached an informal understanding with President Gorbachev and then with President Yeltsin that both sides would gather and destroy thousands of tactical nukes. But the agreement included no inventories of the stockpiles, no outside monitoring, no verification of any kind. It was one of those trust-me deals that, in the hindsight of Sept. 11, amount to an enormous black hole in our security.

Did I say earlier there are about 15,000 Russian warheads? That number includes, alongside the scrupulously counted strategic warheads in bombers, missiles and submarines, the commonly used estimate of 8,000 tactical warheads. But that figure is at best an educated guess. Other educated guesses of the tactical nukes in Russia go as low as 4,000 and as high as 30,000. We just don't know. We don't even know if the Russians know, since they are famous for doing things off the books. "They'll tell you they've never lost a weapon," said Kenneth Luongo, director of a private antiproliferation group called the Russian-American Nuclear Security Advisory Council. "The fact is, they don't know. And when you're talking about warhead counting, you don't want to miss even one."

And where are they? Some are stored in reinforced concrete bunkers like the one at Zhukovka. Others are deployed. (When the submarine Kursk sank with its 118 crewmen in August 2000, the Americans' immediate fear was for its nuclear armaments. The standard load out for a submarine of that class includes a couple of nuclear torpedoes and possibly some nuclear depth charges.) Still others are supposed to be in the process of being dismantled under terms of various formal and informal arms-control agreements. Some are in transit. In short, we don't really know.

The other worrying thing about tactical nukes is that their anti-use devices are believed to be less sophisticated, because the weapons were designed to be employed in the battlefield. Some of the older systems are thought to have no permissive action links at all, so that setting one off would be about as complicated as hot-wiring a car.

Efforts to learn more about the state of tactical stockpiles have been frustrated by reluctance on both sides to let visitors in. Viktor Mikhailov, who ran the Russian Ministry of Atomic Energy until 1998 with a famous scorn for America's nonproliferation concerns, still insists that the United States programs to protect Russian nuclear weapons and material mask a secret agenda of intelligence-gathering. Americans, in turn, sometimes balk at reciprocal access, on the grounds that we are the ones paying the bills for all these safety upgrades, said the former Senator Sam Nunn, co-author of the main American program for securing Russian nukes, called Nunn-Lugar.

People in the field talk of a nuclear 'conex' bomb, using the name of those shack-size steel containers—2,000 of which enter America every hour on trains, trucks and ships. Fewer than 2 percent are cracked open for inspection.

"We have to decide if we want the Russians to be transparent—I'd call it cradle-to-grave transparency with nuclear material and inventories and so forth," Nunn told me. "Then we have to open up more ourselves. This is a big psychological breakthrough we're talking about here, both for them and for us."

The Garage Bomb One of the more interesting facts about the atom bomb dropped on Hiroshima is that it had never been tested. All of those spectral images of nuclear coronas brightening the desert of New Mexico— those were to perfect the more complicated plutonium device that was dropped on Nagasaki. "Little Boy," the Hiroshima bomb, was a rudimentary gunlike device that shot one projectile of highly enriched uranium into another, creating a critical mass that exploded. The mechanics were so simple that few doubted it would work, so the first experiment was in the sky over Japan.

The closest thing to a consensus I heard among those who study nuclear terror was this: building a nuclear bomb is easier than you think, probably easier than stealing one. In the rejuvenated effort to prevent a terrorist from striking a nuclear blow, this is where most of the attention and money are focused.

A nuclear explosion of any kind "is not a sort of high-probability thing," said a White House official who follows the subject closely. "But getting your hands on enough fissile material to build an improvised nuclear device, to my mind, is the least improbable of them all, and particularly if that material is highly enriched uranium in metallic form. Then I'm really worried. That's the one."

To build a nuclear explosive you need material capable of explosive nuclear fission, you need expertise, you need some equipment, and you need a way to deliver it.

Delivering it to the target is, by most reckoning, the simplest part. People in the field generally scoff at the mythologized suitcase bomb; instead they talk of a "conex bomb," using the name of those shack-size steel containers that bring most cargo into the United States. Two thousand containers enter America every hour, on trucks and trains and especially on ships sailing into more than 300 American ports. Fewer than 2 percent are cracked open for inspection, and the great majority never pass through an X-ray machine. Containers delivered to upriver ports like St. Louis or Chicago pass many miles of potential targets before they even reach customs.

"How do you protect against that?" mused Habiger, the former chief of our nuclear arsenal. "You can't. That's scary. That's very, very scary. You set one of those off in Philadelphia, in New York City, San Francisco, Los Angeles, and you're going to kill tens of thousands of

people, if not more." Habiger's view is "It's not a matter of *if*; it's a matter of *when*"—which may explain why he now lives in San Antonio.

The Homeland Security office has installed a plan to refocus inspections, making sure the 2 percent of containers that get inspected are those without a clear, verified itinerary. Detectors will be put into place at ports and other checkpoints. This is good, but it hardly represents an ironclad defense. The detection devices are a long way from being reliable. (Inconveniently, the most feared bomb component, uranium, is one of the hardest radioactive substances to detect because it does not emit a lot of radiation prior to fission.) The best way to stop nuclear terror, therefore, is to keep the weapons out of terrorist hands in the first place.

Fabricating a nuclear weapon is not something a lone madman—even a lone genius—is likely to pull off in his hobby room.

The basic know-how of atom-bomb-building is half a century old, and adequate recipes have cropped up in physics term papers and high school science projects. The simplest design entails taking a lump of highly enriched uranium, about the size of a cantaloupe, and firing it down a big gun barrel into a second lump. Theodore Taylor, the nuclear physicist who designed both the smallest and the largest American nuclear-fission warheads before becoming a remorseful opponent of all things nuclear, told me he recently looked up "atomic bomb" in the World Book Encyclopedia in the upstate New York nursing home where he now lives, and he found enough basic information to get a careful reader started. "It's accessible all over the place," he said. "I don't

mean just the basic principles. The sizes, specifications, things that work."

Most of the people who talk about the ease of assembling a nuclear weapon, of course, have never actually built one. The most authoritative assessment I found was a paper, "Can Terrorists Build Nuclear Weapons?" written in 1986 by five experienced nuke-makers from the Los Alamos weapons laboratory. I was relieved to learn that fabricating a nuclear weapon is not something a lone madman—even a lone genius—is likely to pull off in his hobby room. The paper explained that it would require a team with knowledge of "the physical, chemical and metallurgical properties of the various materials to be used, as well as characteristics affecting their fabrication; neutronic properties; radiation effects, both nuclear and biological; technology concerning high explosives and/or chemical propellants; some hydrodynamics; electrical circuitry; and others." Many of these skills are more difficult to acquire than, say, the ability to aim a jumbo jet.

The schemers would also need specialized equipment to form the uranium, which is usually in powdered form, into metal, to cast it and machine it to fit the device. That effort would entail months of preparation, increasing the risk of detection, and it would require elaborate safeguards to prevent a mishap that, as the paper dryly put it, would "bring the operation to a close."

Still, the experts concluded, the answer to the question posed in the title, while qualified, was "Yes, they can."

David Albright, who worked as a United Nations weapons inspector in Iraq, says Saddam Hussein's unsuccessful crash program to build a nuclear weapon in 1990 illustrates how a single bad decision can mean a huge setback. Iraq had extracted highly enriched uranium from research-reactor fuel and had, maybe, barely enough for a bomb. But the manager in charge of casting the metal was so afraid the stuff would spill or get contaminated that he decided to melt it in tiny batches. As a result, so much of the uranium was wasted that he ended up with too little for a bomb.

"You need good managers and organizational people to put the elements together," Albright said. "If you do a straight-line extrapolation, terrorists will all get nuclear weapons. But they make mistakes."

On the other hand, many experts underestimate the prospect of a do-it-yourself bomb because they are thinking too professionally. All of our experience with these weapons is that the people who make them (states, in other words) want them to be safe, reliable, predictable and efficient. Weapons for the American arsenal are designed to survive a trip around the globe in a missile, to be accident-proof, to produce a precisely specified blast.

But there are many corners you can cut if you are content with a big, ugly, inefficient device that would make a spectacular impression. If your bomb doesn't need to fit in a suitcase (and why should it?) or to endure the stress of a missile launch; if you don't care whether the explosive power realizes its full potential; if you're willing to accept some risk that the thing might go off at the wrong time or might not go off at all, then the job of building it is immeasurably simplified.

"As you get smarter, you realize you can get by with less," Albright said. "You can do it in facilities that look like barns, garages, with simple machine tools. You can do it with 10 to 15 people, not all Ph.D.'s, but some engineers, technicians. Our judgment is that a gun-type device is well within the capability of a terrorist organization."

All the technological challenges are greatly simplified if terrorists are in league with a country—a place with an infrastructure. A state is much better suited to hire expertise (like dispirited scientists from decommissioned nuclear installations in the old Soviet Union) or to send its own scientists for M.I.T. degrees.

Thus Tom Cochran said his greatest fear is what you might call a bespoke nuke—terrorists stealing a quantity of weapons-grade uranium and taking it to Iraq or Iran or Libya, letting the scientists and engineers there fashion it into an elementary weapon and then taking it away for a delivery that would have no return address.

That leaves one big obstacle to the terrorist nuke-maker: the fissile material itself.

To be reasonably sure of a nuclear explosion, allowing for some material being lost in the manufacturing process, you need roughly 50 kilograms—110 pounds—of highly enriched uranium. (For a weapon, more than 90 percent of the material should consist of the very unstable uranium-235 isotope.) Tom Cochran, the master of visual aids, has 15 pounds of depleted uranium that he keeps in a Coke can; an eight-pack would be plenty to build a bomb.

Only 41 percent of Russia's weapon-usable material has been secured… So the barn door is still pretty seriously ajar. We don't know whether any horses have gotten out.

The world is awash in the stuff. Frank von Hippel, a Princeton physicist and arms-control advocate, has calculated that between 1,300 and 2,100 metric tons of weapons-grade uranium exists—at the low end, enough for 26,000 rough-hewed bombs. The largest stockpile is in Russia, which Senator Joseph Biden calls "the candy store of candy stores."

Until a decade ago, Russian officials say, no one worried much about the safety of this material. Vik-tor Mikhailov, who ran the atomic energy ministry and now presides over an affiliated research institute, concedes there were glaring lapses.

"The safety of nuclear materials was always on our minds, but the focus was on intruders," he said. "The system had never taken account of the possibility that these carefully screened people in the nuclear sphere could themselves represent a danger. The system was not designed to prevent a danger from within."

Then came the collapse of the Soviet Union and, in the early 90's, a few frightening cases of nuclear materials popping up on the black market.

If you add up all the reported attempts to sell highly enriched uranium or plutonium, even including those that have the scent of security-agency hype and those where the material was of uncertain quality, the total amount of material still falls short of what a bomb-maker would need to construct a single explosive.

But Yuri G. Volodin, the chief of safeguards at Gosatomnadzor, the Russian nuclear regulatory agency, told me his inspectors still discover one or two instances of attempted theft a year, along with dozens of violations of the regulations for storing and securing nuclear material. And as he readily concedes: "These are the detected cases. We can't talk about the cases we don't know." Alexander Pikayev, a former aide to the Defense Committee of the Russian Duma, said: "The vast majority of installations now have fences. But you know Russians. If you walk along the perimeter, you can see a hole in the fence, because the employees want to come and go freely."

The bulk of American investment in nuclear safety goes to lock the stuff up at the source. That is clearly the right priority. Other programs are devoted to blending down the highly enriched uranium to a diluted product unsuitable for weapons but good as reactor fuel. The Nuclear Threat Initiative, financed by Ted Turner and led by Nunn, is studying ways to double the rate of this diluting process.

Still, after 10 years of American subsidies, only 41 percent of Russia's weapon-usable material has been secured, according to the United States Department of Energy. Russian officials said they can't even be sure how much exists, in part because the managers of nuclear facilities, like everyone else in the Soviet industrial complex, learned to cook their books. So the barn door is still pretty seriously ajar. We don't know whether any horses have gotten out.

And it is not the only barn. William C. Potter, director of the Center for Nonproliferation Studies at the Monterey Institute of International Studies and an expert in nuclear security in the former Soviet states, said the American focus on Russia has neglected other locations that could be tempting targets for a terrorist seeking bomb-making material. There is, for example, a bomb's worth of weapons-grade uranium at a site in Belarus, a country with an erratic president and an anti-American orientation. There is enough weapons-grade uranium for a bomb or two in Kharkiv, in Ukraine. Outside of Belgrade, in a research reactor at Vinca, sits sufficient material for a bomb—and there it sat while NATO was bombarding the area.

"We need to avoid the notion that because the most material is in Russia, that's where we should direct all of our effort," Potter said. "It's like assuming the bank robber will target Fort Knox because that's where the most gold is. The bank robber goes where the gold is most accessible."

Weapons of Mass Disruption The first and, so far, only consummated act of nuclear terrorism took place in Moscow in 1995, and it was scarcely memorable. Chechen rebels obtained a canister of cesium, possibly from a hospital they had commandeered a few months before. They hid it in a Moscow park famed for its weekend flea market and called the press. No one was hurt. Authorities

treated the incident discreetly, and a surge of panic quickly passed.

The story came up in virtually every conversation I had in Russia about nuclear terror, usually to illustrate that even without splitting atoms and making mushroom clouds a terrorist could use radioactivity—and the fear of it—as a potent weapon.

The idea that you could make a fantastic weapon out of radioactive material without actually producing a nuclear bang has been around since the infancy of nuclear weaponry. During World War II, American scientists in the Manhattan Project worried that the Germans would rain radioactive material on our troops storming the beaches on D-Day. Robert S. Norris, the biographer of the Manhattan Project director, Gen. Leslie R. Groves, told me that the United States took this threat seriously enough to outfit some of the D-Day soldiers with Geiger counters.

No country today includes radiological weapons in its armories. But radiation's limitations as a military tool—its tendency to drift afield with unplanned consequences, its long-term rather than short-term lethality—would not necessarily count against it in the mind of a terrorist. If your aim is to instill fear, radiation is anthrax-plus. And unlike the fabrication of a nuclear explosive, this is terror within the means of a soloist.

If your aim is to instill fear, radiation is anthrax-plus. And unlike the fabrication of a nuclear explosive, this is terror within the means of a soloist.

That is why, if you polled the universe of people paid to worry about weapons of mass destruction (W.M.D., in the jargon), you would find a general agreement that this is

probably the first thing we'll see. "If there is a W.M.D. attack in the next year, it's likely to be a radiological attack," said Rose Gottemoeller, who handled Russian nuclear safety in the Clinton administration and now follows the subject for the Carnegie Endowment. The radioactive heart of a dirty bomb could be spent fuel from a nuclear reactor or isotopes separated out in the process of refining nuclear fuel. These materials are many times more abundant and much, much less protected than the high-grade stuff suitable for bombs. Since Sept.11, Russian officials have begun lobbying hard to expand the program of American aid to include protection of these lower-grade materials, and the Bush administration has earmarked a few million dollars to study the problem. But the fact is that radioactive material suitable for terrorist attacks is so widely available that there is little hope of controlling it all.

The guts of a dirty bomb could be cobalt-60, which is readily available in hospitals for use in radiation therapy and in food processing to kill the bacteria in fruits and vegetables. It could be cesium-137, commonly used in medical gauges and radiotherapy machines. It could be americium, an isotope that behaves a lot like plutonium and is used in smoke detectors and in oil prospecting. It could be plutonium, which exists in many research laboratories in America. If you trust the security of those American labs, pause and reflect that the investigation into the great anthrax scare seems to be focused on disaffected American scientists.

Back in 1974, Theodore Taylor and Mason Willrich, in a book on the dangers of nuclear theft, examined things a terrorist might do if he got his hands on 100 grams of plutonium—a thimble-size amount. They calculated that a killer who dissolved it, made an aerosol and introduced it into the ventilation system of an office building could deliver a lethal dose to the entire floor area of a large skyscraper. But plutonium dispersed outdoors in the open air, they estimated, would be far less effective. It would blow away in a gentle wind.

The Federation of American Scientists recently mapped out for a Congressional hearing the consequences of various homemade dirty bombs detonated in New York or Washington. For example, a bomb made with a single footlong pencil of cobalt from a food irradiation plant and just 10 pounds of TNT and detonated at Union Square in a light wind would send a plume of radiation drifting across three states. Much of Manhattan would be as contaminated as the permanently closed area around the Chernobyl nuclear plant. Anyone living in Manhattan would have at least a 1-in-100 chance of dying from cancer caused by the radiation. An area reaching deep into the Hudson Valley would, under current Environmental Protection Agency standards, have to be decontaminated or destroyed.

Frank von Hippel, the Princeton physicist, has reviewed the data, and he pointed out that this is a bit less alarming than it sounds. "Your probability of dying of cancer in your lifetime is already about 20 percent," he said. "This would increase it to 20.1 percent. Would you abandon a city for that? I doubt it."

Indeed, some large portion of our fear of radiation is irrational. And yet the fact that it's all in your mind is little consolation if it's also in the minds of a large, panicky population. If the actual effect of a radiation bomb is that people clog the bridges out of town, swarm the hospitals and refuse to return to live and work in a contaminated place, then the impact is a good deal more than psychological. To this day, there is bitter debate about the actual health toll from the Chernobyl nuclear accident. There are researchers who claim that the people who evacuated are actually in worse health over all from the trauma of relocation, than those who stayed put and marinated in the residual radiation. But the fact is, large swaths of developed land around the Chernobyl site still lie abandoned,

much of it bulldozed down to the subsoil. The Hart Senate Office Building was closed for three months by what was, in hindsight, our society's inclination to err on the side of alarm.

There are measures the government can take to diminish the dangers of a radiological weapon, and many of them are getting more serious consideration. The Bush administration has taken a lively new interest in radiation-detection devices that might catch dirty-bomb materials in transit. A White House official told me the administration's judgment is that protecting the raw materials of radiological terror is worth doing, but not at the expense of more catastrophic threats.

"It's all over," he said. "It's not a winning proposition to say you can just lock all that up. And then, a bomb is pretty darn easy to make. You don't have to be a rocket scientist to figure about fertilizer and diesel fuel." A big fertilizer bomb of the type Timothy McVeigh used to kill 168 people in Oklahoma City, spiced with a dose of cobalt or cesium, would not tax the skills of a determined terrorist.

"It's likely to happen, I think, in our lifetime," the official said. "And it'll be like Oklahoma City plus the Hart Office Building. Which is real bad, but it ain't the World Trade Center."

The Peril of Power Plants Every eight years or so the security guards at each of the country's 103 nuclear power stations and at national weapons labs can expect to be attacked by federal agents armed with laser-tag rifles. These mock terror exercises are played according to elaborate rules, called the "design basis threat," that in the view of skeptics favor the defense. The attack teams can include no more than three commandos. The largest vehicle they are permitted is an S.U.V. They are allowed to have an accomplice inside the plant, but only one. They are not allowed to improvise. (The mock assailants at one Department of Energy

lab were ruled out of order because they commandeered a wheelbarrow to cart off a load of dummy plutonium.) The mock attacks are actually announced in advance. Even playing by these rules, the attackers manage with some regularity to penetrate to the heart of a nuclear plant and damage the core. Representative Edward J. Markey, a Massachusetts Democrat and something of a scourge of the nuclear power industry, has recently identified a number of shortcomings in the safeguards, including, apparently, lax standards for clearing workers hired at power plants.

One of the most glaring lapses, which nuclear regulators concede and have promised to fix, is that the design basis threat does not contemplate the possibility of a hijacker commandeering an airplane and diving it into a reactor. In fact, the protections currently in place don't consider the possibility that the terrorist might be willing, even eager, to die in the act. The government assumes the culprits would be caught while trying to get away.

A nuclear power plant is essentially a great inferno of decaying radioactive material, kept under control by coolant. Turning this device into a terrorist weapon would require cutting off the coolant so the atomic furnace rages out of control and, equally important, getting the radioactive matter to disperse by an explosion or fire. (At Three Mile Island, the coolant was cut off and the reactor core melted down, generating vast quantities of radiation. But the thick walls of the containment building kept the contaminant from being released, so no one died.)

One way to accomplish both goals might be to fly a large jetliner into the fortified building that holds the reactor. Some experts say a jet engine would stand a good chance of bursting the containment vessel, and the sheer force of the crash might disable the cooling system—rupturing the pipes and cutting off electricity that pumps the water through the core. Before nearby residents had begun to evacuate, you could have a meltdown that would spew a volcano of radioactive isotopes into the air, causing fatal radiation sickness for those exposed to high doses and raising lifetime cancer rates for miles around.

This sort of attack is not as easy, by a long shot, as hitting the World Trade Center. The reactor is a small, low-lying target, often nestled near the conspicuous cooling towers, which could be destroyed without great harm. The reactor is encased in reinforced concrete several feet thick, probably enough, the industry contends, to withstand a crash. The pilot would have to be quite a marksman, and somewhat lucky. A high wind would disperse the fumes before they did great damage.

Invading a plant to produce a meltdown, even given the record of those mock attacks, would be more complicated, because law enforcement from many miles around would be on the place quickly, and because breaching the containment vessel is harder from within. Either invaders or a kamikaze attacker could instead target the more poorly protected cooling ponds, where used plutonium sits, encased in great rods of zirconium alloy. This kind of sabotage would take longer to generate radiation and would be far less lethal.

Discussion of this kind of potential radiological terrorism is colored by passionate disagreements over nuclear power itself. Thus the nuclear industry and its rather tame regulators sometimes sound dismissive about the vulnerability of the plants (although less so since Sept. 11), while those who regard nuclear power as inherently evil tend to overstate the risks. It is hard to sort fact from fear-mongering.

Nuclear regulators and the industry grumpily concede that Sept. 11 requires a new estimate of their defenses, and under prodding from Congress they are redrafting the so-called design basis threat, the one plants are required to defend against. A few members of Congress have proposed installing ground-to-air missiles at nuclear plants, which most experts think is a recipe for a disastrous mishap.

"Probably the only way to protect against someone flying an aircraft into a nuclear power plant," said Steve Fetter of the University of Maryland, "is to keep hijackers out of cockpits."

Being Afraid For those who were absorbed by the subject of nuclear terror before it became fashionable, the months since the terror attacks have been, paradoxically, a time of vindication. President Bush, whose first budget cut $100 million from the programs to protect Russian weapons and material (never a popular program among conservative Republicans), has become a convert. The administration has made nuclear terror a priority, and it is getting plenty of goading to keep it one. You can argue with their priorities and their budgets, but it's hard to accuse anyone of indifference. And resistance—from scientists who don't want security measures to impede their access to nuclear research materials, from generals and counterintelligence officials uneasy about having their bunkers inspected, from nuclear regulators who worry about the cost of nuclear power, from conservatives who don't want to subsidize the Russians to do much of anything—has become harder to sustain. Intelligence gathering on nuclear material has been abysmal, but it is now being upgraded; it is a hot topic at meetings between American and foreign intelligence services, and we can expect more numerous and more sophisticated sting operations aimed at disrupting the black market for nuclear materials. Putin, too, has taken notice. Just before leaving to meet Bush in Crawford, Tex., in November, he summoned the head of the atomic energy ministry to the Kremlin on a Saturday to discuss nuclear security. The subject is now on the regular agenda when Bush and Putin talk.

These efforts can reduce the danger but they cannot neutralize the fear, particularly after we have been so vividly reminded of the hostility some of the world feels for us, and of our vulnerability.

Fear is personal. My own—in part, because it's the one I grew up with, the one that made me shiver through the Cuban missile crisis and "On the Beach"—is the horrible magic of nuclear fission. A dirty bomb or an assault on a nuclear power station, ghastly as that would be, feels to me within the range of what we have survived. As the White House official I spoke with said, it's basically Oklahoma City plus the Hart Office Building. A nuclear explosion is in a different realm of fears and would test the country in ways we can scarcely imagine.

A mushroom cloud of irradiated debris would blossom more than two miles into the air. Then highly lethal fallout would begin drifting back to earth, riding the winds into the Bronx or Queens or New Jersey.

As I neared the end of this assignment, I asked Matthew McKinzie, a staff scientist at the Natural Resources Defense Council, to run a computer model of a one-kiloton nuclear explosion in Times Square, half a block from my office, on a nice spring workday. By the standards of serious nuclear weaponry, one kiloton is a junk bomb, hardly worthy of respect, a fifteenth the power of the bomb over Hiroshima.

A couple of days later he e-mailed me the results, which I combined with estimates of office workers and tourist traffic in the area. The blast and searing heat would gut buildings for a block in every direction, incinerating pedestrians and crushing people at their desks. Let's say 20,000 dead in a matter of seconds. Beyond this, to a distance of more than a quarter mile, anyone directly exposed to the fireball would die a gruesome death from radiation sickness within a day—anyone, that is, who survived the third-degree burns. This larger circle would be populated by about a quarter million people on a workday. Half a mile from the explosion, up at Rockefeller Center and down at Macy's, unshielded onlookers would expect a slower death from radiation. A mushroom cloud of irradiated debris would blossom more than two miles into the air, and then, 40 minutes later, highly lethal fallout would begin drifting back to earth, showering injured survivors and dooming rescue workers. The poison would ride for 5 or 10 miles on the prevailing winds, deep into the Bronx or Queens or New Jersey.

A terrorist who pulls off even such a small-bore nuclear explosion will take us to a whole different territory of dread from Sept. 11. It is the event that preoccupies those who think about this for a living, a category I seem to have joined.

"I think they're going to try," said the physicist David Albright. "I'm an optimist at heart. I think we can catch them in time. If one goes off, I think we will survive. But we won't be the same. It will affect us in a fundamental way. And not for the better."

Bill Keller is a Times columnist and a senior writer for the magazine.

From the *New York Times* Magazine, May 26, 2002, pp. 22, 24-29, 51, 54-55, 57. © 2002 by Bill Keller. Distributed by The New York Times Special Features. Reprinted by permission.

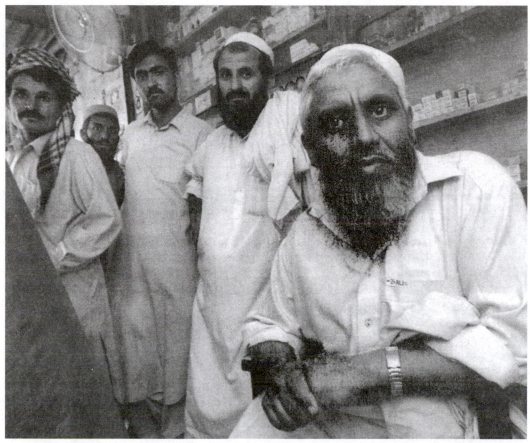

ROBERT HARRISON—STAFF

PROUD FATHER: In a Pakistan pharmacy, pictured above, Amirul Haq (r.), says he is 'satisfied' that his Muslim son was killed in the Kashmir. He's 'against America, because it doesn't care about those who die in Pakistan.'

'Why do they hate us?'

By Peter Ford
Staff writer

asked President Bush in his speech to Congress last Thursday night. It is a question that has ached in America's heart for the past two weeks. Why did those 19 men choose to wreck the icons of US military and economic power?

Most Arabs and Muslims knew the answer, even before they considered who was responsible. Retired Pakistani Air Commodore Sajad Haider—a friend of the US—understood why. Radical Egyptian-born cleric and US enemy Abu Hamza al-Masri understood. And Jimmy Nur Zamzamy, a devout Muslim and advertising executive in Indonesia, understood.

They all understood that this assault was more precisely targeted than an attack on "civilization." First and foremost, it was an attack on America.

In the United States, military planners are deciding how to exact retribution. To many people in the Middle East and beyond, where US policy has bred widespread anti-Americanism, the carnage of Sept. 11 *was* retribution.

And voices across the Muslim world are warning that if America doesn't wage its war on terrorism in a way that the Muslim world considers just, America risks creating even greater animosity.

Mr. Haider is a hero of Pakistan's 1965 war against India, and a sworn friend of America. But he and his neighbors in one of Islamabad's toniest districts are clear about why their warm feelings toward the US are not widely shared in Pakistan.

In his dim office in a north London mosque, Abu Hamza al-Masri sympathizes with the goals of Osama bin Laden, fingered by US officials as the prime suspect behind the Sept. 11 attacks. Abu Hamza has himself directed terrorist operations abroad, according to the British police, although for lack of evidence, they have never brought him to trial.

Mr. Zamzamy, a 30-something advertising executive in Jakarta, knew what was behind the attack, too. Trying to give his ads some zip and still stay within the bounds of his Muslim faith, he is keenly aware of the tensions between Islam and American-style global capitalism.

The 19 men—who US officials say hijacked four American passenger jets and flew them on suicide missions that left more than 7,000 people dead or missing—were all from the Middle East. Most of the hijackers have been identified as Muslims.

The vast majority of Muslims in the Middle East were as shocked and horrified as any American by what they saw happening on their TV screens. And they are frightened of being lumped together in the popular American imagination with the perpetrators of the attack.

But from Jakarta to Cairo, Muslims and Arabs say that on reflection, they are not surprised by it. And they do not share Mr. Bush's view that the perpetrators did what they did because "they hate our freedoms."

Rather, they say, a mood of resentment toward America and its behavior around the world has become so commonplace in their countries that it was bound to breed hostility, and even hatred.

And the buttons that Mr. bin Laden pushes in his statements and interviews—the injustice done to the Palestinians, the cruelty of continued sanctions against Iraq, the presence of US troops in Saudi Arabia, the repressive and corrupt nature of US-backed Gulf governments—win a good deal of popular sympathy.

The resentment of the US has spread through societies demoralized by their recent history. In few of the world's 50 or so Muslim countries have governments offered their citizens either prosperity or democracy. Arab nations have lost three wars against their arch-foe—and America's closest ally—Israel. A sense of failure and injustice is rising in the throats of millions.

Three weeks ago, a leading Arabic newspaper, Al-Hayat, published a poem on its front page. A long lament about the plight of the Arabs, addressed to a dead Syrian poet, it ended:

"Children are dying, but no one makes a move.
Houses are demolished, but no one makes a move.
Holy places are desecrated, but no one makes a move....
I am fed up with life in the world of mortals.
Find me a hole near you. For a life of dignity is in those holes."

It sounds as if it could have been written by a desperate and hopeless man, driven by frustration to seek death, perhaps martyrdom. A young Palestinian refugee planning a suicide bomb attack, maybe. In fact, it was written by the Saudi Arabian ambassador to London, a member of one of the wealthiest and most influential families in the kingdom that is Washington's closest Arab ally.

Against the background of that humiliated mood, America's unchallenged military, economic, and cultural might be seen as an affront even if its policies in the Middle East were neutral. And nobody voices that view.

From one end of the region to the other, the perception is that Israel can get away with murder—literally—and that Washington will turn a blind eye. Clearly, the US and Israel have compelling reasons for their actions. But little that US diplomats have done in recent years to broker a peace deal between Israel and the Palestinians has persuaded Arabs that the US is a fair-minded and equitable judge of Middle Eastern affairs.

Over the past year, Arab TV stations have broadcast countless pictures of Israeli soldiers shooting at Palestinian youths, Israeli tanks plowing into Palestinian homes, Israeli helicopters rocketing Palestinian streets. And they know that the US sends more than $3 billion a year in military and economic aid to Israel.

"You see this every day, and what do you feel?" asks Rafiq Hariri, the portly prime minister of Lebanon, who is not an excitable man. "It hurts me a lot. But for hundreds of thousands of Arabs and Muslims, it drives them crazy. They feel humiliated."

RESENTMENT RISES, AND A RADICAL IS BORN

Ask Sheikh Abdul Majeed Atta why Palestinians may not like the United States, and he does not immediately answer. Instead, he pads barefoot across the red swirls of his living room carpet and reaches for three framed photographs on the floor beside a couch.

The black-and-white prints show dusty, rock-strewn hills dotted with tiny tents and cinderblock houses: the early days of Duheisheh refugee camp, south of Bethlehem in the West Bank. It was where Mr. Atta was born, and where his family has lived for more than half a century. Atta's family village was destroyed in the struggle between Palestinian Arabs and Jews after Britain divided Palestine between them in 1948. For 10 years his family of

13 lived in a tent. The year Atta was born, the United Nations gave them a one-room house.

It doesn't matter to Atta that the United States was not directly involved in "the catastrophe," as Palestinians refer to the events of 1948. Washington averted its eyes when it could have helped, he says, and since then has been firmly on Israel's side.

Heavyset, solid, with a neatly trimmed full beard, Atta is the preacher at a nearby mosque. He looks the part of the community leader, always meticulously turned out in crisp shirts and pressed trousers, gold-rimmed reading glasses tucked into a pocket.

In the past year of the Palestinian-Israeli conflict, Atta has joined Hamas, the radical group responsible for recently sending most of the suicide bombers into Israeli towns. Frustration at watching the rising Palestinian death toll at the hands of the Israeli army played a large part in his decision, he says.

His resentment at Israel, though, dates back to his infancy, and the stories he heard of his village, Ras Abu Amar, which he never knew. That village is still alive for him, just as millions of Palestinians in the West Bank and Gaza Strip, and throughout the Middle East cherish photos, house keys, and deeds to homes that no longer exist or which have housed Israelis for generations.

Today he lives in his own house in Duheisheh, a sprawling tangle of densely packed concrete buildings that crowd snaking, narrow alleys. But he still dreams of the home he never knew, and recalls who took it from him, and remembers who they rely on for their strength.

What happened on Sept. 11 "was an awful thing, a tragedy, and since we live a continuous tragedy, we felt like this touched us," he adds. "But when we see something like this in Israel or the US, we feel a contradiction. We see it's a tragedy, but we remember that these are the people behind our tragedy."

"Even small children know that Israel is nothing without America," says Atta. "And here America means F-16, M-16, Apache helicopters, the tools Israelis use to kill us and destroy our homes."

SUPERPOWER SWAGGER

Such weapons are very much the visible face of American policy in the Middle East, where military might has held the balance of power for 50 years. Thousands of US soldiers stationed in the Gulf, and billions of US dollars each year in military aid to Israel, Egypt, and other allies, have shored up Washington's interests in the strategically crucial, oil-rich region.

That military presence and power looks like swagger to some in the Muslim world, even far from the flashpoints. "Now America is ready with its airplanes to bomb this poor nation [Afghanistan], and most people in Indonesia don't like arrogance," says Imam Budi Prasodjo, an Indonesian sociologist and talk-show host.

"You are a superpower, you are a military superpower, and you can do whatever you want. People don't like that, and this is dangerous," he adds.

"America should spread its culture, rather than weapons or tanks," adds Mohammed el-Sayed Said, deputy director of Cairo's influential Al Ahram think tank. "They need to act like any respectable commander or leader of an army. They can't just project an image of contempt for those they wish to lead."

Ten years ago, at the head of a broad coalition of Western and Arab countries, the United States used its superpower status to kick the Iraqi army out of Kuwait. Since then, however, Washington has found itself alone—save for loyal ally Britain—in its determination to keep bombing Iraq, and to keep imposing strict economic sanctions that the United Nations says are partly responsible for the deaths of half a million Iraqi children.

'We wish the American people could see what their governments are doing in the rest of the world.'

—Saniya Ghussein, whose daughter, Raafat, was killed in the 1986 US bombing raid against Libya

---

Those deaths, and those bombs (which US and British planes drop regularly, but without fanfare), are felt keenly among fellow Arabs. And Saniya Ghussein knows all about bombs.

A DAUGHTER DIES, AND PARENTS WAIT FOR US APOLOGY

In the middle of the night of April 16, 1986, the deafening sound of anti-aircraft guns woke Saniya Ghussein with a sudden start. "My God," she thought, "there's a war being fought above my house."

She slipped out of bed and ran into the bedroom where her husband Bassem and their 7-year-old daughter Kinda had fallen asleep earlier in the evening. "Bassem, the Americans are here," she said urgently. "It looks like they're going to hit us."

She checked on her other daughter, Raafat. She had been suffering from her annual bout of hay fever, and the 18-year-old art student was in the television room next to the humidifier so she could breathe easier.

Raafat was still sleeping, completely oblivious of all the commotion going on around her, due to the medication she had taken earlier. There was little Saniya felt she could do. She climbed back into bed and pulled the sheets tight around her.

Bassem lay awake on the bed, listening to the appalling noise in the night sky above.

A Palestinian-born Lebanese national, Bassem had worked in Libya as an engineer for Occidental, the American oil giant, for 20 years, helping exploit the country's

ROBERT HARRISON—STAFF

'When Bush talked of a crusade...it was not a slip of the tongue.'
—Sajad Haider, retired
Pakistani air force officer

massive oil reserves. He and his family lived in the up-market Ben Ashour neighborhood of Tripoli, the Libyan capital, on the ground floor of a two-story apartment block.

Bassem never heard the explosion. Instead, he watched in astonishment as the window frame suddenly flew into the room, and the roof collapsed on top of him and his daughter.

Kinda was screaming in the darkness near him. Bassem tried to move, but was pinned by the rubble. He groped in the blackness for Kinda. "Don't worry," he said, squeezing his daughter's hand. "Daddy's here, don't cry, it will be okay."

The blast had knocked Saniya unconscious. She woke to hear Bassem calling from the next room and Kinda screaming. She stumbled in the darkness, barefoot across the rubble and glass shards, choking on the fumes from the missile blast, as she called her daughter's name "Raafat! Raafat!" for several minutes. But there was no response, and Saniya knew with a terrible certainty that her daughter was dead.

"Bassem," she cried. "Raafat has gone."

Pinned beneath the rubble, Bassem heard his wife's words, and he felt a deep sense of anger and resentment well up inside him. His life and that of his family had been shattered, and nothing would ever be the same again.

It took them eight hours to dig Raafat out from under the ruins of the house. "Our pain and agony, which I cannot describe, started at that moment," Saniya says.

Raafat was one of an estimated 55 victims of an air raid mounted by US warplanes against a series of targets in Tripoli and another Libyan city, Benghazi.

The attacks were in retaliation for the bombing of a disco in Berlin, Germany, 10 days earlier in which 200 people were injured, 63 of them US soldiers; one soldier

and one civilian were killed. The Reagan administration blamed Libyan leader Muammar Qaddafi.

Bassem and Saniya Ghussein are not natural anti-Americans. Bassem studied in the US before going to work for Esso and then Occidental. He sent Raafat to an American Catholic school, and on family trips to the US, Saniya would take Raafat to Disney World in Florida. "We did all the typical American things," she says.

But since that terrible night 16 years ago, neither Bassem nor Saniya have stepped foot in America. They returned to Beirut in 1994 when Bassem retired.

In 1989, the Libyan government enlisted the help of Ramsey Clark, an attorney general during the Carter administration, to file a lawsuit against President Ronald Reagan and British Prime Minister Margaret Thatcher for the civilian deaths during the air raids. "When Clark came to collect our documents and evidence, I asked him if he thought we had a case," Bassem recalls. "He said 'Oh, definitely. This was murder.' "

But US district court judge Thomas Penfield Jackson disagreed. He dismissed the suit, and fined Clark for presenting a "frivolous" case that "offered no hope whatsoever of success."

Twelve years later, the court's decision still rankles with Bassem. "I will only return to America when I know someone will listen to me and say: 'yes, it was our fault your daughter died, and I am sorry.' So long as they think my daughter's death is 'frivolous,' I won't go back," Bassem says.

The Ghusseins have no sympathy for religious extremism and thoroughly condemn the Sept. 11 suicide bombings in New York and Washington. Yet they both maintain that the devastating attack was a result of America's "arrogant" policies in the Middle East and elsewhere. "We wish the American people could see what their governments are doing in the rest of the world," Saniya says.

A FEELING OF BETRAYAL AMONG FRIENDS

On the other side of Asia, in Pakistan, Air Commodore Haider would sympathize with the Ghusseins' wish. He has always been a friend of the United States, and not just because he enjoyed the 10 years he spent in Washington as his country's military attaché. Like most other members of the ruling elite in Pakistan, in the armed forces, in business, and in the political parties, he sees America as a natural ally.

But not a reliable one. The prevailing mood in Pakistan of anger and suspicion toward the United States springs from a deeply rooted perception that the US has been a fickle friend, Haider says, and not just to Pakistan, but to other nations in the Muslim world.

If there was a moment of betrayal for Haider, it was the 1965 war between India and Pakistan, largely over the future of Kashmir. As Indian tanks advanced on the Pakistani metropolis of Lahore, Haider was head of a squadron of F-86 Sabre jets sent to destroy them. India's Soviet allies helped with money, arms, and diplomatic support. But at a crucial moment, Pakistan's ally, the US, refused to send more weapons. As it turned out, Pakistan was able to defeat the Indian attack on Lahore and elsewhere without US help. Haider's squadron decimated the column of Indian tanks that had reached to within six miles of Lahore. But the lesson lingered: America cannot be trusted.

"There is a feeling of being betrayed, it's a feeling of being let down, and you can only be let down by somebody you care for," says Haider, out for an evening stroll in a tony Islamabad neighborhood.

"They said you will be the bulwark of America and of the free world against Communism. But then they dropped a friend for no good reason."

Today, Haider sees a "convergence of interests" between the United States and Pakistan in the fight against terrorism. But he says that President Bush will need to watch his language when he talks about the Muslim world. "When Bush talked of a Crusade… it was not a slip of the tongue. It was a mindset. When they talk of terrorism, the only thing they have in mind is Islam."

Ultimately, Haider does see a way for America and Muslim nations to become lasting friends, but only if the US begins to give as much weight to the interests of Muslim nations as it does to Israel.

"When you deny justice to people, which you have been doing for several decades in Palestine, and they are intelligent, sensitive people, they are going to find something to do," warns Haider. "They might take shelter in Islam, in fatalism, and some will come to despise you."

AN EGYPTIAN 'INSPIRED' TO JOIN AFGHAN FIGHTERS

Sheikh Abu Hamza al-Masri, the radical Muslim cleric who runs a mosque in a shabby district of north London, has certainly come to despise America.

Abu Hamza says he used to admire the West when he was a young man—so much so that he dropped out of university in his native Alexandria, Egypt, to study in Britain. And he clearly had nothing against the British government when he took a job as a civil engineer at Sandhurst, the British equivalent of West Point, after he graduated.

But as he immersed himself more and more in religious studies, and came into contact with more and more Arab mujahideen, who had travelled from the mountains of Afghanistan to England for medical treatment, he began to change his outlook.

"When you see how happy they are, how anxious to just have a new limb so they can run again and fight again, not thinking of retiring, their main ambition is to get killed in the cause of God… you see another dimension in the verses of the Koran," says Abu Hamza.

How the world views a US military response

In your opinion, once the identity of the terrorists is known, should the American government launch a military attack on the country or countries where the terrorists are based, or should the American government seek to extradite the terrorists to stand trial?

	Launch attack	Try the terrorists	Don't know
Israel	77%	19%	4%
India	72	28	0
United States	54	30	16
Korea	38	54	9
France	29	67	4
Czech Republic	22	64	14
Italy	21	71	8
South Africa	18	75	7
United Kingdom (excluding N. Ireland)	18	75	7
Germany	17	77	6
Bosnia	14	80	6
Colombia	11	85	4
Pakistan	9	69	22
Greece	6	88	6
Mexico	2	94	3

Source: Gallup International surveys Sept. 14 to 17.

Inspired by their example, he took his family to Afghanistan in 1990, to work there as a civil engineer, building roads, tunnels, and "anything I could do." And he also fought with the mujahideen against Afghan President Mohammad Najibullah (seen as a Russian stand-in supported by the Soviets), until he blew both his hands off and lost the sight in his left eye, in a mine explosion.

What transformed him and his comrades-in-arms from anti-Soviet to anti-American militants, he says, was the way Washington abandoned them at the end of the war in Afghanistan, and sought to disarm and disperse them.

"It was when the Americans took the knife out of the Russians and stabbed it in our back, it's as simple as that," says Abu Hamza. "It was a natural turn, not a theoretical one.

"In the meantime, they were bombarding Iraq and occupying the [Arabian] peninsula," he says, referring to the US troops stationed in Saudi Arabia after the Gulf War, "and then with the witch-hunt against the mujahideen, all of it came together, that was a full-scale war, it was very clear."

Abu Hamza would rather see Islamic militants fight corrupt or secular Arab governments before they take on America (indeed, the Yemeni government has sought his extradition from Britain for plotting to overthrow the government in Sana). But he is in no doubt that the American government brought the events of Sept. 11 on its own head.

"The Americans wanted to fight the Russians with Muslim blood, and they could only justify that by triggering the word 'jihad,' " he argues. "Unfortunately for everybody except the Muslims, when that button is pushed, it does not come back that easy. It only keeps going on and on until the Muslim empire swallows every empire existing."

Can he understand the motivation behind the assault on New York and Washington? "The motivation is everywhere," he says, with the current US administration. "When a president stands up before the planet and says an American comes first, he is only preaching hatred. When a president stands up and says we don't honor our missile treaty with the Russians, he is only preaching arrogance. When he refuses to condemn what's happening in Palestine, he is only preaching tyranny.

"American foreign policy has invited everybody, actually, to try to humiliate America, and to give it a bloody nose," he adds.

IN JAKARTA, COUNTERING AMERICAN CULTURE WITHOUT VIOLENCE

You wouldn't catch Rizky "Jimmy" Nur Zamzamy justifying violence that way, though he professes just as deep an attachment to Islam as Abu Hamza.

Mr. Zamzamy, a rangy young Indonesian advertising executive in a pink shirt, is sitting in a Western-style cafe in Jakarta, his cellphone at the ready, and his fried chicken growing cold as he explains how he tries to be a good Muslim by right action, not fighting.

That, he feels, is the best way of countering what he sees as the corrupting influence of American culture and morals on traditional Indonesian ways of life in the largest Muslim country in the world.

Until a few years ago, Zamzamy led a regular secular life, hanging out in bars and dating women. Then he met a Muslim teacher who became his spiritual guide. Now he follows Islamic teachings and donates most of his $1,300 monthly salary to his "guru" to be spent on building mosques and helping the poor.

He says he has made sure that none of the money goes to extremist groups that use violence in the name of Islam, such as the Laskar Jihad group, locked in bloody battle with Christians in the Maluku region of Indonesia.

Two years ago, in line with his growing religious beliefs, he quit the advertising agency he had worked for and set up his own company along Islamic lines: He won't take banks or alcoholic-beverage producers as clients, for example, and he does no business on Friday, the Muslim holy day.

But he is relaxed about those who don't share his beliefs: He does not insist that his wife wear a headscarf, for example, and he is not uncomfortable sitting alongside the rich young Jakartans in the cafe who are flirting and drinking. They must make their own choices, he says.

And though he does not like the sexual overtones of American pop culture, he knows that "you can't hide from American culture." By living his life according to Islamic precepts, he says, "I am fighting America in my own way. But I don't agree with violence."

AMBIVALENCE ABOUT AMERICA

All over the Muslim world, young people like Zamzamy are juggling their sense of Islamic identity with the trappings of a globalized, secular society.

In a classroom of Al Khair University, set in a concrete office park in Islamabad, Nabil Ahmed, a business student, and his classmates are fuming over their president's betrayal of the Pakistani people by pledging to support what they fear will turn into a crusade against Muslims.

Ahmed and his friends are well-dressed, middle-class boys, and represent neither the old-money security of Pakistan's elite nor dirt-poor peasants who make up the bulk of Pakistan's angry conservative masses. They are the silent majority of Pakistan, with their feet firmly planted in both the East and the West. On weekdays, they listen to Whitney Houston and Michael Bolton, wear Dockers and Van Heusen shirts. On weekends, many switch to traditional salwar kameez outfits and go with their fathers to the mosque to pray.

'It is [the] double standard that creates hatred.'

—Nabil Ahmed, a business student at Al Khair University in Islamabad

---

They have much to gain from a Western style of life, and most have plans to move to the United States for a few years to make some money before returning home to Pakistan. Yet despite their attraction to the West, they are wary of it too.

"Most of us here like it both ways, we like American fashion, American music, American movies, but in the end, we are Muslims," says Ahmed. "The Holy Prophet said that all Muslims are like one body, and if one part of the body gets injured, then all parts feel that pain. If one Muslim is injured by non-Muslims in Afghanistan, it is the duty of all Muslims of the world to help him."

Like his friends, Ahmed feels that America has double standards toward its friends and enemies. America attacks Iraq if it invades Kuwait, but allows Israel to bulldoze Palestinian homes in the West Bank and Gaza Strip.

ROBERT HARRISON—STAFF

AFGHAN REFUGEES: These boys are among some 60,000 displaced Afghans at Jalozai refugee camp near Peshawar, Pakistan, along its border with Afghanistan. The camp is crowded, and Pakistan has recently forbidden the UNHCR to register any more refugees.

It ostracizes a Muslim nation like Sudan for oppressing its Christian minority, but allows Russia to bomb its Muslim minority into submission in Chechnya.

And while the US supported many "freedom fighter" movements in the past few decades, including the contra movement in Nicaragua, America labels Pakistan and Afghanistan as terrorist states because they support militant Muslim groups fighting in the Indian state of Kashmir and elsewhere.

'The Americans wanted to fight the Russians with Muslim blood.'

—Sheikh Abu Hamza al-Masri,
a radical Muslim cleric who runs a mosque in London

"There is only one way for America to be a friend of Islam," says Ahmed. "And that is if they consider our lives to be as precious as their own. "If Americans are concerned about the 6,500 deaths in the World Trade Center, let them talk also about the deaths in Kashmir, in Palestine, in Chechnya, in Bosnia. It is this double standard that creates hatred."

Ahmed's ambivalence about America—his desire to live and work there, his admiration for its values, but his anger at its behavior around the world—is broadly shared across the Muslim world and Arab world.

"I think they hate us because of what we do, and it seems to contradict who we say we are," says Bruce Lawrence, a professor of religion at Duke University, referring to people in the Middle East. "The major issue is that our policy seems to contradict our own basic values."

That seems clear enough to Muslims who sympathize with the Palestinians, and who say that Washington should force Israel to abide by United Nations resolutions to withdraw from the occupied territories. "The Americans say September 11th was an attack on civilization," says Mr Hariri, the Lebanese prime minister. "But what does civilized society mean if not a society that lives according to the law?"

It also seems clear to citizens of monarchical states in the Gulf, where elections are unknown and women's rights severely restricted. "Since the Cold War ended, America has talked about promoting democracy," says John Esposito, head of the Center for Muslim-Christian Understanding at Georgetown University in Washington. "But we don't do anything about it in repressive regimes in the Middle East, so you can understand widespread anti-Americanism there."

At the same time, the state-run media—which is all the media there is across much of the Middle East—often fan the flames of anti-American and anti-Israel sentiment because that helps focus citizens' minds on something other than their own government's shortcomings.

In Sana, the Yemeni capital, where queues of visa-seekers line up daily outside the US embassy, the ambivalence about America is clear. "When you go there, you really

50 YEARS OF US POLICY IN THE MIDDLE EAST

1947–48

UN votes to partition Palestine into two states—one for Jews, one for Palestinian Arabs. Arab states invade; 300,000 Palestinians flee Jewish-controlled areas. Jewish forces prevail, declaring Israeli independence. US recognizes Israel.

1953

CIA helps Iran's military stage a coup, deposing elected PM Mohammad Mossadeq, whom US sees as communist threat. US oversees installation of Shah Mohammad Reza Pavlavi as ruler of Iran.

1956

Israel attacks Egypt for control of Suez Canal. Britain and France veto US-sponsored UN resolution calling for halt to military action. British forces attack Egypt.

1960

Iran, Iraq, Kuwait, Saudi Arabia, and Venezuela form Organization of Petroleum Exporting Nations (OPEC).

1966

US sells its firs jet bombers to Israel, breaking with a 1956 decision not to sell arms to the Jewish state.

1967

Six-Day War. Israel launches preemptive strike against Arab neighbors, capturing Jerusalem, the Sinai Peninsula, the Gaza Strip, and the Golan Heights. Kuwait and Iraq cut oil supplies to US, UN adopts Resolution 242, calling on Israel to withdraw from captured territory. Israel refuses.

1968

First major hijacking by Arab militants occurs on El Al flight from Rome to Tel Aviv, marking decades of hostage-takings, hijackings, and assassinations as a strategy by Arab militant groups.

1969

Mummar Qaddafi comes to power in Libyan coup and orders US Air Force to evacuate Tripoli.

1972

Eight Arab commandos of Palestinian group Black September kill 11 Israeli athletes at the Munich Olympic Games.

1973

Egypt and Syria attack Israel over its occupation of the Golan Heights and the Sinai Peninsula. US gives $2.2 billion in emergency aid to Israel, turning tide of battle to Israel's favor. Arab states cut US oil shipments.

1974

UN General Assembly recognizes right of Palestinians to independence.

1976

The UN votes on a resolution accusing Israel of war crimes in occupied Arab territories. US casts lone "no" vote. US Ambassador to Lebanon Francis Meloy and an adviser are shot to death in Beirut. US closes Embassy there.

1978

Egypt and Israel sign US-brokered Camp David peace treaty. Eighteen Arab countries impose an economic boycott on Egypt. Egyptian president Anwar Sadat and Israeli Prime Minister Menachem Begin receive Nobel Peace Prize.

1979

Ayatollah Ruhollah Khomeini leads grass-roots Islamic revolution in Iran, deriding the US as "the great Satan." Iranian students storm US Embassy in Tehran, taking 66 Americans hostage for next 15 months. US imposes sanctions. Protesters attack US Embassies in Libya and Pakistan.

1981

Israel bombs Iraqi nuclear reactor. Muslim militants opposed to Egypt's peace treaty with Israel assassinate Egyptian President Sadat.

1982

Israel invades Lebanon to expel the Palestine Liberation Organization, facilitate election of friendly government, and form 25-mile security zone along Israel's border. Defense Minister Ariel Sharon permits Lebanese Christian militiamen to enter the Sabra and Shatila refugee camps outside Beirut. The ensuing three-day massacre kills 600 or more civilian refugees. US and other nations deploy peacekeeping troops in Lebanon.

1983

A truck bomb explodes in US Marines' barracks in Beirut, Lebanon, killing 241 soldiers. US forces withdraw.

1986

Us bombs Libya in retaliation for the bombing of a Berlin nightclub frequented by US servicemen. The airstrike kills 15 people, including the infant daughter of leader Muammar Qaddafi. All Arab nations condemn the attack.

1987

Start of the Palestinian intifada, or uprising, in the West Bank and Gaza Strip.

1990

Iraq invades Kuwait. Saddam Hussein links pullout to Israel's withdrawal from occupied territories. UN imposes sanctions that continue to hobble Iraq's economy in effort to force Iraqi compliance with weapons resolutions.

(continued)

50 YEARS *(continued)*

1991

US and coalition launch attacks against Iraq from Saudi Arabia. Gulf War ends after some three months, but US deployment continues even now, with 17,000 to 24,000 US troops in region at any time.

1993

World Trade Center in New York is bombed, killing six. US Special Forces, deployed as peacekeepers in Somalia, attempt to capture warlord Mohamed Farah Aidid. Eighteen US servicemen are killed. Israeli PM Yitzhak Rabin and Palestinian leader Yasser Arafat sign historic peace declaration in White House ceremony with President Clinton.

1994

Jordan and Israel sign peace treaty. Yasser Arafat, Yitzhak Rabin, and Foreign Minister Shimon Peres receive Nobel Peace Prize for 1993 agreement.

1995

US announces trade ban against Iran, reinforcing sanctions in effect since 1979. Rabin is assassinated, two years after peace deal with Palestinians. In Riyadh, Saudi Arabia, a car bomb explodes outside an office housing US military personnel. Seven are killed, including five Americans. Three Islamist groups claim responsibility.

1996

A truck bomb explodes outside a US military barracks in Khobar, Saudi Arabia, killing 19 US airmen. UN reports that sanctions cause 4,500 Iraqi children under 5 to die each month.

1997

Egyptian Islamic Group massacres 62 people, mostly foreign tourists, in Luxor, Egypt. The group claims it is retaliation for US imprisonment of Sheikh Omar Abdel al-Rahman, who is later convicted in 1993 World Trade Center bombing.

1998

Bombs explode outside US Embassies in Kenya and Tanzania, killing 224 people. US launches cruise-missile attacks on sites in Sudan and Afghanistan allegedly linked to Osama bin Laden. US indicts bin Laden for committing acts of terrorism against Americans abroad.

1999

Islamic militants, traced to bin Laden, are arrested for plot to bomb tourist sites during millennium celebrations.

2000

Camp David negotiations fail. Sharon visits Temple Mount in Jerusalem, sparking current Palestinian uprising. USS Cole bombing in Yemen's Aden harbor kills 17 American sailors. Bin Laden denies responsibility, but applauds the act.

2001

Hijackers crash two planes into World Trade Center in New York, one into Pentagon, and one in Pennsylvania. More than 7,000 people are dead or missing.

Compiled by Julie Finnin Day

SOURCES: "THE MIDDLE EAST" (CONGRESSIONAL QUARTERLY), NEWS REPORTS.

---

love the United States," says Murad al-Murayri, a US-trained physicist. "You are treated like a human being, much better than in your own country. But when you go back home, you find the US applies justice and fairness to its own people, but not abroad. In this era of globalization, that cannot stand."

Nor has the mood that has gripped Washington over the past two weeks done much to reassure skeptics, says François Burgat, a French social scientist in Yemen.

"When Bush says 'crusade', or that he wants bin Laden 'dead or alive', that is a *fatwa* (religious edict) without any judicial review," he cautions. "It denies all the principles that America is supposed to be."

A *fatwa* is something Amirul Haq, a Pakistani shopkeeper whose son died two years ago in a jihad in Kashmir, understands better than judicial review. "When I heard that my son died, I was satisfied," he says.

It's a sentiment shared by Azad Khan, too. On a hot Sunday afternoon in Mardan, Pakistan, Mr. Khan and his family have laid out a feast in a small guesthouse next to the local mosque. They are celebrating because they have just heard that Mr. Khan's 20-year-old son, Saeed, has

been killed in a gun battle with Indian troops in the part of Jammu and Kashmir state that is under Indian control. With his death, Saeed has become another *shahid*, a martyr and heroic defender of the Muslims against the enemies of Islam. According to the Koran, *shahideen* are not actually dead; they are still alive, they just can't be seen. And through acts of bravery, a *shahid* guarantees that his whole family will go to heaven.

"It is not a thing to be mourned. We are happy," says Khan, sitting down to a meal of chicken and mutton, rice and bread, along with leaders of the group with which Saeed had fought. "I told him to take part in jihad [holy war] because he is the son of a Muslim," Khan says. "And just as we fight in Kashmir, if we need to fight against the United States in Afghanistan we are ready, because we are Muslims. It is our duty to fight against any infidels who are threatening our Muslim brothers."

It's not likely that many Pakistanis, or other Muslims, will actually go to Afghanistan to fight the Americans—assuming American soldiers land there. Khan's militant views are not shared by most of his countrymen.

But in a broader sense, and in the longer term, many people in the Middle East fear that the coming war against terrorism—unless it is waged with the utmost caution—could unleash new waves of anti-American sentiment.

Jamal al-Adimi, a US-educated Yemeni lawyer, speaks for many when he warns that "if violence escalates, you bring seeds and water for terrorism. You kill someone's brother or mother, and you will just get more crazy people."

Trying to root out terrorism without re-plowing the soil in which it grows—which means rethinking the policies that breed anti-American sentiment—is unlikely to succeed, say ordinary Middle Easterners and some of their leaders.

On the practical level, Hariri points out, "launching a war is in the hands of the Americans, but winning it needs everybody. And that means everybody should see that he has an interest in joining the coalition" that Washington is building.

On a higher level, argues Bassam Tibi, a professor of international relations at Gottingen University in Germany, and an expert on political Islam, "we need value consensus between the West and Islam on democracy and human rights to combat Islamic fundamentalism. We can't do it with bombs and shooting—that will only exacerbate the problem."

Reported by staff writers Scott Baldauf in Islamabad, Pakistan; Cameron W. Barr in Amman, Jordan; Peter Ford in London; Nicole Gaouette in Jerusalem; Robert Marquand in Beijing; Scott Peterson in Sana, Yemen; Ilene R. Prusher in Tokyo; as well as contributors Nicholas Blanford in Beirut, Lebanon; Sarah Gauch in Cairo; and Simon Montlake in Jakarta, Indonesia.

CHANGING COURSE ON CHINA

Elizabeth Economy

"Relations between China and the United States are perhaps the best they have been since 1989…What accounts for this seemingly dramatic transformation?"

In the immediate aftermath of his election in November 2000, President George W. Bush proclaimed China a strategic competitor and asserted that US policy in Asia should be reoriented toward American allies in the region, including Taiwan. Consensus withinthe administration on how to implement this new policy, however, remained elusive.

Indeed, for much of the early tenture of the Bush White House, the administration seemed divided on how best to approach China. The US trade representative and members of the State Department preached the virtues of engagement with the mainland while the Pentagon formulated its own policy to enhance US relations with Taiwan. Some within the administration went so far as to place the mantle of the former Soviet Union on China, calling the People's Republic the next great threat to US security. Within months after Bush took office, an ugly altercation over a US spy plane, a sizeable arms sale to Taiwan, and aggressive talk of American missile defense increased tensions markedly.

Nearly three years later, relations between China and the United States are perhaps the best they have been since 1989. President Bush appears to be following in the footsteps of his predecessors in recognizing both the importance of China to US foreign policy interests and the benefits of a more proactive approach to the mainland. Apparent divisions within the Bush administration over how to approach China have resolved themselves, at least temporarily, in favor of a more engagement-oriented policy.

What accounts for this seemingly dramatic transformation in US policy toward China? Above all else, the evolution in policy reflects a new set of strategic realities that have confronted America since September 11, 2001. The events of September 11 caused the Bush adminstration to reorient its international priorities, both diminishing the centrality of China in US foreign policy as a potential long-term threat and offering a new and important opportunity for the two countries to cooperate.

Soon after September 11, when the Bush administration began its campaign to enforce Iraqi compliance with United Nations weapons inspections, US officials actively courted the Chinese leadership. The administration first sought China's support for stepped-up inspections of Iraq's weapons arsenal, then for a UN resolution authorizing military action against Iraq. In the process, the administration turned its agenda with China on its head. More recently, the nuclear crisis on the Korean peninsula has reinforced the importance of China to US strategic interests while opening a new avenue for cooperation. For the United States, China has become an essential partner in meeting its new geostrategic challenges.

China's leaders, in turn, have recognized an opportunity to use these new strategic realities to meet their own fundamental objectives: stability in the Sino-American relationship to ensure continued economic modernization, and the maintenance of China's own domestic security. As Chinese President Hu Jintao has argued, the importance of the United States to China's economic development requires a flexible and accommodating posture that keeps US–China relations on an even keel.

This does not mean that significant policy differences between the two countries have disappeared, or that this partnership will continue indefinitely at its current level of mutual accommodation. The Sino-American relationship remains fragile and continues to require high-level intervention to ensure its stability. Critical differences still mark the manner in which the two countries approach international relations generally, as well as specific issues such as Taiwan, missile defense, and human rights. In addition, President Bush does not make China policy in a vacuum. Differences within the administration over relations with China remain. An active congressional lobby on China and Taiwan, while quiescent for most of 2002 and 2003, may again become energized in an effort to redefine US-China policy.

THE FIRST SIX MONTHS

During the presidential campaign, candidate Bush set the tone for a distinctly new China policy. He promised to refocus US attention in the region away from the mainland and toward Japan, South Korea, and Taiwan. The Bush team stressed that these East Asian powers were democratic and capitalist and thus natural allies of the United

States. Candidate Bush also referred to China as a strategic competitor, emphasizing its human rights abuses and role in the proliferation of missile technology.

> For the United States, China has become an essential partner in meeting its new geostrategic challenges.

While this rhetoric received significant media attention, subtle hints also signaled a degree of continuity with the previous administration's policies. Candidate Bush argued that the United States had to remain deeply engaged on the trade front with China, noting that the development of an entrepreneurial class and the advent of the Internet were cornerstones in a process of long-term political liberalization in China and that American farmers would benefit from China's entry into the World Trade Organization. Bush even reiterated the classic engagement line: "I think if we make China an enemy, they'll wind up being an enemy."

The first testing ground for the new administration and its approach to China came less than three months after Bush assumed the office. In late March 2001, a US Navy EP-3 surveillance aircraft flying near China's Hainan Island collided with a Chinese fighter jet. The Chinese pilot was lost at sea and the EP-3's flight crew was forced to make an emergency landing on Hainan. It was not until April 4, five days after the accident, that diplomatic exchanges occurred.

This was a tense and difficult time, with many members of Congress calling for tough action and some policy analysts predicting military conflict between China and the United States. Once China opened the door to US diplomats, however, President Bush moved quickly to effect a resolution. On April 5, Secretary of State Colin Powell issues a statement of "regret" that progressed to "sorry" and "very sorry" by April 7. On April 11, the US flight crew departed Hainan to return to the United States. In the aftermath of the incident, the consensus in the United States was that President Bush had handled a difficult situation well by preventing the conflict from escalating. Still, some conservative policy analysts critized the administration for appeasing Chinese misbehavior. They argued too that the strategy President Bush adopted represented a victory for Secretary of State Powell and National Security Adviser Condoleezza Rice over Vice President Dick Chaney and Secretary of Defense Donald Rumsfeld.

Still, the EP-3 incident, as well as the initial difficulty in reaching and negotiating with senior Chinese military and party officials, provided the rationale that China hawks within the administration needed to take a series of significant steps. The Defense Department broke off all informal and regular military-to-military contacts, noting that such contacts would have to be approved on a case-by-case basis. The administration announced an arms sale package to Taiwan worth as much as $4 billion that included up to eight diesel submarines and four guided-missile destryoers. It cited as justification China's continued missile buildup opposite Taiwan and the mainland's refusal to renounce the right to use force to reunify with Taiwan. For the first time, the Defense Department allowed Taiwanese military officials to participate in courses at its Hawaii-based think tank, the Asia-Pacific Center for Security Studies.

Meanwhile, discussions of the Bush administration's plans for missile defense, which the Chinese regarded as extremely threatening, filled the news media. And when asked whether he would use military force to defend Taiwan in case of a Chinese attack, the president said, "Whatever it takes to help Taiwan defend itself." This assertion caused great consternation in mainland China, where it was perceived by some as dropping the American commitment to "strategic ambiguity" (whereby the United States had refused to discuss various scenarios in which conflict between the mainland and Taiwan might emerge in order to preserve the full range of options for US action).

Yet even as the security relationship clearly deteriorated, Bush did not permit the EP-3 incident to become entangled with other areas of the Sino-American relationship, such as China's impending entry into the World Trade Organization. Morever, the manner in which the EP-3 incident itself was handled demonstrated that, whatever the administration's rhetoric, the White House was committed to keeping its relationship with China on track and to preventing tensions in the security realm from spilling over into other arenas. In fact, during the summer of 2001 Secretary of State Powell pursued a noticeably more engagement-oriented approach than others within the administration, visiting the People's Republic and elucidating a new foundation for US dialogue with China dubbed the "Three C's": candid, cooperative, and constructive. Thus, for much of the early part of the Bush administration's tenure, China policy proceeded along two distinct tracks: one directed by the Department of Defense and vice president's office and a second navigated primarily by the State Department.

AFTER SEPTEMBER 11

The devasting terrorist attcks on the World Trade Center in New York and on the Pentagon in Virginia caused the United States to reorient radically its foreign policy priorities. The central focus shifted from promotion of free trade, democracy, and stability to a global war against terrorism. No longer was the Bush administration preoccupied with defining the next Soviet-like menace; it needed to identify terrorist cells throughout the world.

For Sino-American relations, this transformation is US foreign policy had two important implications. First, China policy became simply one of many issues rather than a top preoccupation for the administration, Congress, and the media. In a world in which America's physical integrity had been so violently breached, concern over the potential economic and security threat posed by China greatly diminished. Second, September 11 provided a clear opportunity for China to establish common interest with the United States on the latter's number one priority: combating terrorism.

China initially hedged its bets, calling for the United States to support China in battling terrorists in Xinjiang, Tibet, and Taiwan. Just one week after the attacks, the government spokesman Zhu Bangzao remarked that "the United States has asked us to help it fight terrorism. Equally, we have reasons for asking the United States to lend us its support and understanding in our struggle against terrorism and separatism. There can be no double standards. We are not suggesting any horse trading; but China and the United States has a common interest in opposing the Taiwanese independence movement which constitutes the main threat to stability in the [Taiwan] Strait."

China soon amended its request to ask for assistance only in fighting the terrorist threat in Xinjiang Autonomous Region, perhaps realizing that equating Al Qaeda with separatist movements in Taiwan and Tibet would be poorly received by many in the United States. Still, China reiterated its concerns that America establish concrete proof of Osama Bin Laden's guilt, that any military strike it might carry out accord with UN rules, and that any action taken be in the long-term interest of world peace and development. Since September 11, the Bush administration has generally given China high marks for its help in tracking down terrorist financing, cooperating on law enforcement, and providing humanitarian aid to Afghanistan.

But, even as China and the United States began to forge new bonds on the issue of global terrorism, Taiwan remained a sticking point. In March 2002, for example, the US Taiwan Business Council hosted in the United States Taiwanese Defense Minister Tang Yiau-ming and other military officials—including US Deputy Secretary of Defense Paul Wolfowitz—thereby violating an undeclared prohibition against meetings between senior Taiwanese and senior US officials. During the meeting, moreover, Wolfowitz repeated President Bush's earlier controversial statement that the United States would do "whatever it takes" to defend Taiwan from military strikes by China. Wolfowitz's affirmation that "China is not an enemy" was lost in the firestorm that his other rhetoric provoked.

At the same time, the Defense Department's Nuclear Posture Review, a summary of strategic planning for America's nuclear weapons over the next decade, was leaked to the media. The document noted that a conflict between China and Taiwan could lead to the use of nuclear weapons by the United States. This added to the growing tensions in Sino-American relations. Congress, too, always an active player on Taiwan issues, became energized by the Bush administration's more proactive Taiwan policy. On April 9, 2002—the twenty-third anniversary of the Taiwan Relations Act—the House of Representatives established the Taiwan Caucus. One of the forum's primary goals was to promote US military ties with Taiwan. (The 1979 act guarantees continued trade and cultural relations with the island and provides assurances for Taiwan's security.)

Beijing was quick to respond to the Defense Department's increased attention to Taiwan. In reply to Wolfowitz's remarks, Foreign Ministry spokeswoman Zhang Qiyue stated, "The comments of the senior US defense official seriouly violated the clear-cut promises laid out in the three joint communiques [that have framed Sino-American relations] and moreover rudely interfered in China's internal political affairs." In addition, Beijing denied the destroyer the USS *Curtis Wilbur* a port call in Hong Kong for early April, and threatened to cancel the impending visit of then Vice President Hu Juntao later in the month. Thus, only weeks before Hu's visit to the United States, the Pentagon's priority on enhancing ties with Taiwan at the expense of relations with the mainland threatened yet again to derail the Sino-American relationship.

SETTING A NEW AGENDA

While the war against terrorism opened the door to the creation of a new foundation for Sino-American relations, the most significant factor in the evolving relationship was the Bush administration's almost singular focus on Iraq. In working to secure backing for a US-led attack on Iraq, the Bush administration arrived at an entirely new agenda with China. Beginning in April 2002 with the visit of then Vice President Hu Jintao, the White House moved quietly and effectively to set the stage for a newfound unity in US-China relations.

The first sign of this emerged during Hu's visit. Both before and after the visit, the White House downplayed the contentious issues of human rights, Taiwan, and weapons proliferation. Although these concerns came up during a session between Vice President Chaney and Vice President Hu, the thrust of the meetings was positive. One high-ranking but unidentified White House official said: "Mr. Hu is bright, amenable, and very pleasant …she is a pleasure to do business with." Another senior administration official offered even higher praise: "He can come across as warm and even flexible, yet gives nothing away. I can see the day when Mr. Bush feels he can pick up the phone and call Mr. Hu. I don't think he has ever quite felt that way with [President] Jiang Zemin. They are of different eras."

Just one month after Hu's visit, Pentagon hawks, who had defined China as the next Soviet Union, were flying to the People's Republic to discuss conditions for restoring military-to-military contacts. Senior administration officials no longer identified China as a strategic competitor but rather as Secretary of State Powell had—as a participant in a "candid, cooperative, and constructive" dialogue. Deputy Secretary of Defense Wolfowitz began to speak of the need to engage China.

In August 2002 the Bush administration labeled the East Turkestan Islamic Movement—a small group committed to the independence of China's Xinjiang region—a terrorist organization. Beijing, which had long claimed connections between Al Qaeda and ethnic Uighur separatists in Xinjiang, welcomed the move. Perhaps most telling, the Bush administration also indicated in August that it would grant President Jiang Zemin his long-sought invitation to President Bush's ranch in Crawford, Texas, an invitation previously extended only to President Bush's closest perceived allies, such as British Prime Minister Tony Blair and Russian President Vladimir Putin. Moreover, the administration virtually ignored the release of a long-anticipated report by the congressional US-China Security Review Commission in fall 2002, which widely condemned China for its indirect sponsorship of rogue states through the sale of missle technology and called for a range of new sanctions. The report was an embarrassment for an administration intent on making a new friend in China.

China responded positively to th US steps to improve relations. In late August 2002, the Chinese government issued a new set of regulations governing the export of missile technology and reacted only moderately to strong separist rhetoric by Taiwan's leader, Chen Shui-bian, and a visit to the Pentagon by a senior Taiwanese defense official. In October Beijing also announced regulations to control material and technology for biological weapons and took steps to improve the situation of a few political dissidents, including AIDS activist Wan Yanhai, who was permitted to establish his own non-governmental organization to combat AIDS.

On the issue of greatest importance to the United States—reining in any Iraqi program to develop weapons of mass destruction—China supported the United States in its initial effort to press for more aggressive sanctions. It voted in favor of UN Security Council Resolution 1441, which warned Iraq that it would face "serious consequences" if it did not comply with UN weapons inspections. Prime minister Zhu Rongji insisted that "Iraq must cooperate unconditionally with the United Nations" on weapons inspections "At the same time," he continued, "we must respect the sovereignty and territorial integrity of Iraq. If arms inspections do not take place, if there is not clear proof and if there is no authorization from the Security Council, there cannot be a military attack on Iraq."

As the United States pressed for an additional resolution authorizing the automatic use of force in the face of Iraqi noncompliance, China stood with France and Russia in opposition to the proposal. But it did so much less forcefully and publicly than its Security Council partners; there was nothing to be gained from unnecessary antagonizing the United States, particularly in the midst of such a positive overall bilateral relationship. In return, China

escaped the far more vocal criticism the United States directed at France and Russia.

Iraq engendered a new US approach toward China, but it has been the crisis on the Korean peninsula that has served to reinforce the importance of a US-China partnership. In October 2002, North Korea admitted that it had not forsaken its nuclear program, triggering a crisis in US-North Korean relations and setting off shock waves throughout the rest of East Asia. The United States and China had long professed "peace and stability on the Korean peninsula" as a common policy goal. Yet rarely had the policy been put to such a test, and the two countries soon found themselves articulating signficiantly different approaches to the resolution of the crisis.

China, like Japan and South Korea, favored a strategy that engaged North Korea, offering economic incentives for compliance. The United States argued for a much tougher policy, possibly involving sanctions and not ruling out the use of force. China also supported North Korea's desire to resolve the issue through bilateral US-North Korea negotiations; the United States insisted on multilaterial talks. As North Korea pushed the envelope, withdrawing from the nuclear Non-Proliferation Treaty, the United States sought a resolution within the UN security Council condemning North Korea for its actions. But China resisted resorting to the United Nations, believing that doing so would only isolate North Korea further and put China in the awkward position of having to ally itself openly with either North Korea or the United States.

Despite their different policy prescriptions, the United States and China continued to work together behind the scenes at the behest of the United States. Some Chinese scholars within elite policy circles also began to suggest that China needed to reevaluate its position, arguing that a nuclear North Korea posed a threat to China. In February 2003, China shut down an oil supply line to North Korea for three days—ostensibly due to "technical problems" but more likely in an effort to signal to North Korea that its belligerent approach was costing it China's support. Soon thereafter, China brokered an agreement between the United States adn North Korea to hold three-way talks in which China would be the third participant. While the negotiations produced no tangible results, the United States and China had moved their relationship to a new level of partnership in the process and agreed to continue to work together to resolve the crisis.

MARGINALIZED: HUMAN RIGHTS, TAIWAN, TRADE

` After 1989, the issues that defined the Sino-American relationship were human rights, Taiwan and trade. These concerns no longer dominate the bilateral agenda. Securing stability on the Korean peninsula, jointly fighting terrorism, and meeting US objectives in Iraq, as well as China's desire for continued stability to ensure economic growth, are now the dominant factors in the relationship.

This transformation in the bilateral agenda has been marked by substantive changes in policy. For the first time since 1989, the United States in April 2003 did not pursue a resolution at the UN Human Rights Commission in Geneva condemning China for its human rights practices. The United States has been noticeably quiet about human rights violations that otherwise would have given rise to comment from the White House, including China's decision to prevent labor activists in Liaoning province from meeting with US diplomats and foreign reporters, and the continued imprisonment without council of long-time American resident Yang Jianli, a prominent democracy activist who illegally entered China to observe mass labor protests in northeast China. Whereas previously the Bush administration might have exploited Beijing's initial mismanagement of the SARS crisis as an opportunity to criticize the regime, at least publicly, the Bush team instead congratulated the Chinese leadership on its handling of what was feared might become a global epidemic.

With regard to Taiwan, the Bush administration continues to act as the island's primary interlocutor in international forums over the objections of the People's Republic: supporting Taiwan's accession to the World Health Organization (blocked by China) in the midst of the SARS crisis, for instance, and backing Taiwan as it wrangles with Beijing over its formal name within the World Trade Organization. The United States continues to promote military relations with Taiwan through exchanges and the sale of advanced weaponry.

Yet the administration has taken steps to reassure Beijing of its commitment to a one China policy, moving beyond previous US administration to state definitively that America does not support Taiwan independence. Beijing, in turn, has begun to develop policy proposals that might persuade the Bush administration to reduce its commitment to Taiwan. In October 2002, for example, President Jiang proposed withdrawing some of the several hundred missiles targeted at Taiwan in exchange for US cutbacks on arms sales. Although the administration has yet to reply, Bush has reportedly asked his staff to develop a formal response. At the same time, within some quarters of China's elite, growing economic integration and personal links between the mainland and Taiwan have begun to foster a new sense of confidence concerning the eventuality of reunification.

In meetings between US and Chinese officials, Taiwan remains an important matter, but there is little evidence of the rancor that marked bilateral exchanges on the issue prior to September 11. In June 2003 at the Group of Eight summit in Evian, France, Hu Jintao—who replaced Jiang as president this March—stated that he appreciated President Bush's declaration not to support Taiwanese independence.

Unlike human rights and Taiwan, trade issues typically have had an ameliorative effective on Sino-American relations. China's accession to the World Trade Organization in December 2001 was celebrated both in China and the United States as signaling signficant future trade and economic opportunities. America is China's largest export market, with overall exports from China to the United States last year totaling more than $100 billion.

Still, many in the American business community and within the US trade representative's office have begun to express concern with a growing trade deficit that, according to US Trade Representative Robert Zoellick, will exceed $100 billion in 2003. In several critical areas such as agriculture, telecommunications, and finance, Zoellick and various China watchers have noted an increasing number of bureaucratic impediments to foreign access to China's markets.

In addition, while successive US administrations had felt comfortable with Prime Minister Zhu Rongji and his committment to effective implementation of China's World Trade Organization obligations, his retirement in April 2003, as well as the elimination of the Ministry of Foreign Trade and Economic Cooperation, has left some uncertainly in the trade relationship. In May 2003, the United States sanctioned a major Chinese conglomerate, NORINCO, for transferring dual-use technology to an Iranian company known to produce missiles. (The Chinese government's muted response lent weight to the US claim.) Thus, although trade remains one of the pillars of the Sino-American relationship, it is possible that friction will increase over time, particularly if China's WTO implementation proves problematic. Indeed, the United States is the third-largest source of foreign direct investment (FDI) in China after Hong Kong and the Virgin Islands, through which FDI from a number of countries flows. (In 2002, total US FDI contracted totaled more than $7 billion.)

WHAT NEXT?

Even as China and the United States continue to strengthen their bilateral relationship by working closely on issues of global security and downplaying traditional rifs, their recent accommodation may well prove ephemeral. Fundamental divisions persist. The United States coninued to desire, and work toward, evolution in China's political system. Many in China still perceive America as a signficant obstacle to China's growing status as a regional, if not global power.

The uncertainty generated by the recent leadership transition in China and the potential for a new policy agenda in Beijing could also open the door to significant change in China's approach to the United States. Former President Jiang Zemin continues to exercise power beind the scenes, often in conflict with Hu Jintao. While unlikely, it is not impossible that policy toward the United States could fall victim to elite power politics as one side or the other attempts to play an anti-US/nationalism card in hopes of currying popular support. President Hu's new emphassi on slowing down the pace of economic reform to redress the vast social inequities that emerged over the course of Jiang's tenure might also contribute to a slowdown in the implementation of China's trade commitments and a consequent downturn in Sino-American relations.

Important opportunities remain, however, for strengthening the foundation of the relationship and helping it endure beyond the current, potentially transitory accommodation that has resulted from new geostrategic realities. Cooperative ventures in areas where common interest exists naturally, such as public health and the environment, should be fostered. Even more critically, the administration should seek opportunities to advance Sino-American relations in the area most difficult to negotiate: security, human rights, and Taiwan.

In the security realm, the Department of Defense should move quickly to reestablish military-to-military relations. Although the EP-3 incident reinforced the importance of maintaining open channels of communication and fostering military exchange, the Pentagon has yet to follow through on President Bush's directive to restart these exchanges. American military officers must still obtain approval for every meeting or exchange with their Chinese counterparts. This is inefficient and does little to help develop the open dialogue and personal relationships that can prove effective for crisis management as well as longer-term understanding of each country's security priorities and appraches. Military contracts are particularly critical as the United States and China continue to work closely on a highly sensitivie security issue such as North Korea.

In working to secure backing of a US-led attack on Iraq, the Bush administration arrived at an entirely new agenda with China.

With regard to human rights, the new accommodation in Sino-American relations offers the Bush administration the opportunity to call directly on President Hu to embrace a more aggressive program of rights protection. The United States is already cooperating, both directly and indirectly, with the Chinese government on law enforcement and the development of the rule of law. Recent high-profile cooperation in dismantling a major drug-trafficking network in Asia and North America, for example, may mark a breakthrough for future cooperation on law enforcement. Various US and NGO efforts focus on the training of Chinese judges and lawyers, cooperation on intellectual property enforcement, and support for more open media in the hopes of assisting enforcement. The State Department's Bureau of Human Rights, De-

mocracy and Labor in particular has mounted a variety of ambitious initiatives designed to advance the cause of human rights, rule of law, development of NGOS, and freedom of expression.

The administration should also renew the dialogue on human rights at the highest level. While the long-stalled dialogue on human rights between the State Department and the Chinese Foreign Ministry resumed formally in December 2002, it has failed to produce follow-up visits or any tangible results, ostensibly because of personnel shuffling within China. The United States should make it clear to Beijing that dialogue and action on human rights are as important a part of the bilateral relationship as discussions on restoring direct military-to-military relations or providing additional assistance to combat terrorism.

Finally, the administration appears to be successfully navigating the often treacherous waters of cross-strait politics. As the White House presses forward with various security initiatives in Asia, including the restationing of troops in South Korea and Japan and the development of missile defense, administration officials should take the opportunity to consult and even cooperate with the People's Republic. If the United States pursues a missile defense architecture that permits targeting of Chinese missiles, for example, early consultation with China to establish mutually acceptable limits on the number of Chinese missiles deployed in response would be extremely useful in avoiding a potential regional arms race and a downward spiral in US-China relations. In the spirit of the new Sino-American relationship, China as indicated that, while it is concerned about the development of theater missile defense, it is willing to conduct "constructive dialogue" on the issue. The only aspect of a potential US missile defense system that China considers non-negotiable is the inclusion of Taiwan behind the shield.

For the foreseeable future, a strong and stable bilateral relationship serves both Chinese and American interests. China considers US support central to its own continued economic development and security. The United States considers China's support necessary for addressing challenges in North Korea and Iraq, and to combat terrorism. As long as these strategic realities continue to dominate each country's domestic and foreign policy agendas, positive relations between China and the United States are likely to remain a high priority for the leaders of both countries. By developing long-term strategies on key contentious issues, the leaders could improve chances that this stability will continue.

ELIZABETH ECONOMY is a senior fellow and director of Asia studies at the Council on Foreign Relations.

THE KOREA CRISIS

North Korea is not crazy, near collapse, nor about to start a war. But it is dangerous, not to mention dangerously misunderstood. Defusing the threat that North Korea poses to its neighbors and the world will require less bluster, more patience, and a willingness on the part of the United States to probe and understand the true sources of the North's conduct.

By Victor D. Cha and David C. Kang

"North Korea Belongs in the Axis of Evil'"

No. The only link between North Korea and Iran and Iraq, the other two members of the "axis of evil" identified by President George W. Bush in his 2002 State of the Union speech, is financial. North Korea has sold missile technology to Iran, as it has to a number of countries, including U.S. allies Pakistan and Egypt. Unlike the original Axis powers, Japan, Germany, and Italy, which were joined formally by the Tripartite Pact of 1940, North Korea, Iran, and Iraq do not coordinate or work together beyond the sale of goods to one another. Furthermore, North Korea does not share any religious, ideological, or strategic goals with Iran and Iraq. North Korea's concerns focus solely on the peninsula and do not extend to the Middle East. Although it does nasty things like sell drugs and make counterfeit money, North Korea has not engaged in terrorism in the last 16 years, and there has never been any link, nor any suggested, between North Korea and al Qaeda.

Iran, Iraq, and North Korea do share some common traits, the main one being an adversarial relationship with the United States. They are also authoritarian, have allegedly supported or sponsored terrorism, and have programs to develop weapons of mass destruction. However, using those latter criteria, several other countries could fit in the axis. Why not U.S. allies Pakistan or Saudi Arabia, for example?

"Kim Jong Il Is Crazy, Unpredictable, and Undeterrable"

Wrong. Kim Jong Il is as rational and calculating as he is brutal. Dictators generally want to survive, and Kim is no exception. He has not launched a war, because he has good reason to think he would face fatal opposition from the United States and South Korea. In fact, like his father Kim Il Sung, Kim has clearly shown he is deterrrable: North Korea has not started a war in five decades.

Dictators do not survive without sophisticated political skills. Kim has maintained power despite intelligence assessments that his leadership would not survive the death of his father in July 1994. And he has endured despite famine, floods, economic collapse, nuclear crises, the loss of two major patrons in Russia and China, and U.S. pressure. There has been no palace or military coup, no extensive social unrest, no obvious chaos in the military, and no wholesale purge of various officials. Moreover, Kim's decision to proceed with North Korea's tentative and measured economic reforms is further proof that, however morally repugnant he may be, he is also quite capable of assessing costs and benefits.

But his rationality does not make him any less dangerous. Under Kim's rule, North Korea has engaged in a coercive bargaining strategy designed to ratchet up a crisis with the United States. Provocations such as test-firing missiles, shadowing spy planes, and walking away from treaties can grab attention and even force the United States and its allies to provide inducements persuading North Korea back from the brink. A risky approach, perhaps—but rational, too. If you have little to negotiate with, it makes sense to leverage the status quo for maximum bargaining advantage.

"North Korea Poses a Direct Nuclear Threat to the United States"

Calm down. What sparked the current crisis over North Korea's nuclear intentions were revelations last October that Pyongyang has pursued a secret program to produce highly enriched uranium that could be used to make nuclear bombs. That effort violated the Agreed Framework negotiated between North Korea and the Clinton administration in 1994, under which the North had agreed to freeze its nuclear program and accept international inspections in return for fuel oil shipments and, eventually, two "proliferation-resistant" nuclear reactors. The October revelations prompted a stiff U.S. response that included a cutoff of fuel oil deliveries. North Korea, meanwhile, has kicked out international inspectors, withdrawn from the Non-Proliferation Treaty, and begun to restart its nuclear reactors—prompting fears that the country will soon have much more than the one or two bombs' worth of nuclear material typically cited by U.S. intelligence analysts.

But lost amid all the alarm and bluster is the reality that the logic of deterrence will prevail even if North Korea develops and deploys a nuclear force. North Korea pursues nuclear weapons not for leverage but for the same reason that other highly vulnerable nations arm themselves: to deter an adversary, in this case a superpower that is armed with nuclear weapons. But even if the North develops nuclear weapons, the threat of a devastating U.S. response will prevent it from ever using them—after all, unlike shadowy terrorist cells, nations cannot hide from a retaliatory strike.

What about the North's missile threat? Some analysts claim that North Korea already possesses a long-range nuclear missile capability. That's false. The longest-range missile currently deployed by North Korea is the No Dong missile, which can carry a 1,500-pound payload approximately 800 miles. However, North Korea has reportedly tested the No Dong only once. The untested Taepo Dong 2 can potentially carry a several hundred–pound payload between 6,000 and 9,000 miles—far enough to reach the West Coast of the United States. But without adequate testing, such a nuclear missile would be highly unreliable.

The fact is, North Korea could blow up terrorist bombs in downtown Seoul every week if it had the desire to do so. It could smuggle a nuclear device into Japan, given the extensive network of Koreans in that country with ties to the North. For that matter, why should North Korea develop an expensive ballistic missile to shoot at the United

States when smuggling a nuclear weapon in a shipping container would be so much easier? The primary value of the North's missiles is as a military deterrent, not as an offensive weapon.

The only nuclear threat to the United States from North Korea is indirect, in the potential transfer of such capabilities to third parties. Pyongyang has shown no aversion to selling weapons to anyone with the hard currency or barter to pay for them. North Korean nuclear weapons or fissile material hidden in tens of thousands of underground caves would likely go undetected even by the most intrusive inspections. But a transfer of nuclear material would be a risky proposition for a regime that values survival above all else. Given the preemptive mind-set of a post–September 11 United States, the North would have to be confident that any transfer would escape U.S. detection and therefore the threat of a massive U.S. retaliation.

"North Korea Does Not Honor International Agreements"

Mostly true. Heralded for a half century as an outlaw state, North Korea has maintained some of its international commitments. It is a member of the Conference on Disarmament, Biological Weapons Convention, and Geneva Protocol. After the attacks of September 11, 2001, the country signed on to two U.N. antiterrorism protocols. During the negotiations for the Agreed Framework, the United States required the North to improve relations with South Korea. Pyongyang eventually responded by agreeing to a summit (just prior to Kim Il Sung's death). As of this writing, the North has also honored its 1999 ballistic missile-testing moratorium for four years.

But the North also has a history of engaging in strategic deception"—signing agreements to convey reliability but purposefully cheating on them to its own advantage. The history of inter-Korean relations, for example, is littered with pacts that Pyongyang has not honored, including the 1992 denuclearization declaration in which North Korea agreed to forgo developing nuclear and nuclear-reprocessing facilities. The United States may have been slow to implement the Agreed Framework, but the North is blatantly breaking the framework's spirit, if not letter, with its covert uranium enrichment program. But perhaps the best evidence of strategic deception occurred in June 1950: On the eve of the Korean War, North Korea put forth a major peace initiative to the South.

"North Korea's Political and Economic Collapse Is Imminent"

Don't bet on it. Observers have predicted an imminent North Korean collapse since the fall of the Berlin Wall in 1989. The country's economic situation is desperate, but signs of political collapse are absent. The best indicator of regime stability is that social control, however vicious, remains solid. Although the flow of refugees from the North is increasing, there is no widespread internal migration, and few observable signs of protest.

Some evidence suggests that North Korea is serious about normal political and economic relations with South Korea and the rest of the world. By December 2002, North Korea had cleared land mines from sections of the demilitarized zone (DMZ) separating North and South Korea. Tracks are being connected on the Kyonggui Railway, which would run from South Korea through the western corridor of the DMZ into North Korea. Pyongyang has also begun work on a four-lane highway on the eastern corridor as well. In July 2002, the central government formally abandoned the centrally planned economy and allowed prices and wages to be set by the market. The government has also created three special economic zones to exploit tourism and investment and amended its laws on foreign ownership, land leases, and taxes and tariffs. Although these reforms have been halting and only marginally successful, they are also becoming increasingly hard to reverse.

Though the regime appears resilient, there are two sources of potential fissures. First, the decidedly mixed results of several recent initiatives by Kim—among them, his decisions to lift price controls and to acknowledge North Korea's kidnapping of Japanese nationals in the 1970s—have exposed "Dear Leader" to potential disgruntlement in the top ranks. Second, the process of reform could create cracks in the regime's foundation. As Montesquieu observed, revolutions don't occur when the people's conditions are at rock bottom but when reform creates a spiral of expectations that spurs people to action against the old stultifying system.

"China Has the Most Influence on North Korea"

Yes, but good luck getting the Chinese to use it. If a state's influence on North Korea is merely a function of the North's material dependence on it, then China holds the trump cards. Seventy to 90 percent of North Korea's annual energy supplies, roughly 30 percent of its total outside assistance, and 38 percent of its imports reportedly come from China. Beijing played a quiet but critical role in inter-Korean dialogue leading up to the June 2000 summit. It also influenced Kim Jong Il and his decision to tentatively reform the North Korean economy by hosting Kim in Shanghai in 2002 and backing the creation of special economic zones.

Notwithstanding this close history, Chinese Foreign Ministry officials deny any influence on North Korea, complaining that "North Korea doesn't listen to us, it doesn't listen to anyone." But China's protestations largely reflect its unwillingness to put real pressure on its neighbor. China's traditional stake in North Korea has rested in keeping the regime afloat as a geostrategic buffer against U.S. influence on China's border. It also has no desire to provoke a regime collapse that would send millions of North Korean refugees flooding across the border.

A different set of Chinese interests may now come to the fore. Beijing opposes nuclear weapons on the Korean peninsula and delivered a dressing-down to North Korean embassy officials in Beijing in January about the country's cheating on the Agreed Framework. For China, nothing good comes from a nuclear North Korea. Such an outcome could prompt Japan to move from merely developing missile defense capabilities to acquiring ballistic missiles or nuclear weapons. And Taiwan might also cross the nuclear threshold if the country's leaders see North Korea successfully guaranteeing its security this way. While the rest of Asia provides China's economic lifeblood, Beijing continues to throw good money, food, and fuel down a rat hole in North Korea with little prospect of major reform.

Chinese policy may change under a new cadre of leaders such as President Hu Jintao who are less wedded to the Cold War relationship with Pyongyang, which used to be characterized as close as "lips and teeth." Or more likely, Beijing's hesitation to intervene may be tactical, as it waits for the United States to do the heavy lifting with North Korea (despite Bush's rhetoric to the contrary) and then swoops in to help close the deal and maximize its influence.

"The DMZ Is the Scariest Place in the World"

Yes, if looks could kill. When former U.S. President Bill Clinton called the border between the two Koreas the world's scariest place, he was referring to the massive forward deployment of North Korean forces around the DMZ and the shaky foundations of the 50-year-old armistice—not peace treaty—that still keeps the peace between the two former combatants. Since the end of the Korean War in 1953, there have been more than 1,400 incidents across the DMZ, resulting in the deaths of 899 North Koreans, 394 South Koreans, and 90 U.S. soldiers. Tensions have been so high that in 1976 the United States mobilized bombers and an aircraft carrier battle group to trim one tree in the DMZ. The deployments and operational battle plans on both sides suggest that if a major outbreak of violence were to start, a rapid escalation of hostilities would likely ensue.

In practice, however, no such outbreak has occurred. North Korea has faced both a determined South Korean military, and more important, U.S. military deployments that at their height comprised 100,000 troops and nuclear-tipped Lance missiles and even today include 37,000 troops, nuclear-capable airbases, and naval facilities that guarantee U.S. involvement in any Korean conflict.

The balance of power has held because any war would have disastrous consequences for both sides. Seoul and Pyongyang are less than 150 miles apart—closer than New York is to Washington, D.C. Seoul is 30 miles from the DMZ and easily within reach of North Korea's artillery tubes. Former Commander of U.S Forces Korea Gen. Gary Luck estimated that a war on the Korean peninsula would cost $1 trillion in economic damage and result in 1 million casualties, including 52,000 U.S. military casualties. As one war gamer described, the death toll on the North Korean side would be akin to a "holocaust," and Kim Jong Il and his 1,000 closest generals would surely face death or imprisonment. As a result, both sides have moved cautiously and avoided major military mobilizations that could spiral out of control.

Ironically enough, as for the DMZ itself, although bristling with barbed wire and sown with land mines, it has also become a remarkable nature preserve stretching across the peninsula that is home to wild birds and a trove of other rare species.

"The Clinton Administration's Policies Toward North Korea Failed"

No. The North's breach of the Agreed Framework may make Clinton's policies look ineffective, but consider the counterfactual proposition. If Clinton had not succeeded in freezing North Korea's main nuclear facilities at Yongbyon for nine years, North Korea would today have enough plutonium for at least 30 nuclear weapons rather than one or two bombs' worth.

Clinton's engagement with North Korea also provided a useful test of North Korean intentions and expectations. Previously, the United States had little sense of the North's interest in swapping its proliferation threat for external assistance. True, the debate between hawks and doves over this question still has a "he said, she said" quality to it: Hawks see North Korea's violations of the Agreed Framework as evidence of the North's lack of interest in such a deal; doves see those same violations as a reaction to the U.S. failure to fulfill the framework and still believe Pyongyang will give up nukes in return for outside support of economic reform. But now there is a baseline or "data" for a debate that previously took place at a theological and ideological level. Before Clinton, there was also no way to use leverage on a country with which the United States had next to no contact for five decades. Since 1994, the North has gained food, fuel, economic assistance, and diplomatic relations not just with South Korea but also Japan, the European Union, Australia, Canada, and others. Ironically, Clinton's carrots have become Bush's sticks, enabling the latter to pursue a harder-line policy by threatening to withhold what was once previously promised.

"The Bush Administration Caused the Current Crisis"

No. Bush's "axis of evil" speech and his professed loathing of Kim may have exacerbated the current crisis, but they certainly did not cause it.

First, North Korea started its covert uranium enrichment program for nuclear weapons long before Bush took office. As far back as 1997, Pakistani nuclear scientists were shuttling to Pyongyang, providing technology for uranium enrichment in return for North Korean missile systems.

Second, prior to the October 2002 revelations and despite Bush's occasional negative statements on North Korea, the United States had offered a string of consistent assurances at lower levels that it would pursue some form of engagement. These assurances included the creation of a package of new incentives and the expressed willingness to meet "any time, any place, and without preconditions." In addition, the Bush administration abandoned several initiatives cited as attempts to derail North Korean engagement—revision of the Agreed Framework and a push for conventional force reductions—after they proved to be nonstarters with U.S. allies. Compared with the Clinton administration's effusive advances to North Korea, Bush's aggressive posturing was portrayed by some media as a dramatic shift, but the U.S. predisposition for engagement remained. The North Koreans' response? They refused to engage in direct bilateral dialogue with the United States, accusing Washington of high-handedness.

Third, there is no denying the harder turn in both U.S. statements and policy after October 2002. North Korea's perception of the preemptive language in the Bush administration's new national security strategy and nuclear posture review could only have heightened North Koreans' worst fears. But Bush's unconditional refusal to talk with North Korea didn't create the crisis. The administration believes North Korea stands so far outside the nonproliferation regime that negotiating its return would be tantamount to blackmail. Should Pyongyang first make compliance gestures, however, then the United States would be willing to discuss incentives including security assurances, energy, and economic assistance. Sounds like a negotiating position to us.

"The United States Should Pull Its Troops Out of an Ungrateful South Korea"

Not yet. Massive demonstrations, Molotov cocktails hurled into U.S. bases, and American soldiers stabbed on the streets of Seoul have stoked anger in Congress and on the op-ed pages of major newspapers about South Korea. As North Korea appears on the nuclear brink, Americans are puzzled by the groundswell of anti-Americanism. They cringe at a younger generation of Koreans who tell

CBS television's investigative program *60 Minutes* that Bush is more threatening than Kim, and they worry about reports that South Korea's new president, Roh Moo-hyun, was avowedly anti-American in his younger days.

Most Koreans have complicated feelings about the United States. Some of them are anti-American, to be sure, but many are grateful. South Korea has historically been one of the strongest allies of the United States. Yet it would be naive to dismiss the concerns of South Koreans about U.S. policy and the continued presence of U.S. forces as merely emotional. Imagine, for example, how Washingtonians might feel about the concrete economic impact of thousands of foreign soldiers monopolizing prime real estate downtown in the nation's capital, as U.S. forces do in Seoul.

But hasty withdrawal of U.S. forces is hardly the answer to such trans-Pacific anxiety, particularly as the U.S.–South Korean affiance enters uncharted territory. The North Koreans would claim victory, and the United States would lose influence in one of the most dynamic economic regions in the world—an outcome it neither wants nor can afford. In the long term, such a withdrawal would also pave the way for Chinese regional dominance. Some South Koreans might welcome a larger role for China—a romantic and uninformed notion at best. Betting on China, after all, did not make South Korea the 12th largest economy and one of the most vibrant liberal democracies in the world.

The alternatives to the alliance are not appealing to either South Koreans or Americans. Seoul would have to boost its relatively low level of defense spending (which, at roughly 3 percent of gross domestic product, is less than that of Israel and Saudi Arabia, for example). Washington would run the risk of jeopardizing its military presence across East Asia, as a U.S. withdrawal from the peninsula raised questions about the raison d'être for keeping its troops in Japan. A revision in the U.S. military presence in Korea is likely within the next five years, but withdrawal of that presence and abrogation of its alliance are not.

Victor D. Cha is associate professor of government and D.S. Song-Korea foundation chair at Georgetown University's Edmund Walsh School of Foreign Service. David C. Kang is associate professor of government and adjunct associate professor at the Tuck School of Business at Dartmouth College. They are coauthors of Nuclear North Korea: A Debate On Engagement Strategies *(New York: Columbia University Press, October 2003).*

Want to Know More?

On North Korean history, the classic remains Chong-sik Lee and Robert Scalapino's *Communism in Korea* (Berkeley: University of California Press, 1972). The best recent history is Don Obberdorfer's *The Two Koreas* (New York: Basic Books, 2001). On North Korea's leadership, one of the few good works is Dae-Sook Suh's ***Kim Il-Sung: The North Korean Leader*** (New York: Columbia University Press, 1988). The inauguration speech of South Korea's new president, Roh Moo-hyun, can be found on the Web site of the *Korean Information Service.* The Korean Central News Agency Web site offers the North's perspective, served up zany and fresh.

On U.S.–North Korean relations, see Samuel Kim's, ed., ***North Korean Foreign Relations in the Post–Cold War Era*** (New York: Oxford University Press, 1998), Leon Sigal's ***Disarming Strangers: Nuclear Diplomacy With North Korea*** (Princeton: Princeton University Press, 1998), and Michael May's, ed., **"Verifying the Agreed Framework"** (Livermore: Center for Global Security Research, April 2001). For a recent roundtable featuring the views of 28 top Korea experts, see the February 2003 report of the Task Force of U.S. Korea Policy, **"Turning Point in Korea: New Dangers and New Opportunities for the United States,"** available on the Web site of the Center for International Policy. A deja vu–inducing snapshot of U.S.–North Korean relations circa 1979 can be found in journalist Gareth Porter's **"Time to Talk With North Korea"** (FOREIGN POLICY, Spring 1979).

On the history of nuclear and missile proliferation on the Korean peninsula, see James Clay Moltz and Alexandre Mansourov's, eds., ***The North Korean Nuclear Program*** (New York: Routledge, 1999), Peter Hayes's **Pacific Powderkeg** (Lexington: Lexington Books, 1991), and David Albright and Kevin O'Neill's, eds., **Solving the North Korean Nuclear Puzzle** (Washington: Institute for Science and International Security, 2000). David Shambaugh argues that China does not ultimately want to end North Korea's nuclear program in **"China and the Korean Peninsula: Playing for the Long Term"** (The Washington Quarterly, Spring 2003).

On North Korea's economy, see Marcus Noland's *Avoiding the Apocalypse: The Future of the Two Koreas* (Washington: Institute for International Economics., 2000), Nicholas Eberstadt's *The End of North Korea* (Washington: American Enterprise Institute Press, 1999), and Hy-Sang Lee's **North Korea: A Strange Socialist Fortress** (Westport: Praeger, 2001).

For links to relevant Web sites, access to the *FP* Archive, and a comprehensive index of related FOREIGN POLICY articles, go to *www.foreignpolicy.com.*

From *Foreign Policy*, May/June 2003, pp. 20-28. © 2003 by the Carnegie Endowment for International Peace. Reprinted by permission.

India, Pakistan and the Bomb

The Indian subcontinent is the most likely place in the world for a nuclear war

by M. V. Ramana and A. H. Nayyar

As the U.S. mobilized its armed forces in the aftermath of the terrorist attacks of September 11, the world's attention focused on Pakistan, so crucial to military operations in Afghanistan. When Pakistani president Pervez Musharraf pledged total support for a U.S.-led multinational force on September 14, many people's first thought was: What about Pakistan's nuclear weapons? Could they fall into the hands of extremists? In an address to his nation, Musharraf proclaimed that the "safety of nuclear missiles" was one of his priorities. The Bush administration began to consider providing Pakistan with perimeter security and other assistance to guard its nuclear facilities.

The renewed concern about nuclear weapons in South Asia comes a little more than three years after the events of May 1998: the five nuclear tests conducted by India at Pokharan in the northwestern desert state of Rajasthan, followed three weeks later by six nuclear explosions conducted by Pakistan in its southwestern region of Chaghai. These tit-for-tat responses mirrored the nuclear buildup by the U.S. and the former Soviet Union, with a crucial difference: the two cold war superpowers were separated by an ocean and never fought each other openly. Neighboring India and Pakistan have gone to war three times since British India was partitioned in 1947 into Muslim-majority and Hindu-majority states. Even now artillery guns regularly fire over the border (officially, a cease-fire line) in the disputed region of Kashmir.

In May 1999, just one year after the nuclear tests, bitter fighting broke out over the occupation of a mountain ledge near the Kashmiri town of Kargil. The two-month conflict took a toll of between 1,300 (according to the Indian government) and 1,750 (according to Pakistan) lives. For the first time since 1971, India deployed its air force to launch attacks. In response, Pakistani fighter planes were scrambled for fear they might be hit on the ground; air-raid sirens sounded in the capital city of Islamabad. High-level officials in both countries issued at least a dozen nuclear threats. The peace and stability that some historians and political scientists have ascribed to nuclear weapons—because nuclear nations are supposed to be afraid of mutually assured destruction—were nowhere in sight.

Wiser counsel eventually prevailed. The end of the Kargil clash, however, was not the end of the nuclear confrontation in South Asia. The planned deployment of nuclear weapons by the two countries heightens the risks. With political instability a real possibility in Pakistan, particularly given the conflict in Afghanistan, the dangers have never been so near.

Learning to Love the Bomb

BOTH COUNTRIES have been advancing their nuclear programs almost ever since they gained independence from Britain. Understanding this history is crucial in figuring out what to do now, as well as preventing the further proliferation of nuclear weapons. Although the standoff between Pakistan and India has distinct local characteristics, both countries owe much to other nuclear states. The materials used in their bombs were manufactured with Western technology; both countries' justifications for joining the nuclear club drew heavily on cold war thinking. The continued reliance of the U.S. and Russia on thousands of nuclear weapons on hair-trigger alert only adds to the perceived need for nuclear arsenals in India and Pakistan.

While setting up the Indian Atomic Energy Commission (IAEC) in 1948, Jawaharlal Nehru, India's first prime minister, laid out his desire that the country "develop [atomic energy] for peaceful purposes." But at the same time, he recognized that "if we are compelled as a nation to use it for other purposes, possibly no pious sentiments will stop the nation from using it that way." Such ambivalence remained a central feature of India's nuclear policy as it developed.

MAKING NUCLEAR WEAPONS MATERIAL

THE MOST DIFFICULT part of making nuclear weapons is manufacturing the fuel, either plutonium or highly enriched uranium. The starting point is natural uranium, which is 99.3 percent uranium 238 and 0.7 percent uranium 235. Only the latter can sustain a chain reaction. To build a uranium bomb, one needs to increase the uranium 235 content to 80 percent or more. Most modern enrichment facilities, including the ones in Pakistan and India, use high-speed centrifuges [see "The Gas Centrifuge," by Donald R. Olander; SCIENTIFIC AMERICAN, August 1978].

The alternative route involves plutonium. This element is not found in nature. It is produced by irradiating uranium fuel in nuclear reactors, then extracted through a chemical process called reprocessing [see "The Reprocessing of Nuclear Fuels," by William P. Bebbington; SCIENTIFIC AMERICAN, December 1976.] In the most commonly followed reprocessing scheme, the irradiated fuel is chopped up, dissolved in acid and exposed to a solvent called tributyl phosphate mixed with kerosene. The solvent separates out the plutonium and uranium from other fission products. Plutonium is then precipitated out by a reductant, a chemical that changes it to an insoluble form.

—M.V.R. and A.H.N.

1 URANIUM MINING The ore is extracted, crushed into fine particles and leached with acid or alkali to separate out the uranium

2 PROCESSING The uranium is converted to a chemical form suited to either enrichment or fuel fabrication

3a ENRICHMENT Uranium 235 is concentrated, converted to metallic form and fabricated into bomb cores

3b FUEL FABRICATION Natural or slightly enriched uranium is clad in metal casing to make fuel rods

4 REACTOR During a chain reaction, some of the uranium absorbs neutrons and transmutes into plutonium. Afterward the fuel rods cool in a water pool

5 REPROCESSING Plutonium is extracted through a chemical process, converted to metallic form and fabricated into bomb cores

INDIAN PLUTONIUM INVENTORY

Cumulative production (in reactors): 450–722 kg
Consumption (in tests and reactors): 165 kg
Net stock: 285–557 kg (equivalent to 55–110 bombs)

PAKISTANI ENRICHED-URANIUM INVENTORY

Cumulative production (by enrichment): 450–750 kg
Consumption (during tests): 120 kg
Net stock: 330–630 kg (equivalent to 20–40 bombs)

To Indian leaders, the program symbolized international political clout and technological modernity. Over the next two decades, India began to construct and operate nuclear reactors, mine uranium, fabricate fuel and extract plutonium. In terms of electricity produced, these activities often proved uneconomical—hardly, one might think, where a developing nation should be putting its resources. Politicians and scientists justified the nuclear program on the grounds that it promoted self-sufficiency, a popular theme in postcolonial India. Rhetoric aside, India solicited and received ample aid from Canada, the U.S. and other countries.

UNDERGROUND NUCLEAR EXPLOSIONS conducted by India on May 11, 1998, caused the surface immediately above to collapse. Seismic readings suggest that the total explosive yield was between 16 and 30 kilotons, about half of what India claimed.

After India's defeat in the 1962 border war with China, some right-wing politicians issued the first public calls for developing a nuclear arsenal. These appeals became louder after China's first nuclear test in 1964. Countering this bomb lobby were other prominent figures, who argued that the economic cost would be too high. Many leading scientists advocated the bomb. Homi Bhabha, the theoretical physicist who ran the IAEC, claimed that his organization could build nuclear weapons "within 18 months." Citing a Lawrence Livermore National Laboratory report, Bhabha predicted that nuclear bombs would be cheap. He also promised economic gain from "peaceful nuclear explosions," which many American nuclear researchers extolled for, say, digging canals.

In November 1964 Indian prime minister Lal Bahadur Shastri compromised, permitting the commission to explore the technology for such an explosion. It turned out that Bhabha had already been doing some exploring. In 1960 he reportedly sent Vasudev Iya, a young chemist, to France to absorb as much information as he possibly could about how polonium—a chemical element used to trigger a nuclear explosion—was prepared. Bhabha died in 1966, and design work on the "peaceful" device did not begin for another two years. But by the late 1960s, between 50 and 75 scientists and engineers were actively developing weapons. Their work culminated in India's first atomic test—the detonation on May 11, 1974, of a plutonium weapon with an explosive yield of five to 12 kilotons. For comparison, the bomb dropped on Hiroshima had a yield of about 13 kilotons.

Nuclear Tipping Point

THE 1974 TEST was greeted with enthusiasm within India and dismay elsewhere. Western countries cut off cooperative efforts on nuclear matters and formed the Nuclear Suppliers Group, which restricts the export of nuclear technologies and materials to nations that refuse to sign the 1968 Nuclear Non-Proliferation Treaty, including both India and Pakistan.

In the years that followed, the bomb lobby pushed for tests of more advanced weapons, such as a boosted-fission design and a hydrogen bomb. It appears that in late 1982 or early 1983, Prime Minister Indira Gandhi tentatively agreed to another test, only to change her mind within 24 hours. One of the causes for the volte-face is said to have been a conversation with the Indian foreign secretary, whom an American official had confronted with satellite evidence of preparations at the test site. The conversation seems to have convinced Gandhi that the U.S. reaction would create economic difficulties for India. Instead, it is reported, she wanted to "develop other things and keep them ready."

The "other things" she had in mind were ballistic missiles. In 1983 the Integrated Guided Missile Development Program was set up under the leadership of Abdul Kalam, a renowned rocket engineer. This followed an earlier, secret attempt to reverse-engineer a Soviet antiaircraft missile that India had purchased in the 1960s. Although that effort did not succeed, it led to the development of several critical technologies, in particular a rocket engine. Kalam adopted an open management style—as compared with the closed military research program—and involved academic institutions and private firms. Anticipating restrictions on imports, India went on a shopping spree for gyroscopes, accelerometers and motion simulators from suppliers in France, Sweden, the U.S. and Germany.

In 1988 India tested its first short-range surface-to-surface missile. A year later came a medium-range missile; in April 1999, a longer-range missile. The latter can fly 2,000 kilometers, well into the heart of China. Despite this ability, India is unlikely to achieve nuclear parity with China. According to various estimates, China has 400 warheads and an additional 200 to 575 weapons' worth of fissile material. If India's plutonium production reactors have been operating on average at 50 to 80 percent of full power, India has somewhere between 55 and 110 weapons' worth of plutonium. The stockpile could be much larger if commercial reactors earmarked for electricity generation have also been producing plutonium for weapons.

Eating Grass

PAKISTAN'S NUCLEAR PROGRAM drew on a general desire to match India in whatever it does. The country set up its Atomic Energy Commission in 1954, began operating its first nuclear research reactor in 1965 and opened its first commercial reactor in 1970. As scientific adviser to the government, physicist Abdus Salam, who later won the Nobel Prize in Physics, played an important role.

The program was severely handicapped by a shortage of manpower. In 1958 the commission had only 31 scientists and engineers; it was run by Nazir Ahmad, the former head of the Textile Committee. The commission pursued an active program of training personnel by sending more than 600 scientists and engineers to the U.S., Canada and western Europe. With generous help from these countries, some of which also aided India, Pakistan had a few nuclear research laboratories in place by the mid-1960s.

THROUGH THE STREETS OF KARACHI, a mock missile is paraded by Pasban, a youth wing of Pakistan's main fundamentalist party, Jamaat-e-Islami. The parade took place in February 1999 on a day of solidarity with Kasmiris in India-administered Kashmir. Such enthusiasm for nuclear weapons is widespread, though not universal, in both India and Pakistan.

After the 1965 war with India, many Pakistani politicians, journalists and scientists pressed for the development of nuclear weapons. The most prominent was Foreign Minister Zulfikhar Ali Bhutto, who famously declared that if India developed an atomic bomb, Pakistan would follow "even if we have to eat grass or leaves or to remain hungry." After Pakistan's defeat in the December 1971 war, Bhutto became prime minister. In January 1972 he convened a meeting of Pakistani scientists to discuss making bombs.

As the first prong of their two-pronged effort to obtain weapons material, researchers attempted to purchase plutonium reprocessing plants from France and Belgium. After initially agreeing to the sale, France backed down under American pressure. But a few Pakistani scientists did go to Belgium for training in reprocessing technology. Returning to Pakistan, they constructed a small-scale reprocessing laboratory in the early 1980s. Using spent fuel from a plutonium production reactor that opened in 1998, this lab is capable of producing two to four bombs' worth of plutonium annually.

As the second prong, researchers explored techniques for enriching uranium—that is, for concentrating the bomb-usable isotope uranium 235. In 1975 A. Q. Khan, a Pakistani metallurgist who had worked at an enrichment plant in the Netherlands, joined the group. With him came classified design information and lists of component suppliers

Nuclear Arena

For five decades, India and Pakistan have fought an incessant low-level war in Kashmir and engaged in a nuclear arms race. They now possess large and diverse nuclear weapons infastructures. Meanwhile hundreds of millions of people in the region remain impoverished.

Pakistan's Nuclear Establishment§	India's Nuclear Establishment§
REACTORS	
Research/Plutonium Production Reactor, 40–70 MW* **LOCATION:** Khushab **OPENED:** 1998 **FOREIGN PARTNER:** China **MODERATOR:** heavy water [?] **COOLANT:** heavy water **ANNUAL OUTPUT:** 6.6–18 kg of plutonium[†]	**CIRUS, 40 MW*** **LOCATION:** Mumbai **OPENED:** 1960 **FOREIGN PARTNER:** Canada **MODERATOR:** heavy water **COOLANT:** light water **ANNUAL OUTPUT:** 6.6–10.5 kg of plutonium **Dhruva, 100 MW*** **LOCATION:** Mumbai **OPENED:** 1985 **MODERATOR:** heavy water **COOLANT:** heavy water **ANNUAL OUTPUT:** 16–26 kg of plutonium[†] **Fast Breeder Test Reactor, 40 MW*** **LOCATION:** Kalpakkam **OPENED:** 1983 **FOREIGN PARTNER:** France **COOLANT:** liquid sodium **ANNUAL OUTPUT:** 4–6.4 kg of plutonium[†]
PLUTONIUM REPROCESSING	
New Labs **LOCATION:** Rawalpindi **OPENED:** early 1980s **ANNUAL OUTPUT:** 10–20 kg of plutonium	**Trombay** **LOCATION:** Mumbai **COMMISIONED:** 1964 **ANNUAL CAPACITY:** 30–50 tons of spent metallic fuel **PREFRE** **LOCATION:** Tarapur **COMMISSIONED:** 1977 **ANNUAL CAPACITY:** 100 tons of spent oxide fuel **KARP** **LOCATION:** Kalpakkam **COMMISSIONED:** 1997 **ANNUAL CAPACITY:** 100–125 tons of spent oxide fuel
URANIUM ENRICHMENT	
Khan Research Laboratories **LOCATION:** Kahuta **OPENED:** 1984 **ANNUAL OUTPUT:** 57–93 kg of highly enriched uranium	**Rattehalli**[††] **LOCATION:** Mysore **OPENED:** 1990 **ANNUAL PRODUCTION:** unknown
URANIUM MINE	
Dera Ghazi Khan **OPENED:** 1974 **ANNUAL PRODUCTION:** 23–30 tons	**Jadugoda** **OPENED:** 1968 **ANNUAL PRODUCTION:** 200 tons

*thermal power output
[†]running at 50%–90% of capacity
[††]said to produce fuel for a nuclear submarine
§bomb-related facilities; commercial power reactors omitted

in the West, many of which proved quite willing to violate export-control laws [see box "Secrets, What Secrets"]. Success came in 1979 with the enrichment of small quantities of uranium. Since then, Pakistan is estimated to have produced 20 to 40 bombs' worth of enriched uranium. Every year it produces another four to six bombs' worth.

By 1984 designs for aircraft-borne bombs were reportedly complete. Around this time, some American officials started alleging that China had given Pakistan the design for a missile-ready bomb. China and Pakistan have indeed exchanged technology and equipment in several areas, including those related to nuclear weapons and missiles. For example, it is believed that Pakistan has imported short-range missiles from China. But the accusation that China supplied Pakistan with a weapons design has never been substantiated. And understandably, Pakistan's nuclear scientists have denied it.

In spring 1990 events in Kashmir threatened to erupt into another full-scale war. According to a 1993 *New Yorker* article by American journalist Seymour M. Hersh, U.S. satellites detected a convoy of trucks moving out of Kahuta, Pakistan's uranium-enrichment facility, toward an air base where F-16 fighter jets stood ready. Hersh reported that American diplomats conveyed this information to India, which recalled the troops it had amassed at the border. But the overwhelming opinion among scholars who have analyzed these claims is that Pakistan never contemplated the use of nuclear weapons; experts are also skeptical that U.S. satellites ever detected the claimed movement. Nevertheless, the Pakistani bomb lobby has used the allegations to assert that nuclear weapons protect the country from Indian attack. In India, officials have never acknowledged Hersh's story; it would be an admission that Pakistan's nuclear capability had neutralized India's conventional military advantage.

"Now I Am Become Death"

FURTHER BUILDUP of nuclear capabilities in both countries took place against a background transformed by the end of the cold war. Superpower arsenals shrank, and the Comprehensive Test Ban Treaty, which prohibits explosive tests, was negotiated in 1996. But the five declared nuclear states—the U.S., Russia, Britain, France and China—made it clear that they intend to hold on to their arsenals. This ironic juxtaposition strengthened the bomb lobbies in India and Pakistan.

Domestic developments added to the pressure. India witnessed the rise of Hindu nationalism. For decades, parties subscribing to this ideology, such as the Bharatiya Janata Party (BJP), had espoused the acquisition of greater military capability—and nuclear weapons. It was therefore not surprising that the BJP ordered nuclear tests immediately after coming to power in March 1998.

The Indian tests, in turn, provided Pakistani nuclear advocates with the perfect excuse to test. Here again, religious extremists advocated the bomb. Qazi Hussain Ahmad of the Jamaat-e-Islami, one of the largest Islamist groups in Pakistan, had declared in 1993: "Let us wage jihad for Kashmir. A nuclear-armed Pakistan would deter India from a wider conflict." Meanwhile the military sought nuclear weapons to counter India's vastly larger armed forces.

This lobbying was partially offset by U.S. and Chinese diplomacy after India's tests. In addition, some analysts and activists enumerated the ill effects that would result from the economic sanctions that were sure to follow any test. They suggested that Pakistan not follow India's lead—leaving India to face international wrath alone—but to no avail. Three weeks after India's blasts, Pakistan went ahead with its own tests. Bombast notwithstanding, the small size of seismic signals from the tests of both countries has cast doubt on the declared explosive yields. The data released by the Indian weapons establishment to support its claims are seriously deficient; for example, a graph said to be of yields of radioactive by-products has no units on the axes. Independent scientists have not been able to verify that the countries set off as many devices as they profess.

Whatever the details, the tests have dramatically changed the military situation in South Asia. They have spurred the development of more advanced weapons, missiles, submarines, antiballistic missile systems, and command-and-control systems. In August 1999 the Indian Draft Nuclear Doctrine called for the deployment of a triad of "aircraft, mobile land-missiles and sea-based assets" to deliver nuclear weapons. Such a system would cost about $8 billion. This past January the Indian government declared that it would deploy its new long-range missile. A month later the Pakistani deputy chief of naval staff announced that Pakistan was thinking about equipping at least one of its submarines with nuclear missiles.

Critical Mass

DEPLOYMENT INCREASES the risk that nuclear weapons will be used in a crisis through accident or miscalculation. With missile flight times of three to five minutes between the two countries, early-warning systems are useless. Leaders may not learn of a launch until they look out their window and see a blinding flash of light. They will therefore keep their fingers close to the button or authorize others, geographically dispersed, to do so.

Broadly speaking, there are two scenarios. The first postulates that India crosses some threshold during a war—its troops reach the outskirts of Lahore or its ships impose a naval blockade on Karachi—and Pakistan responds with tactical nuclear weapons as a warning shot. The other scenario supposes that under the same circumstances, Pakistan decides that a warning shot would not work and instead attacks an Indian city directly. In 1998 one of us (Ramana) conducted the first scientific study of how much damage a modest, 15-kiloton bomb dropped on Bombay

INDIAN MISSILES

AGNI ("Fire") 1
TYPE: Solid-fueled, first stage; liquid-fueled, second stage
RANGE: 1,500–2,000 km
WARHEAD: 1,000 kg
DEVELOPMENT STAGE: Suspended

AGNI II
TYPE: Solid-fueled, two-stage
RANGE: 2,000 km
WARHEAD: 1,000 kg
DEVELOPMENT STAGE: Tested April 1999

PRITHVI ("Earth") I
TYPE: Liquid-fueled, single-stage; engine based on Russian SA-2 air defense missile
RANGE: 150 km
WARHEAD: 1,000 kg
DEVELOPMENT STAGE: Deployed

PRITHVI II
TYPE: Liquid-fueled, single-stage
RANGE: 250 km
WARHEAD: 500 kg
DEVELOPMENT STAGE: Tested January 1996

PRITHVI III
TYPE: Liquid-fueled, single-stage naval missile
RANGE: 350 km (?)
WARHEAD: Unknown
DEVELOPMENT STAGE: Under development

SAGARIKA ("Born on the Ocean")
TYPE: Submarine-launched cruise/ballistic missile
RANGE: 300 km (?)
WARHEAD: Unknown
DEVELOPMENT STAGE: Under development

PAKISTANI MISSILES

HATF ("Armor") I
TYPE: Solid-fueled, single-stage; based on French sounding rocket
RANGE: 60–80 km
WARHEAD: 500 kg
DEVELOPMENT STAGE: Tested January 1989

HATF II
TYPE: Solid-fueled, single-stage
RANGE: 280–300 km
WARHEAD: 500 kg
DEVELOPMENT STAGE: Tested January 1989

HATF III
TYPE: Solid-fueled, single-stage
RANGE: Up to 600 km
WARHEAD: 250 kg
DEVELOPMENT STAGE: Tested July 1997

M-11
TYPE: Solid-fueled, single-stage
RANGE: 290 km
WARHEAD: 500 kg
DEVELOPMENT STAGE: Allegedly imported from China; in storage?

GHAURI (name refers to 12th century Afghan king)
TYPE: Liquid-fueled, single-stage; similar to North Korean missile
RANGE: 1,500 km
WARHEAD: 700 kg
DEVELOPMENT STAGE: Tested April 1998; serial production started November 1998

SHAHEEN ("Eagle")
TYPE: Solid-fueled, single-stage
RANGE: 600-750 km
WARHEAD: 1,000 kg
DEVELOPMENT STAGE: Tested April 1999

SHAHEEN II
TYPE: Solid-fueled, two-stage
RANGE: 2,400 km
DEVELOPMENT STAGE: Under development

would cause: over the first few months, between 150,000 and 850,000 people would die.

The Indian military is already preparing for these eventualities. This past May it carried out its biggest exercises in more than a decade, called Operation Complete Victory. Tens of thousands of troops, backed by tanks, aircraft and attack helicopters, undertook drills close to the border with Pakistan. The stated aim was to train the armed forces to operate in an "environment of chemical, biological and nuclear assault" and "to teach the enemy a lesson once and for all." In one significant exercise, the military had to "handle a warlike situation wherein an enemy aircraft is encountered carrying a nuclear warhead." Abdul Kalam, head of India's missile program, said that India's nuclear weapons "are being tested for military operations … for training by our armed forces."

Even before September 11, South Asia had all the ingredients for a nuclear war: possession and continued development of bombs and missiles, imminent deployment of nuclear weapons, inadequate precautions to avoid unauthorized use of these weapons, geographical proximity, ongoing conflict in Kashmir, militaristic religious extremist movements, and leaders who seem sanguine about the dangers of nuclear war.

The responses of India and Pakistan to the events of September 11 and the U.S.-led attack on targets in Afghanistan reflect the strategic competition that has shaped much of their history. India was quick to offer air bases and logistical support to the U.S. military so as to isolate Pakistan. Attempting to tie its own problems in Kashmir with the global concern about terrorism, Indian officials even threatened to launch attacks on Pakistani supply lines and alleged train-

Secrets, What Secrets?

Terrorists might exploit Pakistan's cavalier attitude toward nuclear information

by David Albright

Over the years, successive Pakistani governments have assured the West that they had a secure grip on the country's nuclear weapons, materials and technology. But nuclear analysts have never been entirely comforted by these assertions. Many people in the Pakistani nuclear weapons program and the military could well be sympathetic to radical Islamist or anti-American causes. What is especially worrisome is that the historical development of Pakistan's program has heightened the risk of illegal assistance and other security violations.

From its inception, the program has relied on illicit procurement and deliberate deception. It has fostered extensive contacts with the world of shady middlemen and companies whose allegiance to Western export controls depends on the price one is willing to pay. In the organizational culture of such a program, disaffected individuals could find plenty of justifications and opportunities to transfer classified information or sensitive items. Others might be disinclined to report on the suspicious actions of colleagues. Some might even feel ownership over parts of the program and believe it is their right to sell their contributions for personal benefit.

Such problems affect India less, because it started its nuclear weapons program earlier than Pakistan did. India obtained much of its nuclear infrastructure from foreign suppliers before Western governments understood the extent to which developing countries were misusing civilian nuclear assistance to make nuclear explosives. To be sure, Pakistan is not alone in dealing with an organizational culture that scorns security guidelines. The German civilian gas-centrifuge program was notorious for its weak security. In the late 1980s German nuclear experts secretly assisted Iraq.

A key component of Pakistan's program, the production of highly enriched uranium for bombs, was born in an act of industrial espionage. In the mid-1970s the father of that effort, A. Q. Khan, worked at a Dutch engineering firm and was given the task of translating classified designs and specifications for gas centrifuges. He gained access to a wide variety of sensitive information. On his return to Pakistan, Khan founded the Engineering Research Laboratories, now known as the Dr. A. Q. Khan Research Laboratories, to transform this knowledge into a bomb factory.

According to a declassified 1983 U.S. State Department memorandum, the enrichment program disguised its activities by providing false statements about the final use of items imported from Western countries. Pakistan once described its gas-centrifuge plant as a synthetic butter factory. In a 1999 interview in the Egyptian newspaper *Al-Ahram,* Khan said that his program purchased items through offshore front companies in Japan, Singapore and elsewhere. Those companies took a cut of 15 to 25 percent of the purchase price.

Khan and his colleagues took a Robin Hood approach to classified information. In the late 1980s they published a series of technical articles in Western journals about gas centrifuges. The intention was to demonstrate Pakistan's self-sufficiency in centrifuges and thereby signal that the country

was ready to make a bomb. One paper stated its purpose thus: to "provide useful and practical information, as technical information on balancing of centrifuge rotors is hardly available because most of the work is shrouded in the clouds of the so-called secrecy." These articles aided other countries, such as South Africa, in their own nuclear programs.

One Pakistani article is the only publicly available study on bellows built from maraging steel, a superstrong type of steel. For years, Urenco—a British, German and Dutch enrichment consortium—considered the mere mention of these bellows a violation of its secrecy rules.

How much further did the Pakistani nuclear scientists go in spreading the art of bomb making? The U.N. arms inspections in Iraq came across a one-page Iraqi intelligence document, marked TOP SECRET, that contained an offer of nuclear weapons assistance from the Pakistanis. According to the document, an intermediary approached Iraqi intelligence in October 1990—two months after the Iraqi invasion of Kuwait and three months before the U.S.-led counterattack—with the following proposition: Khan would give Iraq bomb designs, help to procure materials through a company in Dubai and provide other services. In return, Iraq would pay handsomely.

Arms inspectors were unable to find the middleman, and Pakistan and Khan have denied any involvement. Nevertheless, the Iraqis took this offer as genuine—and apparently rejected it. Khidhir Hamza, a former weapons scientists who left Iraq in 1994 and worked with me in the late 1990s, says he knew of this offer at the time and believes Iraq would not have pursued it, for fear that Khan would gain too much knowledge about, and control over, Iraq's nuclear programs. Khan already had a track record of misleading the Iraqis, having used a contract for a petrochemical facility as a cover to obtain maraging steel.

In March of this year the government of Pakistan removed Khan as head of the nuclear laboratory and offered him a position as a special science and technology adviser. The move is widely viewed as an attempt to rein him in. This past summer, however, reports emerged that the laboratory has kept up its ties with North Korea's ballistic-missile program, reviving fears of nuclear cooperation. Pakistani officials have denied any connection.

No evidence links elements in the Pakistani government with any terrorist group, but the Pakistani government has had extensive contact with the Taliban. It is conceivable that terrorists could exploit these connections to gain access to sensitive nuclear items. The culture within the nuclear program increases this risk.

David Albright is a physicist, president of the Institute for Science and International Security in Washington, D.C., and a former U.N. weapons inspector in Iraq.

ing camps for militants fighting in Kashmir. Pakistan, for its part, realizing both the geopolitical advantage it possessed and the dangers of civil instability, deliberated before agreeing to provide support to fight the Taliban. The diplomatic machinations, war in Afghanistan and violence in Kashmir may well have worsened the prospects for peace on the subcontinent. The lifting of American sanctions, which had been imposed in the 1990s, freed up resources to invest in weapons.

The limitations of Western nonproliferation policy are now painfully obvious. It has relied primarily on supply-side export controls to prevent access to nuclear technologies. But Pakistan's program reveals that these are inadequate. Any effective strategy for nonproliferation must also involve demand-side measures—policies to assure countries that the bomb is not a requisite for true security. The most important demand-side measure is progress toward global nuclear disarmament. Some people argue that global disarmament and nonproliferation are unrelated. But as George Perkovich of the W. Alton Jones Foundation in Charlottesville, Va., observed in his masterly study of the Indian nuclear program, that premise is "the grandest illusion of the nuclear age." It may also be the most dangerous.

MORE TO EXPLORE

Fissile Material Production Potential in South Asia. A.H. Nayyar, A.H. Toor and Zia Mian in *Science and Global Security*, Vol. 6, No. 2, pages 189–203; 1997.

The Making of the Indian Atomic Bomb: Science, Secrecy and the Postcolonial State. Itty Abraham. Zed Books, 1998.

India's Nuclear Bomb: The Impact on Global Proliferation. George Perkovich. University of California Press, 1999.

Out of the Nuclear Shadow. Edited by Smitu Kothari and Zia Mian. Zed Books, 2001.

For articles by M. V. Ramana, visit
www.geocities.com/CollegePark/5409/nuclear.html

For extensive information on both countries' nuclear weapons, visit:
www.isis-online.org/publications/southasia/index.html
www.ceip.org/files/Publications/trackingTOC.asp
www.fas.org/nuke/hew/

SCIENTIFIC AMERICAN ARTICLES ON NUCLEAR PROLIFERATION

Since the dawn of the nuclear age, SCIENTIFIC AMERICAN *has published articles on nuclear weapons policy. A sampling:*

The Hydrogen Bomb: II. Hans A. Bethe. April 1950.

The Proliferation of Nuclear Weapons. William Epstein. April 1975.

Stopping the Production of Fissile Materials for Weapons. Frank von Hippel, David H. Albright and Barbara G. Levi. September 1985.

The Future of American Defense. Philip Morrison, Kosta Tsipis and Jerome Wiesner. February 1994.

The Real Threat of Nuclear Smuggling. Phil Williams and Paul N. Woessner. January 1996.

Iran's Nuclear Puzzle. David A. Schwarzbach. June 1997.

M. V. RAMANA and *A. H. NAYYAR* are physicists and peace activists who have worked to bridge the divide between India and Pakistan. Ramana, a research staff member in Princeton University's Program on Science and Global Security (www.princeton.edu/~globsec), is a founding member of the Indian Coalition for Nuclear Disarmament and Peace. He was born and raised in southern India and has written extensively on the region's classical music. Nayyar, a physics professor at Quaid-e-Azam University in Islamabad, is co-founder of the Pakistan Peace Coalition. He also runs a project to provide education to underprivileged children.

From *Scientific American*, December 2001, pp. 72-83. © 2001 by Scientific American, Inc. All rights reserved. Reprinted by permission.

UNIT 6

Cooperation

Unit Selections

Key Points to Consider

- Itemize the products you own that were manufactured in another country.

- What recent contacts have you had with people from other countries? How was it possible for you to have these contacts?

- Do you use the Internet to access people or Web sites in other countries?

- Identify nongovernmental organizations in your community that are involved in international cooperation (e.g., Rotary International).

- What are the prospects for international governance? How do trends in this direction enhance or threaten American values and constitutional rights?

- What new strategies for cooperation can be developed to fight terrorism, international narcotics trafficking and other criminal threats?

- How can conflict and rivalry be transformed into meaningful cooperation?

 Links: www.dushkin.com/online/
These sites are annotated in the World Wide Web pages.

Carnegie Endowment for International Peace
http://www.ceip.org

Commission on Global Governance
http://www.sovereignty.net/p/gov/gganalysis.htm

OECD/FDI Statistics
http://www.oecd.org/statistics/

U.S. Institute of Peace
http://www.usip.org

person can write a letter to another, and assuming it is properly addressed, the sender can be relatively certain that the letter will be delivered to just about any location in the world. This is true even though the sender pays for postage only in the country of origin and not in the country where it is delivered. A similar pattern of international cooperation is true when a traveler boards an airplane and never gives a thought to the issue of potential language and technical barriers, even though the flight's destination is halfway around the world.

Many of the most basic activities of our lives are the direct result of multinational cooperation. International organizational structures, for example, have been created to monitor threats to public health and to scientifically evaluate changing weather conditions. The flow of mail, the safety of airlines, and the monitoring of changing atmospheric conditions are just some of the examples of individual governments recognizing that their self-interest directly benefits from cooperation (in some cases by giving up some of their sovereignty through the creation of international governmental organizations, or IGOs).

Transnational activities are not limited to the governmental level. There are now tens of thousands of international nongovernmental organizations (INGOs). The activities of INGOs range from staging the Olympic Games to organizing scientific meetings to actively discouraging the hunting of seals. The number of INGOs along with their influence has grown tremendously in the past 50 years.

During the same period in which the growth in importance of IGOs and INGOs has taken place, there also has been a parallel expansion of corporate activity across international borders. Most U.S. consumers are as familiar with Japanese or German brand-name products as they are with items made in their own country. The multinational corporation (MNC) is an important non-state actor. The value of goods and services produced by the biggest MNCs is far greater than the gross domestic product (GDP) of many countries. The international structures that make it possible to buy a Swedish automobile in Sacramento or a Swiss watch in Singapore have been developed over many years. They are the result of governments negotiating treaties that create IGOs to implement the agreements (e.g., the World

Trade Organization). As a result, corporations engaged in international trade and manufacturing have created complex transnational networks of sales, distribution, and service that employ millions of people.

To some observers these trends indicate that the era of the nation-state as the dominant player in international politics is passing. Other experts have observed these same trends and have concluded that the state system has a monopoly of power and that the diverse variety of transnational organizations depends on the state system and, in significant ways, perpetuates it.

In many of the articles that appear elsewhere in this book, the authors have concluded their analysis by calling for greater international cooperation to solve the world's most pressing problems. The articles in this section provide examples of successful cooperation. In the midst of a lot of bad news, it is easy to overlook the fact that we are surrounded by international cooperation and that basic day-to-day activities in our lives often directly benefit from it.

STRATEGIES FOR WORLD PEACE:

THE VIEW OF THE UN SECRETARY-GENERAL

By Kofi A. Annan

Today, in Afghanistan, a girl will be born. Her mother will hold her and feed her, comfort her and care for her, just as any mother would anywhere in the world. In these most basic acts of human nature, humanity knows no divisions.

But to be born a girl in today's Afghanistan is to begin life centuries away from the prosperity that one small part of humanity has achieved. It is to live under conditions that many of us would consider inhuman. Truly, it is as if it were a tale of two planets.

I speak of a girl in Afghanistan, but I might equally well have mentioned a baby boy or girl in Sierra Leone. No one today is unaware of this divide between the world's rich and poor. No one today can claim ignorance of the cost that this divide imposes on the poor and dispossessed who are no less deserving of human dignity, fundamental freedoms, security, food, and education than any of us. The cost, however, is not borne by them alone. Ultimately, it is borne by all of us—North and South, rich and poor, men and women of all races and religions.

Today's real borders are not between nations, but between powerful and powerless, free and fettered, privileged and humiliated. Today, no walls can separate humanitarian or human-rights crises in one part of the world from national-security crises in another.

Scientists tell us that the world of nature is so small and interdependent that a butterfly flapping its wings in the Amazon rain forest can generate a violent storm on the other side of the earth. This principle is known as the "Butterfly Effect." Today, we realize, perhaps more than ever, that the world of human activity also has its own "Butterfly Effect"—for better or for worse.

The Universal Bond of Tragedy

We have entered the third millennium through a gate of fire. If today, after the horror of September 11, we see better and we see further, we will realize that humanity is indivisible. New threats make no distinction among races, nations, or regions. A new insecurity has entered every mind, regardless of wealth or status. A deeper awareness of the ties that bind us all—in pain as in prosperity—has gripped young and old.

In the early beginnings of the twenty-first century—a century already violently disabused of any hopes that progress toward global peace and prosperity is inevitable—this new reality can no longer be ignored. It must be confronted.

The twentieth century was perhaps the deadliest in human history, devastated by innumerable conflicts, untold suffering, and unimaginable crimes. Time after time, a group or a nation inflicted extreme violence on others, often driven by irrational hatred and suspicion or by unbounded arrogance and thirst for power and resources. In response to these cataclysms, the leaders of the world came together at mid-century to unite the nations as never before.

A forum was created—the United Nations—where all nations could join forces to affirm the dignity and worth of every person and to secure peace and development for all peoples. Here, states could unite to strengthen the rule of law, recognize and address the needs of the poor, restrain man's brutality and greed, conserve the resources and beauty of nature, sustain the equal rights of men and women, and provide for the safety of future generations.

We thus inherit from the twentieth century the political, scientific, and technological power that—if only we have the will to use them—give us the chance to vanquish poverty, ignorance, and disease.

Redefining the United Nations

In the twenty-first century, I believe the mission of the United Nations will be defined by a new, more profound awareness of the sanctity and dignity of every human life, regardless of race or religion. This will require us to look beyond the framework of states and beneath the surface of nations or communities. We must focus, as never before, on improving the conditions of the individual men and women who give the state or nation its richness and character.

Over the past five years, I have often recalled that the United Nations' Charter begins with the words, "We the peoples." What is not always recognized is that "We the peoples" are made up of individuals whose claims to the most fundamental rights have too often been sacrificed in the supposed interests of the state.

The United Nations' priorities for the future are to promote democracy, prevent conflict, and lessen the burden of global poverty.

A genocide begins with the killing of one man not for what he has done but because of who he is. A campaign of "ethnic cleansing" begins with one neighbor turning on another. Poverty begins when even one child is denied his or her fundamental right to education. What begins with the failure to uphold the dignity of one life all too often ends with a calamity for entire nations.

In this new century, we must start from the understanding that peace belongs not only to states or peoples but to each member of those communities. The sovereignty of states must no longer be used as a shield for gross violations of human rights. Peace must be made real and tangible in the daily existence of every individual in need. Peace must be sought, above all, because it is the condition that enables every member of the human family to live a life of dignity and security.

Focusing on Diversity And Dialogue

From this vision of the role of the United Nations in the next century flow three key priorities for the future: eradicating poverty, preventing conflict, and promoting democracy.

Only in a world that is rid of poverty can all men and women make the most of their abilities. Only where individual rights are respected can differences be channeled politically and resolved peacefully. Only in a democratic environment, based on respect for diversity and dialogue, can individual self-expression and self-government be secured and freedom of association be upheld.

The idea that there is one people in possession of the truth, one answer to the world's ills, or one solution to humanity's needs has done untold harm throughout history—especially in the last century. Today, however, even amid continuing ethnic conflict around the world, there is a growing understanding that human diversity is both the reality that makes dialogue necessary and the basis for that dialogue.

We understand, as never before, that each of us is fully worthy of the respect and dignity essential to our common humanity. We recognize that we are the products of many cultures, traditions, and memories; that mutual respect allows us to study and learn from other cultures; and that we gain strength by combining the foreign with the familiar.

In every great faith and tradition one can find the values of tolerance and mutual understanding. The Qur'an, for example, tells us, "We created you from a single pair of male and female and made you into nations and tribes, that you may know each other." Confucius urged his followers, "When the good way prevails in the State, speak boldly and act boldly. When the State has lost the way, act boldly and speak softly." In the Jewish tradition, the injunction to "love thy neighbor as thyself" is considered to be the very essence of the Torah.

This thought is reflected in the Christian Gospel, which also teaches us to love our enemies and pray for those who wish to persecute us. Hindus are taught, "Truth is one, the sages give it various names." And in the Buddhist tradition, individuals are urged to act with compassion in every facet of life.

Each of us has the right to take pride in our particular faith or heritage. But the notion that what is "ours" is necessarily in conflict with what is "theirs" is both false and dangerous. It has resulted in endless enmity and conflict, leading men to commit the greatest of crimes in the name of a higher power.

It need not be so. People of different religions and cultures live side by side in almost every part of the world, and most of us have overlapping identities that unite us with very different groups. We can love what we are without hating what—and who—we are not. We can thrive in our own tradition, even as we learn from others, and come to respect their teachings.

This will not be possible, however, without freedom of religion, of expression, of assembly, and basic equality under the law. Indeed, the lesson of the past century has been that where the dignity of the individual has been trampled or threatened—where citizens have not enjoyed the basic right to choose their government or the right to change it regularly—conflict has too often followed, with innocent civilians paying the price in lives cut short and communities destroyed.

PEACE HAS NO PARADE: KOFI A. ANNAN ON THE NOBEL PEACE PRIZE

In 1960, the Nobel Peace Prize was awarded for the first time to an African—Albert Luthuli, one of the earliest leaders of the struggle against apartheid in South Africa. In 1961, the Prize was first awarded to a secretary-general of the United Nations—posthumously, because Dag Hammarskjöld had already given his life for peace in Central Africa. For me, as a young African beginning his career in the United Nations a few months later, those two men set a standard that I have sought to follow throughout my working life.

This award belongs not just to me. My own path to service at the United Nations was made possible by the sacrifice and commitment of my family and many friends from all continents—some of whom have passed away—who taught me and guided me. To them, I offer my most profound gratitude.

In a world filled with weapons of war and all too often words of war, the Nobel Committee has become a vital agent for peace. Sadly, a prize for peace is a rarity in this world. Most nations have monuments or memorials to war, bronze salutations to heroic battles, archways of triumph. But peace has no parade, no pantheon of victory.

Only by understanding and addressing the needs of individuals for peace, for dignity, and for security can we at the United Nations hope to live up to fulfill the vision of our founders. This is the broad mission of peace that United Nations staff members carry out every day in every part of the world.

—*Kofi A. Annan*

The obstacles to democracy have less to do with culture or religion than with the desire of those in power to maintain their position at any cost. This is neither a new phenomenon nor one confined to any particular part of the world. People of all cultures value their freedom of choice and feel the need to have a say in decisions affecting their lives.

The United Nations, whose membership comprises almost all the states in the world, is founded on the principle of the equal worth of every human being. It is the nearest thing we have to a representative institution that can address the interests of all states and all peoples. Through this universal, indispensable instrument of human progress, states can serve the interests of their citizens by recognizing common goals and pursuing them in unity. No doubt, that is why the Nobel Committee says that it "wishes, in its centenary year, to proclaim that the only negotiable route to global peace and cooperation goes by way of the United Nations."

Our era of global challenges leaves us no choice but to cooperate at the global level. When states undermine the rule of law and violate the rights of their individual citizens, they become a menace not only to their own people but also to their neighbors and the world. What we need today is better governance—legitimate, democratic governance that allows each individual to flourish and each state to thrive.

A Test of Common Humanity

I began with a reference to the girl born in Afghanistan today. Even though her mother will do all in her power to protect and sustain her, there is a one-in-four risk that she will not live to see her fifth birthday. Whether she does is just one test of our common humanity—of our belief in our individual responsibility for our fellow men and women. But it is the only test that matters.

If we remember this girl, then our larger aims—to fight poverty, prevent conflict, or cure disease—will not seem distant or impossible. Indeed, those aims will seem very near and very achievable—as they should. Beneath the surface of states and nations, ideas and language, lies the fate of individual human beings in need. Answering their needs will be the mission of the United Nations in the century to come.

Kofi A. Annan is secretary-general of the United Nations, #S-8000, New York, New York 10017. Web site www.un.org. This article is drawn from his Nobel Lecture delivered on December 10, 2001, in Oslo, Norway.

EUROPE DRAFTING ITS CONSTITUTION

ELAINE SCIOLINO

[On Friday, despite deep disagreements within the delegation, Valéry Giscard d'Estaing, the former French president who is the convention's president, told the final plenary session in Brussels that the convention had adopted a historic first draft. The forum rose for the union's anthem, Beethoven's "Ode to Joy," and toasted their endeavor with Champagne.]

It will be much less than a United States of Europe. But it will be more than the distillation of five decades of treaties into one document.

For 16 months, Europe's most important and exclusive club has struggled to draft its first constitution. The process has been awkward and unpredictable, ambitious and timid, as delegates from the 15 member nations of the European Union and the 10 that are to join next year fight to protect their countries' national interests even as they agree to cede bits of sovereignty.

Philadelphia it ain't.

The founding fathers came together in 1787 for a Constitutional Convention to forge a document that created a national identity and institutionalized the sovereignty of the American people in one nation-state. The 105 delegates who made up the Convention on the Future of Europe tried to do something much more modest: codifying common ground among long-established states that will give their union more of a logical structure—and perhaps more power—as they expand eastward.

"Until now, Europe was mainly associated with a common market," Ana Palacio, Spain's foreign minister and a delegate representing her government, said in an interview. "Now Europe will be more and more a place of citizenship. Now people will associate Europe with a constitution."

Indeed, one article in the draft constitution states, "Every national of a member state shall be a citizen of the union." When the union expands, that means a mega-Europe of 450 million citizens, larger than any population mass except for China and India, and an economy of more than $9 trillion, close to that of the United States.

The proposed constitution also states that European Union law will have primacy over that of member states. It simplifies voting rules and spells out areas like trade policy in which the union will have full authority and other areas to be shared with the member states, including justice, transportation and economic and social policy.

It will also set up a new structure for an organization that was created for only 6 states and will soon have 25, with two permanent presidents, one foreign minister, a stronger administrative arm and a Parliament with expanded power to pass more legislation.

But for many participants in the process, including Giuliano Amato, a former Italian prime minister and a constitutional law expert who is one of two vice presidents of the convention, the proposed constitution is lacking because it fails to create a common foreign and security policy.

"I'm not entirely satisfied," he said in an interview. "Too many member states are defending themselves instead of sharing power at the European level to make things better. It's each state beyond the constitution. That's why I'm not even sure we are entitled to call it a constitution."

With over 400 articles, the constitution is very much a work in progress. Mr. Giscard d'Estaing will present it to a summit meeting of the member heads of state in Greece next week. Then, in October, it will go into intergovernmental review, in which each member state has the right to demand changes. Each parliament—including those of next year's 10 newcomers—must ratify the document before it comes into force. Some countries, like Ireland and Denmark, will have national referendums—as required by their constitutions.

Even the pope has weighed in, lobbying—thus far successfully—for a specific reference in the text to God and Europe's Christian heritage. After all, the union's debt to the "civilizations of Greece and Rome" and later "by the philosophical currents of the Enlightenment" are mentioned.

One of the main challenges to forming a more perfect European Union is one that the American founding fathers confronted: how to find a way for big states and small states to share power. France and other big states would like a strong president from a large country who would reflect their views, an idea that is anathema to the smaller states. Spain has vowed to fight to retain complex voting rules that give it power disproportionate to its population. (Spain has 27 votes in the union, only 2 fewer than Germany, which has more than twice its population.)

Britain, which is skeptical about creating anything that looks like a European state, is demanding the absolute right for any member nation to veto decisions on foreign policy and taxation. Sometimes the big-small divide is trumped by history. Germany, for example, is more inclined to create a federal structure that would more closely resemble a United States of Europe.

Another issue yet to be resolved is how to make the union more accountable to its citizens by opening the decision-making process to public scrutiny. "Right now, if my prime minister goes to Brussels and makes decisions behind closed doors, I as a parliamentarian cannot hold him to account because I only know the outcome, I don't know the process," said Gisela Stuart, a member of the British delegation and of the European Parliament. "It's the same with the ministers. They can tell me anything."

The new constitution will introduce a single foreign minister to give the union a single actor on the international stage. It will also create a permanent president, elected by member heads of state, who will serve up to a five-year term to replace an unwieldy system in which the presidency rotates among member states every six months.

Already there is intense speculation that Joschka Fischer, the German foreign minister as well as a convention delegate, is eager for the job of European foreign minister, even though it will probably not be created before 2006. In recent weeks, he suddenly began to talk to Anglophone journalists in English, and friends in Brussels said that he had asked them where one might want to live there.

But there will continue to be two presidents indefinitely—one for the Council of the European Union, which consists of the heads of state of each member country, another for the European Commission, a kind of executive body that is more federal in nature and tends to take the smaller states more seriously.

"You have an animal with two heads," said Mr. Amato, who favored a proposal to merge the two presidencies in 2015. "Can an animal with two heads survive for long?"

Mr. Giscard d'Estaing answered yes. "We still have seven monarchies in the system," he said in an interview. "Some went through violent revolutionary uprisings, like France. Some were under the Communist rule for 50 years, 70 years. So if we try for an oversimplified system it cannot work."

The draft constitution clearly states that "member states shall actively and unreservedly support the union's common foreign and security policy in a spirit of loyalty and mutual solidarity" and shall "refrain from action contrary to the union's interests or likely to impair its effectiveness."

But that was language picked up from previous treaties and did not prevent the union's deeply painful split on Iraq, which pitted countries like France and Germany against Spain and Britain.

In a setback to those who wanted a more powerful union to help counterbalance the United States when it comes to issues like foreign policy, defense and taxation, each country—even Luxembourg, with a population of 440,000—has the right to veto any decision on foreign policy and defense.

In one of the most ambitious expansions of the union's authority, the draft constitution also would create a European public prosecutor to combat terrorism and cross-border crimes like corruption, fraud and people-trafficking. It simplifies legislative and legal procedures and extends decision-making by majority vote, particularly in areas like justice, law enforcement, immigration, asylum, energy and the annual European Union budget.

The draft document also gives the union a "legal personality" that would allow it to sign international treaties. A solidarity clause will require member states to provide mutual assistance in case of terrorist attack. The constitution also explicitly bans slavery (which the original United States Constitution did not) and the death penalty (which was never banned in the American Constitution). There is even an exit clause so that a member state can secede from the union if it chooses.

On defense matters, the constitution pledges enhanced "structured cooperation" for "more demanding tasks," but does not pledge military resources for common purposes. Not surprisingly, no effort was made to coax France and Britain to give up their seats as permanent members of the United Nations Security Council.

Underscoring just how important national differences remain, the constitution will be published in the union's 11 current official languages—21 when the 10 new members are admitted next year. There was no agreement on what to call the new union once it has a constitution, so delegates deleted the space in the draft's preamble where a new name would have appeared.

Even the inclusion of the dreaded word "federal" as a description of the way the union would function was found to be objectionable, particularly by Britain. It was replaced by anodyne phrases like "united in an ever closer fashion."

"The reality is that you have different visions for Europe," Jean-Luc Dehaene, the former Belgian prime minister who is a convention vice president, said in an interview. "So never fight for words. Just because someone doesn't want to name the baby, you don't throw out the baby."

Even in the best of circumstances, the constitution will not come into effect for years. So it will not solve the immediate problem of how to absorb the 10 new countries next year. With the expansion, the population of the European club will increase by 20 percent, but the average wealth per person will fall by about 13 percent because most of the newcomers are relatively poor.

That means that the new union, which started out as a club for the rich, will have to find ways to balance the interests of a country like Luxembourg, which has a per capita gross domestic product of nearly $43,000, with a country like Lithuania, which has a per capita G.D.P. of $3,200.

The constitution also will not do away with the 80,000 pages of European Union laws and regulations that dictate what members can and cannot do in some of the biggest and smallest areas of life. The rules govern such things as how to make cars and cigarettes, how corporations carry out acquisitions, how high a budget deficit a country is allowed to have, who is a dentist, what preservatives can be used to make beer, how many hours a week people can work and when hunters can shoot small birds.

The New Containment

An Alliance Against Nuclear Terrorism

Graham Allison & Andrei Kokoshin

DURING THE Cold War, American and Russian policy-makers and citizens thought long and hard about the possibility of nuclear attacks on their respective homelands. But with the fall of the Berlin Wall and the disappearance of the Soviet Union, the threat of nuclear weapons catastrophe faded away from most minds. This is both ironic and potentially tragic, since the threat of a nuclear attack on the United States or Russia is certainly greater today than it was in 1989.

In the aftermath of Osama bin Laden's September 11 assault, which awakened the world to the reality of global terrorism, it is incumbent upon serious national security analysts to think again about the unthinkable. Could a nuclear terrorist attack happen today? Our considered answer is: yes, unquestionably, without any doubt. It is not only a possibility, but in fact the most urgent unaddressed national security threat to both the United States and Russia.[1]

Consider this hypothetical: A crude nuclear weapon constructed from stolen materials explodes in Red Square in Moscow. A 15-kiloton blast would instantaneously destroy the Kremlin, Saint Basil's Cathedral, the ministries of foreign affairs and defense, the Tretyakov Gallery, and tens of thousands of individual lives. In Washington, an equivalent explosion near the White House would completely destroy that building, the Old Executive Office Building and everything within a one-mile radius, including the Departments of State, Treasury, the Federal Reserve and all of their occupants—as well as damaging the Potomac-facing side of the Pentagon.

Psychologically, such a hypothetical is as difficult to internalize as are the plot lines of a writer like Tom Clancy (whose novel *Debt of Honor* ends with terrorists crashing a jumbo jet into the U.S. Capitol on Inauguration Day, and whose *The Sum of All Fears* contemplates the very scenario we discuss—the detonation of a nuclear device in a major American metropolis by terrorists). That these kinds of scenarios are physically possible, however, is an undeniable, brute fact.

After the first nuclear terrorist attack, the Duma, Congress—or what little is left of them—and the press will investigate: Who knew what, when? They will ask what could have been done to prevent the attack. Most officials will no doubt seek cover behind the claim that "no one could have imagined" this happening. But that defense should ring hollow. We have unambiguous strategic warning today that a nuclear terrorist attack could occur at any moment. Responsible leaders should be asking hard questions now. Nothing prevents the governments of Russia, America and other countries from taking effective action immediately—nothing, that is, but a lack of determination.

The argument made here can be summarized in two propositions: first, nuclear terrorism poses a clear and present danger to the United States, Russia and other nations; second, nuclear terrorism is a largely *preventable* disaster. Preventing nuclear terrorism is a large, complex, but ultimately finite challenge that can be met by a bold, determined, but nonetheless finite response. The current mismatch between the seriousness of the threat on the one hand, and the actions governments are now taking to meet it on the other, is unacceptable. Below we assess the threat and outline a solution that begins with a U.S.-Russian led Alliance Against Nuclear Terrorism.

Assessing the Threat

A COMPREHENSIVE threat assessment must consider both the likelihood of an event and the magnitude of its anticipated consequences. As described above, the impact of even a crude nuclear explosion in a city would produce devastation in a class by itself.[2] A half dozen nuclear explosions across the United States or Russia would shift the course of history. The question is: how likely is such an event?

Security studies offer no well-developed methodology for estimating the probabilities of unprecedented events. Contemplating the possibility of a criminal act, Sherlock Holmes investigated three factors: motive, means and opportunity. That framework can be useful for analyzing the question at hand. If no actor simultaneously has motive, means and opportunity, no nuclear terrorist act will occur. Where these three factors are abundant and widespread, the likelihood of a nuclear terrorist attack increases. The questions become: Is anyone *motivated* to instigate a nuclear attack? Could terrorist groups acquire the *means* to attack the United States or Russia with nuclear weapons? Could these groups find or create an *opportunity* to act?

I. Motive

There is no doubt that Osama bin Laden and his associates have serious nuclear ambitions. For almost a decade they have been actively seeking nuclear weapons, and, as President Bush has noted, they would use such weapons against the United States or its allies "in a heartbeat." In 2000, the CIA intercepted a message in which a member of Al-Qaeda boasted of plans for a "Hiroshima" against America. According to the Justice Department indictment for the 1998 bombings of the American embassies in Kenya and Tanzania, "At various times from at least as early as 1993, Osama bin Laden and others, known and unknown, made efforts to obtain the components of nuclear weapons." Additional evidence from a former Al-Qaeda member describes attempts to buy uranium of South African origin, repeated travels to three Central Asian states to try to buy a complete warhead or weapons-usable material, and discussions with Chechen criminal groups in which money and drugs were offered for nuclear weapons.

Bin Laden himself has declared that acquiring nuclear weapons is a religious duty. "If I have indeed acquired [nuclear] weapons," he once said, "then I thank God for enabling me to do so." When forging an alliance of terrorist organizations in 1998, he issued a statement entitled "The Nuclear Bomb of Islam." Characterized by Bernard Lewis as "a magnificent piece of eloquent, at times even poetic Arabic prose," it states: "It is the duty of Muslims to prepare as much force as possible to terrorize the enemies of God." If anything, the ongoing American-led war on global terrorism is heightening our adversary's incentive to obtain and use a nuclear weapon. Al-Qaeda has discovered that it can no longer attack the United States with impunity. Faced with an assertive, determined opponent now doing everything it can to destroy this terrorist network, Al-Qaeda has every incentive to take its best shot.

Russia also faces adversaries whose objectives could be advanced by using nuclear weapons. Chechen terrorist groups, for example, have demonstrated little if any restraint on their willingness to kill civilians and may be tempted to strike a definitive blow to assert independence from Russia. They have already issued, in effect, a radioactive warning by planting a package containing cesium-137 at Izmailovsky Park in Moscow and then tipping off a Russian reporter. Particularly as the remaining Chechen terrorists have been marginalized over the course of the second Chechen war, they could well imagine that by destroying one Russian city and credibly threatening Moscow, they could persuade Russia to halt its campaign against them.

All of Russia's national security documents—its *National Security Concept*, its military doctrine and the recently-updated *Foreign Policy Concept*—have clearly identified international terrorism as the greatest threat to Russia's national security. As President Putin noted in reviewing Russian security priorities with senior members of the Foreign Ministry in January 2001, "I would like to stress the danger of international terrorism and fundamentalism of any, absolutely any stripe." The illegal drug trade and the diffusion of religious extremism throughout Central Asia, relating directly to the rise of the Taliban in Afghanistan, threaten Russia's borders and weaken the Commonwealth of Independent States. The civil war in Tajikistan, tensions in Georgia's Pankisi Gorge, and the conflicts in South Ossetia, Abkhazia and Nagorno-Karabakh—all close to the borders of the Russian Federation—provide feeding grounds for the extremism that fuels terrorism. Additionally, Russia's geographical proximity to South Asia and the Middle East increases concerns over terrorist fallout from those regions. President Putin has consistently identified the dark hue that weapons of mass destruction (WMD) give to the threat of terrorism. In a December 2001 interview in which he named international terrorism the "plague of the 21st century," Putin stated: "We all know exactly how New York and Washington were hit.… Was it ICBMs? What threat are we talking about? We are talking about the use of mass destruction weapons terrorists may obtain."

Separatist militants (in Kashmir, the Balkans and elsewhere) and messianic terrorists (like Aum Shinrikyo, which attacked the Tokyo subway with chemical weapons in 1995) could have similar motives to commit nuclear terrorism. As Palestinians look to uncertain prospects for independent statehood—and never mind whose leadership actually increased that uncertainty in recent years—Israel becomes an ever more attractive target for a nuclear terrorist attack. Since a nuclear detonation in any part of the world would be extremely destabilizing, it threatens American and Russian interests even if few or no Russians or Americans are killed. Policymakers would therefore be foolish to ignore any group with a motive to use a nuclear weapon against any target.

II. Means

To the best of our knowledge, no terrorist group can now detonate a nuclear weapon. But as Secretary of Defense Donald Rumsfeld has stated, "the absence of evidence is not evidence of absence." Are the means beyond terrorists' reach, even that of relatively sophisticated groups like Al-Qaeda?

Over four decades of Cold War competition, the superpowers spent trillions of dollars assembling mass arsenals, stockpiles, nuclear complexes and enterprises that engaged hundreds of thousands of accomplished scientists and engineers. Technical know-how cannot be un-invented. Reducing arsenals that include some 40,000 nuclear weapons and the equivalents of more than 100,000 nuclear weapons in the form of highly enriched uranium (HEU) and plutonium to manageable levels is a gargantuan challenge.

Terrorists could seek to buy an assembled nuclear weapon from insiders or criminals. Nuclear weapons are known to exist in eight states: the United States, Russia, Great Britain, France, China, Israel, India and Pakistan. Security measures, such as "permissive action links" designed to prevent unauthorized use, are most reliable in the United States, Russia, France and the United Kingdom. These safeguards, as well as command-and-control systems, are much less reliable in the two newest nuclear states—India and Pakistan. But even where good systems are in place, maintaining high levels of security requires constant attention from high-level government officials.

Alternatively, terrorists could try to build a weapon. The only component that is especially difficult to obtain is the nuclear fissile material—HEU or plutonium. Although the largest stockpiles of weapons-grade material are predominantly found in the nuclear weapons programs of the United States and Russia, fissile material in sufficient quantities to make a crude nuclear weapon can also be found in many civilian settings around the globe. Some 345 research reactors in 58 states together contain twenty metric tons of HEU, many in quantities sufficient to build a bomb.[3] Other civilian reactors produce enough weapons-grade nuclear material to pose a proliferation threat; several European states, Japan, Russia and India reprocess spent fuel to separate out plutonium for use as new fuel. The United States has actually facilitated the spread of fissile material in the past—over three decades of the Atoms for Peace program, the United States exported 749 kg of plutonium and 26.6 metric tons of HEU to 39 countries.[4]

Terrorist groups could obtain these materials by theft, illicit purchase or voluntary transfer from state control. There is ample evidence that attempts to steal or sell nuclear weapons or weapons-usable material are not hypothetical, but a recurring fact.[5] Just last fall, the chief of the directorate of the Russian Defense Ministry responsible for nuclear weapons reported two recent incidents in which terrorist groups attempted to perform reconnaissance at Russian nuclear storage sites. The past decade has seen repeated incidents in which individuals and groups have successfully stolen weapons material from sites in Russia and sought to export them—but were caught trying to do so. In one highly publicized case, a group of insiders at a Russian nuclear weapons facility in Chelyabinsk plotted to steal 18.5 kg (40.7 lbs.) of HEU, which would have been enough to construct a bomb, but were thwarted by Russian Federal Security Service agents.

In the mid-1990s, material sufficient to allow terrorists to build more than twenty nuclear weapons—more than 1,000 pounds of highly enriched uranium—sat unprotected in Kazakhstan. Iranian and possibly Al-Qaeda operatives with nuclear ambitions were widely reported to be in Kazakhstan. Recognizing the danger, the American government itself purchased the material and removed it to Oak Ridge, Tennessee. In February 2002, the U.S. National Intelligence Council reported to Congress that "undetected smuggling [of weapons-usable nuclear materials from Russia] has occurred, although we do not know the extent of such thefts." Each assertion invariably provokes blanket denials from Russian officials. Russian Atomic Energy Minister Aleksandr Rumyantsev has claimed categorically: "Fissile materials have not disappeared." President Putin has stated that he is "absolutely confident" that terrorists in Afghanistan do not have weapons of mass destruction of Soviet or Russian origin.

For perspective on claims of the inviolable security of nuclear weapons or material, it is worth considering the issue of "lost nukes." Is it possible that the United States or Soviet Union lost assembled nuclear weapons? At least on the American side the evidence is clear. In 1981, the U.S. Department of Defense published a list of 32 accidents involving nuclear weapons, *many of which resulted in lost bombs*.[6] One involved a submarine that sank along with two nuclear torpedoes. In other cases, nuclear bombs were lost from aircraft. Though on the Soviet/Russian side there is no official information, we do know that four Soviet submarines carrying nuclear weapons have sunk since 1968, resulting in an estimated 43 lost nuclear warheads.[7] These accidents suggest the complexity of controlling and accounting for vast nuclear arsenals and stockpiles.

Nuclear materials have also been stolen from stockpiles housed at research reactors. In 1999, Italian police seized a bar of enriched uranium from an organized crime group trying to sell it to an agent posing as a Middle Eastern businessman with presumed ties to terrorists. On investigation, the Italians found that the uranium originated from a U.S.-supplied research reactor in the former Zaire, where it presumably had been stolen or purchased *sub rosa*.

Finally, as President Bush has stressed, terrorists could obtain nuclear weapons or material from states hostile to the United States. In his now-infamous phrase, Bush called hostile regimes developing WMD and their terrorist allies an "axis of evil." He argued that states such as Iraq, Iran and North Korea, if allowed to realize their nuclear ambitions, "could provide these arms to terrorists, giving them the means to match their hatred." The fear that a hostile regime might transfer a nuclear weapon to terrorists has contributed to the Bush Administration's development of a new doctrine of preemption against such regimes, with Iraq as the likeliest test case. It also adds to American concerns about Russian transfer of nuclear technologies to Iran. While Washington and Moscow continue to disagree over whether any safeguarded civilian nuclear cooperation with Iran is justified, both agree on the dangers a nuclear-armed Iran would pose. Russia is more than willing to agree that there should be no transfers of technology that could help Iran make nuclear weapons.

III. Opportunity

Security analysts have long focused on ballistic missiles as the preferred means by which nuclear weapons would be delivered. But today this is actually the least likely vehicle by which a nuclear weapon will be delivered against Russia or the United States. Ballistic weapons are hard to produce, costly and difficult to hide. A nuclear weapon delivered by a missile also leaves an unambiguous return address, inviting devastating retaliation. As Robert Walpole, a National Intelligence Officer, told a Senate subcommittee in March, "Nonmissile delivery means are less costly, easier to acquire, and more reliable and accurate."[8] Despite this assessment, the U.S. government continues to invest much more heavily in developing and deploying missile defenses than in addressing more likely trajectories by which weapons could arrive.

Terrorists would not find it very difficult to sneak a nuclear device or nuclear fissile material into the United States via shipping containers, trucks, ships or aircraft. Recall that the nuclear material required is smaller than a football. Even an assembled device, like a suitcase nuclear weapon, could be shipped in a container, in the hull of a ship or in a trunk carried by an aircraft. After this past September 11, the number of containers that are X-rayed has increased, to about 500 of the 5,000 containers currently arriving daily at the port of New York/New Jersey—approximately 10 percent. But as the chief executive of CSX Lines, one of the foremost container-shipping companies, put it: "If you can smuggle heroin in containers, you may be able to smuggle in a nuclear bomb."

Effectively countering missile attacks will require technological breakthroughs well beyond current systems. Success in countering covert delivery of weapons will require not just technical advances but a conceptual breakthrough. Recent efforts to bolster border security are laudable, but they only begin to scratch the surface. More than 500 million people, 11 million trucks and 2 million rail cars cross into the United States each year, while 7,500 foreign-flag ships make 51,000 calls in U.S. ports. That's not counting the tens of thousands of people, hundreds of aircraft and numerous boats that enter illegally and uncounted. Given this volume and the lengthy land and sea borders of the United States, even a radically renovated and reorganized system cannot aspire to be airtight.

The opportunities for terrorists to smuggle a nuclear weapon into Russia or another state are even greater. Russia's land borders are nearly twice as long as America's, connecting it to more than a dozen other states. In many places, in part because borders between republics were less significant in the time of the Soviet Union, these borders are not closely monitored. Corruption has been a major problem among border patrols. Visa-free travel between Russia and several of its neighbors creates additional opportunities for weapons smugglers and terrorists. The "homeland security" challenge for Russia is truly monumental.

In sum: even a conservative estimate must conclude that dozens of terrorist groups have sufficient motive to use a nuclear weapon, several could potentially obtain nuclear means, and hundreds of opportunities exist for a group with means and motive to make the United States or Russia a victim of nuclear terrorism. The mystery before us is not how a nuclear terrorist attack could possibly occur, but rather why no terrorist group has yet combined motive, means and opportunity to commit a nuclear attack. We have been lucky so far, but who among us trusts luck to protect us in the future?

Chto Delat?[9]

THE GOOD NEWS about nuclear terrorism can be summarized in one line: no highly enriched uranium or plutonium, no nuclear explosion, no nuclear terrorism. Though the world's stockpiles of nuclear weapons and weapons-usable materials are vast, they are finite. The prerequisites for manufacturing fissile material are many and require the resources of a modern state. Technologies for locking up super-dangerous or valuable items—from gold in Fort Knox to treasures in the Kremlin Armory—are well developed and tested. While challenging, a specific program of actions to keep nuclear materials out of the hands of the most dangerous groups is not beyond reach, *if* leaders give this objective highest priority and hold subordinates accountable for achieving this result.

The starting points for such a program are already in place. In his major foreign policy campaign address at the Ronald Reagan Library, then-presidential candidate George W. Bush called for "Congress to increase substantially our assistance to dismantle as many Russian weapons as possible, as quickly as possible." In his September 2000 address to the United Nations Millennium Summit, Russian President Putin proposed to "find ways to block the spread of nuclear weapons by excluding use of enriched uranium and plutonium in global atomic energy production." The Joint Declaration on the New Strategic Relationship between the United States and Russia, signed by the two presidents at the May 2002 summit, stated that the two partners would combat the "closely linked threats of international terrorism and the proliferation of weapons of mass destruction." Another important result yielded by the summit was the upgrading of the Armitage/Trubnikov-led U.S.-Russia Working Group on Afghanistan to the U.S.-Russia Working Group on Counterterrorism, whose agenda is to thwart nuclear biological and chemical terrorism.

Operationally, however, priority is measured not by words, but by deeds. A decade of Nunn-Lugar Cooperative Threat Reduction Programs has accomplished much in safeguarding nuclear materials. Unfortunately, the job of upgrading security to minimum basic standards is mostly unfinished: according to Department of Energy reports, two-thirds of the nuclear material in Russia remains to be adequately secured.[10] Bureaucratic inertia, bolstered by mistrust and misperception on both sides, leaves these joint programs bogged down on timetables that extend to 2008. Unless implementation improves significantly, they will probably fail to meet even this unacceptably distant target. What is required on both sides is personal, presidential priority measured in commensurate energy, specific orders, funding and accountability. This should be embodied in a new U.S.-Russian led Alliance Against Nuclear Terrorism.

Five Pillars of Wisdom

WHEN IT COMES to the threat of nuclear terrorism, many Americans judge Russia to be part of the problem, not the solution. But if Russia is welcomed and supported as a fully responsible non-proliferation partner, the United States stands to accomplish far more toward minimizing the risk of nuclear terrorism than if it treats Russia as an unreconstructed pariah. As the first step in establishing this alliance, the two presidents should pledge to each other that his government will do everything technically possible to prevent criminals or terrorists from stealing nuclear weapons or weapons-usable material, and to do so on the fastest possible timetable. Each should make clear that he will personally hold accountable the entire chain of command within his own government to assure this result. Understanding that each country bears responsibility for the security of its own nuclear materials, the United States should nonetheless offer Russia any assistance required to make this happen. Each nation—and each leader—should provide the other sufficient transparency to monitor performance.

Archy: "An optimist is a guy that has never had much experience."

—Don Marquis, *Archy and Mehitabel* (1927)

To ensure that this is done on an expedited schedule, both governments should name specific individuals, directly answerable to their respective presidents, to co-chair a group tasked with developing a joint Russian-American strategy within one month. In developing a joint strategy and program of action, the nuclear superpowers would establish a new world-class "international security standard" based on President Putin's Millennium proposal for new technologies that allow production of electricity with low-enriched, non-weapons-usable nuclear fuel.

A second pillar of this alliance would reach out to all other nuclear weapons states—beginning with Pakistan. Each should be invited to join the alliance and offered assistance, if necessary, in assuring that all weapons and weapons-usable material are secured to the new established international standard in a manner sufficiently transparent to reassure all others. Invitations should be diplomatic in tone but nonetheless clear that this is an offer that cannot be refused. China should become an early ally in this effort, one that could help Pakistan understand the advantages of willing compliance.

A third pillar of this alliance calls for global outreach along the lines proposed by Senator Richard Lugar in what has been called the Lugar Doctrine.[11] All states that possess weapons-usable nuclear materials—even those without nuclear weapons capabilities—must enlist in an international effort to guarantee the security of such materials from theft by terrorists or criminals groups. In effect, each would be required to meet the new international security standard and to do so in a transparent fashion. Pakistan is particularly important given its location and relationship with Al-Qaeda, but beyond nuclear weapons states, several dozen additional countries hosting research reactors—such as Serbia, Libya and Ghana—should be persuaded to surrender such material (almost all of it either American or Soviet in origin), or have the material secured to acceptable international standards.

A fourth pillar of this effort should include Russian-American led cooperation in preventing any further spread of nuclear weapons to additional states, focusing sharply on North Korea, Iraq and Iran. The historical record demonstrates that when the United States and Russia have cooperated intensely, nuclear wannabes have been largely stymied. It was only during periods of competition or distraction, for example in the mid-1990s, that new nuclear weapons states realized their ambitions. India and Pakistan provide two vivid case studies. Recent Russian-American-Chinese cooperation in nudging India and Pakistan back from the nuclear brink suggests a good course of action. The failure and subsequent freeze of North Korean nuclear programs offers complementary lessons about the consequences of competition and distraction. The new alliance should reinvent a robust non-proliferation regime of controls on the sale and export of weapons of mass destruction, nuclear material and missile technologies, recognizing the threat to each of the major states that would be posed by a nuclear-armed Iran, North Korea or Iraq.

Finally, adapting lessons learned in U.S.-Russian cooperation in the campaign against bin Laden and the Taliban, this new alliance should be heavy on intelligence sharing and affirmative counter-proliferation, including disruption and pre-emption to prevent acquisition of materials and know-how by nuclear wannabes. Beyond joint intelligence sharing, joint training for pre-emptive actions against terrorists, criminal groups or rogue states attempting to acquire weapons of mass destruction would provide a fitting enforcement mechanism for alliance commitments.

AS FORMER Senator Sam Nunn has noted: "At the dawn of a new century, we find ourselves in a new arms race. Terrorists are racing to get weapons of mass destruction; we ought to be racing to stop them."[12] Preventing nuclear terrorism will require no less imagination, energy and persistence than did avoiding nuclear war between the superpowers over four decades of Cold War. But absent deep, sustained cooperation between the United States, Russia and other nuclear states, such an effort is doomed to failure. In the context of the qualitatively new relationship Presidents Putin and Bush have established in the aftermath of last September 11, success in such a bold effort is within the reach of determined Russian-American leadership. Succeed we must.

Notes

1. This judgment echoes that of a Department of Energy task force on nonproliferation programs with Russia led by Howard Baker and Lloyd Cutler: "The most urgent unmet national security threat to the United States today is the danger that weapons of mass destruction or weapons-usable material in Russia could be stolen and sold to terrorists or hostile nation states and used against American troops abroad or citizens at home." *A Report Card on the Department of Energy's Nonproliferation Programs with Russia*, January 10, 2001.

2. Although biological and chemical weapons can cause huge devastation as well, "the massive, assured, instantaneous, and comprehensive destruction of life and property" of a nuclear weapon is unique. See Matthew Bunn, John P. Holdren and Anthony Wier, "Securing Nuclear Weapons and Materials: Seven Steps for Immediate Action," *Nuclear*

Threat Initiative and the Managing the Atom Project, May 20, 2002, p. 2. This report provides extensive, but not-too-technical detail on many of the points in this essay.

3. See U.S. Department of Energy, *FY 2003 Budget Request: Detailed Budget Justifications—Defense Nuclear Nonproliferation* (Washington, DC: DOE, 2002), p. 172.

4. Summarized in NIS Nuclear Trafficking Database, available at www.nti.org.

5. The Nuclear Threat Initiative maintains a database of cases and reported incidents of trafficking in nuclear and radioactive materials in and from the former Soviet Union. Available at www.nti.org.

6. U.S. Department of Defense, "Narrative Summaries of Accidents involving U.S. Nuclear Weapons: 1950-1980" (April 1981).

7. Joshua Handler, Amy Wickenheiser and William M. Arkin, "Naval Safety 1989: The Year of the Accident," *Neptune Paper No. 4* (April 1989).

8. U.S Senate Subcommittee on International Security, Proliferation, and Federal Services, "Statement of Robert Walpole before the Senate Subcommittee on International Security, Proliferation, and Federal Services," 107th Cong., 1st sess., March 11, 2002.

9. A proverbial Russian refrain: "What is to be done?"

10. Bunn, Holdren and Wier, "Securing Nuclear Weapons and Materials."

11. Speech by Senator Richard Lugar, May 27, 2002, at the Moscow Nuclear Threat Initiative Conference.

12. Sam Nunn, "Our New Security Framework," *Washington Post*, October 8, 2001.

Graham Allison is director of the Belfer Center for Science and International Affairs at Harvard's John F. Kennedy School of Government. Andrei Kokoshin is director of the Institute for International Security Studies of the Russian Academy of Sciences and a former secretary of the Security Council of Russia.

countdown TO ERADICATION

Rotary and its partners move ever closer to a polio-free world

by Anne Stein

In 1985, Rotary International committed to help immunize the world's children against polio, with a target date of a polio-free world by 2005, Rotary's centennial year. Since 1988, when the Global Polio Eradication Initiative was launched, millions of Rotary volunteers have raised funds and assisted in oral polio vaccine delivery, social mobilization, and garnering cooperation from national governments and health ministries—making Rotary the leading private-sector partner in the effort.

The program began with donations from Rotarians of more than US $240 million. By 2005, Rotary's financial contribution will reach an estimated $500 million. Along with its global partners—the World Health Organization (WHO), UNICEF, and the U.S. Centers for Disease Control and Prevention (CDC)—Rotary can take pride in the largest public health initiative in history.

ENDING POLIO Ten polio-endemic nations and a $275 million shortfall stand between failure and success.

According to the most recent data, only 537 cases of polio have been reported worldwide for 2001, although that number will likely increase somewhat as final reports are filed. However, it is sure to be down considerably from 2000, when 2,979 cases were logged. The program partners report an astounding 99 percent decrease in polio cases since 1988, the year WHO resolved to eradicate the disease. But despite the remarkable progress in reaching children and the unprecedented outpouring of time and resources from Rotarians and their partners, there's still much work to do and $1 billion in donor contributions needed to reach the final goal.

So far, $725 million has been pledged or projected, leaving a gap of $275 million. In response, the RI Board of Directors and The Rotary Foundation Trustees have announced a new, $80 million fundraising campaign—Rotary's contribution to closing the gap—that will run through the 2002-03 Rotary year. The campaign, Fulfilling Our Promise: Eradicate Polio, was officially launched in June during the RI Convention in Barcelona by 2001-02 Trustees Chairman Luis V. Giay.

"The very principles of Rotary are on the line," says Past RI Vice President Louis Piconi, chairman of the North American Polio Eradication Fundraising Campaign Committee. "We should never make a promise without fulfilling it; it's our word. To have Rotary be a major factor in eradicating polio will be a very significant milestone for the organization and its history."

"Each of these countries has made tremendous progress, but each has its own unique set of challenges."

About 60 percent of today's Rotarians—including all women members—have joined since the launch of the PolioPlus campaign, observes former RI General Secretary Herbert Pigman, who is now the director of the polio eradication fundraising campaign. "This new campaign gives them an opportunity to play an important part in our great humanitarian cause."

Up to $25 million of the contributions to this campaign can be matched by an equal amount from the Bill & Melinda Gates Foundation. The resulting $50 million can be further augmented by $75 million in assistance from the World Bank. "If we raise this money now, we will be saving all of humankind from this disease for all time, and that's priceless," Giay said in April during a joint news conference with representatives of the CDC, UNICEF, and WHO.

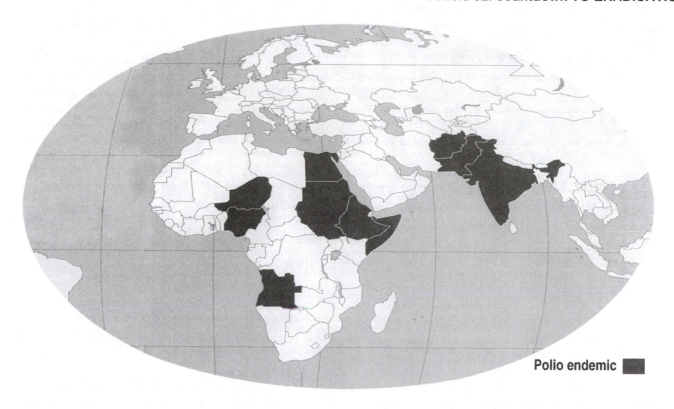

Polio endemic ■

Countdown to 2005: The wild poliovirus remains endemic in 10 countries: Afghanistan, Angola, Egypt, Ethiopia, India, Niger, Nigeria, Pakistan, Somalia and Sudan.

What remains to be done? Quite simply, a lot. Intensified immunization activities must take place in the 10 remaining polio-endemic nations (down from 20 nations in 2000 and 127 in 1985) and in several more countries that are considered high risk for outbreaks. Surveillance must also be maintained to ensure polio-free status. "We need to immunize more than 600 million children over the next four years," explains Piconi. "Rotary's campaign will help to buy and get the vaccine to all of those children, no matter where they are."

Areas of greatest focus are the remaining polio-endemic nations: Afghanistan, India, Niger, Nigeria, and Pakistan, which collectively accounted for more than 85 percent of the 2001 polio cases and are considered high-intensity transmission areas, and Angola, Egypt, Ethiopia, Somalia, and Sudan, which are low-intensity transmission areas, accounting for less than 15 percent of new polio cases last year.

The high risk areas have been identified and are receiving special attention during the immunization drives.

"Each of these countries has made tremendous progress, but each has its own unique set of challenges," said Carol Bellamy, executive director of UNICEF, during the April news conference. "Through the battle to rid the

world of polio, we have managed to reach children living in some of the most remote and challenging circumstances imaginable. Over the coming days and months, we must continue this unprecedented effort, using all of our resources to reach the very last child with polio vaccine. We have a unique opportunity to deliver a truly global victory in an uncertain world."

Deepak Kapur, India's National PolioPlus Committee Chairman and a member of the Rotary Club of Delhi South, says: "The job is almost done [in India]. But the most dangerous word in the dictionary is *almost*. We rejoiced when the thousands of cases fell to a few hundred, and today we have cause to celebrate when only a few cases are left. But this new chapter begins with those few cases."

The good news in India is that the geographic distribution of the virus has decreased. In 2001, India reported 268 cases of polio (a slight increase from 2000), about 95 percent of them in the northern states of Bihar and Uttar Pradesh. Kapur says that a range of projects will be undertaken this year, including mobilizing village volunteers and school students to spread awareness of the campaign; continuing the "video van" project and musical street dramas, which are especially useful in explaining PolioPlus to illiterate populations; and updating local media to enlist their editorial support.

According to Kapur, challenges include generally poor immunization rates for routine childhood diseases, overworked public health officials, public indifference because of repeated immunization drives, and outright distrust of immunization efforts by members of certain populations.

"[PolioPlus India] Executive Director Rajiv Tandon and I went to Moradabad on a field visit. It's considered the 'polio capital of the world since more than half the cases have been genetically traced to the virus prevalent in this area," recounts Kapur. "One of the reasons for proliferation is the significant resistance of members of a particular minority community." The two visited a hut in a slum to persuade a mother to let them administer the polio drops to her infant son. Finally, after making the pair promise not to tell her husband, she allowed her baby to be immunized.

"Her husband was convinced that the immunization program was aimed at curtailing the population of their community by making children impotent," Kapur explains. "As we stepped out of the hut, we saw a polio-crippled girl playing in the mud. The neighbors told us that the little girl had contracted polio a few months ago and was that very mother's daughter." Polio workers in India face this kind of resistance because of incomplete information, misinformation, and rumors, says Kapur.

Afghanistan and Pakistan also pose challenges, but progress has been good. Despite the war and refugee crises in the wake of 11 September, Afghanistan reported only 11 cases of polio in 2001 (down from 120 in 2000), clustered in two main areas. Two rounds of National Immunization Days (NIDs) were held in September and November 2001, reaching 35 million children; this year, NIDs began anew in mid-April, coordinated with Pakistan to ensure coverage of cross-border areas.

"The most urgent need is to bring stability to the country [Afghanistan] and start reconstruction work without delay," says Rotarian Abdul Haiy Khan, chairman of Pakistan's National PolioPlus Committee and a member of the Rotary Club of Karachi. "The Afghans are resilient, and during my visits to Afghanistan to monitor their NIDs, I was impressed with their commitment. I think that once they get the opportunity to freely move around the country and have the assurance of obtaining the needed resources, the polio-eradication job will be done in a short time."

In Pakistan, Khan reports, "we have to deal with a large population, 140 million people with about 30 million of them children under age 5, nearly 30 percent living in densely populated or slum areas with poor sanitary conditions." However, the high-risk areas have been identified and are receiving special attention during the immunization drives. The high-transmission season of wild polio-virus runs from May to September.

"This will be the testing period to prove that we are in control of the situation and can hope to come close to the zero polio-case reporting status," says Khan. "The real challenge includes training and retraining health workers, volunteers and monitors, improving the surveillance system, and providing better transport for health workers

and volunteers so they can reach the farthest, most isolated areas in the hot summer season."

But ingenuity and creativity bring success. Khan cites one story of an Afghan refugee camp on the outskirts of Karachi, where health workers reported that camp residents refused to let their children be immunized. "We spent some time investigating the matter and were pleasantly surprised to find that the refusal was based on a lack of communication. Our health workers didn't speak [their] language and couldn't convey the purpose of house-to-house visits. And the camp housed a large number of unregistered refugees, so camp elders were reluctant to let government officials (health workers) go door-to-door."

The solution? "We simply located two Afghan girls studying in a local medical college, explained the problem, and they came to the camp to talk to the elders. The next day, the two girls were accompanied by more than 100 fellow students, and in two days, working eight hours a day, they immunized 10,000 Afghan children under the age of 5 whom we had been missing for a couple of years."

The funding gap, along with the continuing threat of conflict worldwide and particularly within the 10 polio-endemic nations, brought a recent warning from experts: While polio has been pushed to the lowest level in history, efforts to complete eradication and immunization must be intensified now before global instability closes the window of opportunity.

"I urge the world to finish the job. Eradicate polio while we still have the opportunity."

— Dr. Gro Harlem Brundtland, WHO director-general

"When we began the eradication effort in 1988, polio paralyzed more than 1,000 children each day. In 2001, there were far fewer than 1,000 cases the entire year," said Dr. Gro Harlem Brundtland, director-general of WHO, during the joint April news conference. "But we're not finished yet, and the past year has reminded us that we live in a world where security and access to children cannot be guaranteed. So I urge the world to finish the job. Eradicate polio while we still have the opportunity."

Fortunately, in the African nations of Angola, Sierra Leone, and Somalia, warring forces have laid down their arms and allowed health workers to reach the children. In 2000, 16 West African countries coordinated immunization activities, and more than 80 million children were immunized. In 2001, conflict-affected countries in central Africa, including Angola and the Democratic Republic of the Congo, synchronized their NIDs for the first time and 16 million children were vaccinated.

To stop the transmission of the wild poliovirus in Africa by the end of 2002, NIDs will target northern Nigeria along the border with Niger. Major progress has been achieved in Angola, Egypt, Ethiopia, Somalia, and Sudan, and the goal is to stop transmission of the wild poliovirus by mid-2002—given continued political commitment in the five nations and no further deterioration of security.

Seventeen nations are considered at particularly high risk of reestablished wild poliovirus transmission, including Central African Republic, Chad, Congo Republic, Democratic Republic of the Congo, Indonesia, Madagascar, and Mozambique. These countries either are recently endemic or border endemic areas. They also have low routine-immunization coverage and inadequate surveillance capability or both. These nations require continuing, supplementary immunization activities and the assurance that no child has been missed.

"If we don't finish this off," warns Piconi, "this virus can creep back rather quickly. We need to stay focused and eliminate a disease that's vaccine preventable."

Anne Stein is a health and fitness writer in Evanston, Ill., USA, who has written frequently about the polio-eradication effort for THE ROTARIAN.

AERIAL WAR AGAINST DISEASE

Satellite tracking of epidemics is soaring

BY JOHN PICKRELL

Imagine this: It's the height of summer a decade from now, and stifling heat blankets New York City. A hidden foe has lurked out of sight for many months, and authorities are on high alert. The streets are deserted, Central Park is sealed, helicopters circle overhead, and residents remain inside with their windows firmly closed. Dousing the city, the surrounding suburbs, and perhaps wider areas with protective chemicals is the only defense.

In this fictional scenario, the city isn't under terrorist attack. The concealed enemy is the deadly virus that causes West Nile fever, which first emerged in New York in 1999. Attempts at vaccines (*SN: 3/16/02, p. 164*) haven't borne fruit, and the heat is creating the ideal conditions for drawing virus-carrying mosquitoes out of their winter refuges.

However, officials of the future are prepared to fight the disease, courtesy of an unlikely medical tool: the National Oceanic and Atmospheric Administration's (NOAA) series of polar-orbiting satellites about 850 kilometers above Earth.

The NOAA satellites aren't sending pictures of crows—the most common avian reservoir of West Nile virus in the United States—but they're producing maps of vegetation and weather affecting the mosquitoes that carry the disease to people.

In 2002, satellites already enable people to forecast weather, spy on enemy armies, broadcast television signals, and phone a friend in the Amazon jungle. A growing group of researchers has also been harnessing satellites in the battle against infectious diseases (*SN: 8/2/97, p. 72; 2/2/02, p. 78*).

The amount of data created by environmental satellites is increasing at an exponential rate, and researchers worldwide are catching on to the value of this information. The February *Photogrammetric Engineering and Remote Sensing* details several research projects that use satellite data to reveal patterns of diseases including schistosomiasis, tuberculosis, Ebola, Rift Valley fever, and West Nile fever.

TALA RESEARCH GROUP, UNIVERSITY OF OXFORD.

SATELLITE FEVER—INTREPID's preliminary risk map for West Nile fever. Land-surface temperatures (shading variation) and locations of birds harboring West Nile virus (black dots) help researchers predict where outbreaks are likely to occur.

FOLLOW THE FLORA The rationale behind the use of satellites to fight vector-borne disease lies in the field of landscape epidemiology, which began in Russia in the 1930s. Theory holds that features of the landscape, such as temperature, humidity, rainfall, and plant growth, can help public health officials pinpoint where deadly microbes—or the animals that carry them—dwell. Ideally, that indicates where people are at greatest risk and so where governments should target disease-control measures and medicines.

Researchers first discussed adding satellites to the tools of landscape epidemiology in the early 1970s, with the launch of NASA's Landsat-1 satellite, says Robert Venezia, program manager for public health applications at NASA's headquarters. In recent years, NASA has promoted the use of satellite data for health purposes. The agency is currently working with an international team of universities and U.S.-government bodies under the banner of the International Research Partnership for Infectious Diseases, or INTREPID. The program's researchers are today in the early stages of creating a satellite-borne early-warning system for West Nile virus in the United States, says epidemiologist Simon Hay at the University of Oxford in England. With it, workers will be better prepared, as in the fictional example above.

The scientists are focusing on areas with warm land surfaces, high humidity, and vegetation types likely to harbor mosquitoes. Data on these variables will be combined with other information, such as bird-migration routes, to create risk maps for West Nile fever during each summer.

Ebola—Chlorophyll reflects light in a distinctive way, so satellite images clearly reveal vegetation changes, which act as a surrogate for rainfall patterns.

Two other deadly diseases—Rift Valley fever and Ebola—are also the focus of remote-sensing initiatives. These viral infections, which cause swelling and fever, strike in intermittent outbreaks in Africa. Rift Valley fever was first documented in the 1930s in the area of Kenya from which it takes its name. It kills few people, but it's a major problem in livestock, where it causes spontaneous abortions and many deaths.

Ebola is more deadly to people and more enigmatic. Named after a river in Sudan, it was reported appearing for the first time in 1976. Reappearing periodically, it spreads rapidly and kills up to 88 percent of the people it infects.

Despite the disease's deadliness, scientific understanding of Ebola is limited. Infections first show up in people who have been in tropical forests, but researchers still have no idea which animal species act as vectors or reservoirs for the virus.

Scientists at the NASA-Goddard Space Flight Center in Greenbelt, Md., are approaching the challenges of Ebola and Rift Valley fever in a different way from the strategy employed to study U.S. West Nile virus. They're not only trying to predict future outbreaks of the African diseases but also attempting to better understand their causes.

The only documented Ebola outbreaks have occurred within two limited periods. The disease struck in Congo and Sudan during the 1970s, before satellite data were being collected con-

sistently in Africa, and resurfaced for a further sporadic round of infection in Gabon, Ivory Coast, and Congo during the 1990s.

Compton J. Tucker of Goddard and his colleagues suspect that transitions between wet and dry periods in tropical forests may lie at the heart of Ebola outbreaks. Using images taken by the NOAA and Landsat satellites from 1994 to 1996, the team has studied vegetation changes as a surrogate for rainfall patterns.

That's relatively easy to do from space. Chlorophyll—the green pigment in leaves—reflects light in a distinctive way, so satellite images clearly reveal vegetation. The researchers found that images taken during three separate Ebola outbreaks indicated a transition from below-average to above-average vegetation cover within equatorial Africa's forests. The finding suggests that the disease "may follow when a rare tropical dry period is brought to an abrupt end with a change to very wet conditions," says Tucker.

He notes that it will take additional research to confirm that a wet-to-dry transition is the trigger and to explain why such an event spreads the virus. "It's fortunate for those affected by Ebola that we have so few outbreaks to study," Tucker says, "but it makes our job more difficult."

CREATURE FEATURES The researchers have had more luck nailing down the conditions that promoted Rift Valley fever during the 1980s and 1990s.

From the satellite data, Rift Valley fever appears linked to above-average rainfall in savanna ecosystems. The wet conditions lead to booms in mosquito populations that spread Rift Valley fever to people.

The accuracy with which the researchers can use satellite data to predict the disease's course has already led the team to create a disease-surveillance and early-warning system. Computers analyze daily satellite data for the telltale predictors of mosquito outbreaks in areas prone to the disease, says Tucker. The researchers pass along 2-week summaries of the data to the U.S. Army and the World Health Organization. These groups can warn their personnel and others in threatened areas to take precautions against infection, for instance, by stepping up immunizations. Satellite surveillance programs can now "monitor diseases in both space and time," Hay says.

Schistosomiasis—Researchers turned to data from the Landsat TM satellite to identify areas most likely to harbor snails.

So far, satellite epidemiology has focused mostly on insect-borne diseases because correlations among weather, plant cover, and insect infestations are clearly established. But scientists are starting to use NASA's Landsat satellites—700 km above Earth—to track other disease carriers.

In China, researchers are using satellites to map areas at highest risk from the debilitating disease schistosomiasis. Caused by a parasitic flatworm that spends part of its life in amphibious snails, the illness afflicts 200 million people worldwide. The parasite, *Schistosoma japonica*, burrows into the skin of people who venture into contaminated water. Inside a victim, it migrates to internal organs, such as the lungs or liver, causing serious long-term illness.

In the mid-1990s, Edmund Seto of the University of California, Berkeley was approached by colleagues from China to help find high-risk areas and monitor the spread of the disease as the environment changes during the country's rapid development.

Seto's team of researchers turned to data from the Landsat TM satellite to identify areas most likely to harbor snails. In China, the snail host is divided into two subspecies, one that lives in certain types of mountainous irrigated farmland and the other that favors marshy floodplains of the lower Yangtze River. Both snails are susceptible to environmental change and tightly linked to their favored habitats.

The researchers selected sites in the two habitats and then compared painstakingly collected field data with satellite images of the regions taken with seven visible-light and infrared wavelengths. Different combinations are useful for classifying different types of vegetation and land cover. The team noted whether each 30-meter-by-30-meter pixel in the images matched snail habitat, as determined by the field data.

As the data accumulates, computer programs may be able to recognize snail habitat—and thus, areas of high disease risk—in future images of the same or other areas of China, so health officials can pinpoint risky areas without sending researchers into the field.

As China undergoes major environmental changes, the technology will enable scientists to monitor potential snail habitat. Seto cites the example of the enormous Three Gorges Dam being constructed on the Yangtze River. To be completed in 2003, the dam will create a reservoir covering 395 square miles, alter the course of rivers and streams, and perhaps create large new tracts of snail habitat. "The impacts are huge," says Seto.

POSSUM POSSIBILITIES In New Zealand, veterinary researchers are using remote sensing to spy on a disease culprit that's larger and fluffier than snails or mosquitoes. The brushtail possum carries the bacterium responsible for bovine tuberculosis (TB).

Bovine TB—McKenzie and her colleagues used images from the French SPOT3 satellite to map out vegetation types to predict infected-possum hot spots.

Mycobacterium bovis, a close relative of the germ responsible for tuberculosis in people, preferentially afflicts cattle. It can, however, infect a person who drinks tainted milk, as it commonly did before pasteurization.

Bovine TB still strikes dairy and beef herds, causes major reductions in milk yields, and makes livestock unfit for export to many countries. The disease also continues to pose a threat to farm and slaughterhouse workers.

Eradicating the disease in cattle wouldn't be a problem if possums didn't complicate matters, says Joanna McKenzie, of Massey University in Palmerston North, New Zealand. These marsupials are also susceptible to *M. bovis*.

The brushtail possum is a relentless marauder that invaded New Zealand from Australia. Listed among the world's 100 most-invasive species, it is one of New Zealand's worst pests. It costs the government there $13 million a year in pest-control expenses.

The otherwise nocturnal mammal becomes disorientated during its dying throes from bovine TB, and it then wanders into pastures during daylight, says McKenzie. Inquisitive deer and cattle often come to investigate, sniffing and licking the afflicted animals and, in the process, getting a strong draught of TB bacteria.

The traditional approach to fighting bovine TB in possums is extensive culling of both the healthy and ill animals in wide areas of infection. McKenzie's group hopes to use remote sensing to focus efforts, which would reduce costs and bring hope of eradicating the disease from the country.

The researchers studied a 60-square-kilometer area of New Zealand's North Island. The location has rolling hills, thousands of cattle, and numerous pockets of bovine-TB-infected possums. McKenzie and her colleagues used images from the French SPOT3 satellite to map out vegetation types for each 20-square-meter piece of the overall area. Vegetation type determines food sources and suitable den sites for the possums, says McKenzie.

Because the animals prefer to make dens in large trees or steep slopes, the researchers also took into account topographic data indicating the slope of each pixel. Overlaid with a map of farms within the area, the data enabled McKenzie's team to predict infected possum hot spots.

To test the accuracy of the approach, the researchers and agricultural authorities focused annual possum-eradication campaigns on 30 farms that satellite data had identified as possum hot spots. From 1997 through 1999, the researchers directed intensive trapping and killing programs on those areas, and during the seasons of high risk, farmers kept their grazing cattle out of pastures where they would be likely to encounter sick possums.

After the 3-year trial, the incidence of TB in cattle in the test area was significantly lower than among livestock on a similar group of farms receiving traditional, less-targeted efforts to control possums.

HIGH-TECH INTELLIGENCE Researchers' interest in satellite mapping of infectious diseases has been building steadily, says Hay. "The new generation of satellite sensors is also continually opening up new opportunities," he adds.

One recent advance is the development of radar satellites. Traditionally, satellites rely on passive detection of reflected energy, such as visible light or infrared radiation. Radar satellites provide similar information but generate their own signals and then receive the reflections. So, they can pick up images night and day. The wavelengths that the radar satellites emit also enable them to see through clouds.

Researchers have been finding it tough to keep abreast of technological advances in recent years, with new satellites improving the scope and quality of data. At any given time, NASA has in store about 1 petabyte of data from all its Earth-observing satellites, says Venezia. This breathtaking quantity—one quadrillion bytes—is roughly equivalent to the amount of data stored in a mile-high stack of CD-ROMS.

Developing the human-health applications of all those data is more important today than at any other time in the last 30 years, says Durland Fish of the Yale University School of Medicine. Many "new vector-borne diseases are emerging… and we have very few people trained in how to handle them," he says.

"Thirty years ago we thought we were on top of infectious disease," says Hay. However, with pests' growing resistance to drugs and insecticides, many of the diseases have resurged. Increasingly mobile populations are escalating the problem, and now there is the additional threat of deliberate introduction of disease by terrorists. The numerous satellites encircling Earth may provide powerful medical intelligence.

From *Science News,* April 6, 2002, pp. 218-220 by John Pickrell. © 2001 by Science Service Inc. Reprinted by permission.

UNIT 7
Values and Visions

Unit Selections

Key Points to Consider

- Comment on the idea that it is naive to speak of global issues in terms of ethics.

- What role can governments, international organizations, and the individual play in making the world a more ethical place?

- How is the political role of women changing, and what impacts are these changes having on conflict resolution and community building?

- The consumption of resources is the foundation of the modern economic system. What are the values underlying this economic system, and how resistant to change are they?

- What are the characteristics of leadership?

- In addition to the ideas presented here, what other new ideas are being expressed, and how likely are they to be widely accepted?

 Links: www.dushkin.com/online/
These sites are annotated in the World Wide Web pages.

Human Rights Web
http://www.hrweb.org
InterAction
http://www.interaction.org

The final unit of this book considers how humanity's view of itself is changing. Values, like all other elements discussed in this anthology, are dynamic. Visionary people with new ideas can have a profound impact on how a society deals with problems and adapts to changing circumstances. Therefore, to understand the forces at work in the world today, values, visions, and new ideas in many ways are every bit as important as new technology or changing demographics.

Novelist Herman Wouk, in his book *War and Remembrance,* observed that many institutions have been so embedded in the social fabric of their time that people assumed that they were part of human nature. Slavery and human sacrifice are two examples. However, forward-thinking people opposed these institutions. Many knew that they would never see the abolition of these social systems within their own lifetimes, but they pressed on in the hope that someday these institutions would be eliminated.

Wouk believes the same is true for warfare. He states, "Either we are finished with war or war will finish us." Aspects of society such as warfare, slavery, racism, and the secondary status of women are creations of the human mind; history suggests that they can be changed by the human spirit.

The articles of this unit have been selected with the previous six units in mind. Each explores some aspect of world affairs from the perspective of values and alternative visions of the future.

New ideas are critical to meeting these challenges. The examination of well-known issues from new perspectives can yield new insights into old problems. It was feminist Susan B. Anthony who once remarked that "social change is never made by the masses, only by educated minorities." The redefinition of human values (which, by necessity, will accompany the successful confrontation of important global issues) is a task that few people take on willingly. Nevertheless, in order to deal with the dangers of nuclear war, overpopulation, and environmental degradation, educated people must take a broad view of history. This is going to require considerable effort and much personal sacrifice.

When people first begin to consider the magnitude of contemporary global problems, many often become disheartened and depressed. Some ask: What can I do? What does it matter? Who cares? There are no easy answers to these questions, but people need only look around to see good news as well as bad. How individuals react to the world is not solely a function of so-called objective reality but a reflection of themselves.

As stated at the beginning of the first unit, the study of global issues is the study of people. The study of people, furthermore, is the study of both values and the level of commitment supporting these values and beliefs.

It is one of the goals of this book to stimulate you, the reader, to react intellectually and emotionally to the discussion and description of various global challenges. In the process of studying these issues, hopefully you have had some new insights into your own values and commitments. In the presentation of the allegory of the balloon, the third color added represented the "meta" component, all of those qualities that make human beings unique. It is these qualities that have brought us to this "special moment in time," and it will be these same qualities that will determine the outcome of our historically unique challenges.

Are Human Rights Universal?

Shashi Tharoor

The growing consensus in the West that human rights are universal has been fiercely opposed by critics in other parts of the world. At the very least, the idea may well pose as many questions as it answers. Beyond the more general, philosophical question of whether anything in our pluri-cultural multipolar world is truly universal, the issue of whether human rights is an essentially Western concept—ignoring the very different cultural, economic, and political realities of the other parts of the world—cannot simply be dismissed. Can the values of the consumer society be applied to societies that have nothing to consume? Isn't talking about universal rights rather like saying that the rich and the poor both have the same right to fly first class and to sleep under bridges? Don't human rights as laid out in the international convenants ignore the traditions, the religions, and the socio-cultural patterns of what used to be called the Third World? And at the risk of sounding frivolous, when you stop a man in traditional dress from beating his wife, are you upholding her human rights or violating his?

This is anything but an abstract debate. To the contrary, ours is an era in which wars have been waged in the name of human rights, and in which many of the major developments in international law have presupposed the universality of the concept. By the same token, the perception that human rights as a universal discourse is increasingly serving as a flag of convenience for other, far more questionable political agendas, accounts for the degree to which the very idea of human rights is being questioned and resisted by both intellectuals and states. These objections need to be taken very seriously.

The philosophical objection asserts essentially that nothing can be universal; that all rights and values are defined and limited by cultural perceptions. If there is no universal culture, there can be no universal human rights. In fact, some philosophers have objected that the concept of human rights is founded on an anthropocentric, that is, a human-centered, view of the world, predicated upon an individualistic view of man as an autonomous being whose greatest need is to be free from interference by the state—free to enjoy what one Western writer summed up as the "right to private property, the right to freedom of contract, and the right to be left alone." But this view would seem to clash with the communitarian one propounded by other ideologies and cultures where society is conceived of as far more than the sum of its individual members.

Who Defines Human Rights?

Implicit in this is a series of broad, culturally grounded objections. Historically, in a number of non-Western cultures, individuals are not accorded rights in the same way as they are in the West. Critics of the universal idea of human rights contend that in the Confucian or Vedic traditions, duties are considered more important than rights, while in Africa it is the community that protects and nurtures the individual. One African writer summed up the African philosophy of existence as: "I am because we are, and because we are therefore I am." Some Africans have argued that they have a complex structure of communal entitlements and obligations grouped around what one might call four "r's": not "rights," but respect, restraint, responsibility, and reciprocity. They argue that in most African societies group rights have always taken precedence over individual rights, and political decisions have been made through group consensus, not through individual assertions of rights.

These cultural differences, to the extent that they are real, have practical implications. Many in developing countries argue that some human rights are simply not relevant to their societies—the right, for instance, to political pluralism, the right to paid vacations (always good for a laugh in the sweatshops of the Third World), and, inevitably, the rights of women. It is not just that some societies claim they are simply unable to provide certain rights to all their citizens, but rather that they see the "universal" conception of human rights as little more than an attempt to impose alien Western values on them.

Rights promoting the equality of the sexes are a contentious case in point. How, critics demand, can women's rights be universal in the face of widespread divergences of cultural practice, when in many societies, for example, marriage is not seen as a contract between two individuals but as an alliance between lineages, and when the permissible behavior of womenfolk is central to the society's perception of its honor?

And, inseparable from the issues of tradition, is the issue of religion. For religious critics of the universalist definition of human rights, nothing can be universal that is not founded on transcendent values, symbolized by God, and sanctioned by the guardians of the various faiths. They point out that the cardinal document of the contemporary human rights movement, the Universal Declaration of Human Rights, can claim no such heritage.

Recently, the fiftieth anniversary of the Universal Declaration was celebrated with much fanfare. But critics from countries that were still colonies in 1948 suggest that its provisions reflect the ethnocentric bias of the time. They go on to argue that the concept of human rights is really a cover for Western interventionism in the affairs of the developing world, and that "human rights" are merely an instrument of Western political neocolonialism. One critic in the 1970s wrote of his fear that "Human Rights might turn out to be a Trojan horse, surreptitiously introduced into other civilizations, which will then be obliged to accept those ways of living, thinking and feeling for which Human Rights is the proper solution in cases of conflict."

In practice, this argument tends to be as much about development as about civilizational integrity. Critics argue that the developing countries often cannot afford human rights, since the tasks of nation building, economic development, and the consolidation of the state structure to these ends are still unfinished. Authoritarianism, they argue, is more efficient in promoting development and economic growth. This is the premise behind the so-called Asian values case, which attributes the economic growth of Southeast Asia to the Confucian virtues of obedience, order, and respect for authority. The argument is even a little more subtle than that, because the suspension or limiting of human rights is also portrayed as the sacrifice of the few for the benefit of the many. The human rights concept is understood, applied, and argued over only, critics say, by a small Westernized minority in developing countries. Universality in these circumstances would be the universality of the privileged. Human rights is for the few who have the concerns of Westerners; it does not extend to the lowest rungs of the ladder.

The Case for the Defense

That is the case for the prosecution—the indictment of the assumption of the universality of human rights. There is, of course, a case for the defense. The philosophical objection is, perhaps surprisingly, the easiest to counter. After all, concepts of justice and law, the legitimacy of

government, the dignity of the individual, protection from oppressive or arbitrary rule, and participation in the affairs of the community are found in every society on the face of this earth. Far from being difficult to identify, the number of philosophical common denominators between different cultures and political traditions makes universalism anything but a distortion of reality.

Historically, a number of developing countries—notably India, China, Chile, Cuba, Lebanon, and Panama—played an active and highly influential part in the drafting of the Universal Declaration of Human Rights. In the case of the human rights covenants, in the 1960s the developing world actually made the decisive contribution; it was the "new majority" of the Third World states emerging from colonialism—particularly Ghana and Nigeria—that broke the logjam, ending the East–West stalemate that had held up adoption of the covenants for nearly two decades. The principles of human rights have been widely adopted, imitated, and ratified by developing countries; the fact that therefore they were devised by less than a third of the states now in existence is really irrelevant.

In reality, many of the current objections to the universality of human rights reflect a false opposition between the primacy of the individual and the paramountcy of society. Many of the civil and political rights protect groups, while many of the social and economic rights protect individuals. Thus, crucially, the two sets of rights, and the two covenants that codify them, are like Siamese twins—inseparable and interdependent, sustaining and nourishing each other.

Still, while the conflict between group rights and individual rights may not be inevitable, it would be naïve to pretend that conflict would never occur. But while groups may collectively exercise rights, the individuals within them should also be permitted the exercise of their rights within the group, rights that the group may not infringe upon.

A Hidden Agenda?

Those who champion the view that human rights are not universal frequently insist that their adversaries have hidden agendas. In fairness, the same accusation can be leveled against at least some of those who cite culture as a defense against human rights. Authoritarian regimes who appeal to their own cultural traditions are cheerfully willing to crush culture domestically when it suits them to do so. Also, the "traditional culture" that is sometimes advanced to justify the nonobservance of human rights, including in Africa, in practice no longer exists in a pure form at the national level anywhere. The societies of developing countries have not remained in a pristine, pre-Western state; all have been subject to change and distortion by external influence, both as a result of colonialism in many cases and through participation in modern interstate relations.

You cannot impose the model of a "modern" nation-state cutting across tribal boundaries and conventions on your country, appoint a president and an ambassador to the United Nations, and then argue that tribal traditions should be applied to judge the human rights conduct of the resulting modern state.

In any case, there should be nothing sacrosanct about culture. Culture is constantly evolving in any living society, responding to both internal and external stimuli, and there is much in every culture that societies quite naturally outgrow and reject. Am I, as an Indian, obliged to defend, in the name of my culture, the practice of suttee, which was banned 160 years ago, of obliging widows to immolate themselves on their husbands' funeral pyres? The fact that slavery was acceptable across the world for at least 2,000 years does not make it acceptable to us now; the deep historical roots of anti-Semitism in European culture cannot justify discrimination against Jews today.

The problem with the culture argument is that it subsumes all members of a society under a cultural framework that may in fact be inimical to them. It is one thing to advocate the cultural argument with an escape clause—that is, one that does not seek to coerce the dissenters but permits individuals to opt out and to assert their individual rights. Those who freely choose to live by and to be treated according to their traditional cultures are welcome to do so, provided others who wish to be free are not oppressed in the name of a culture they prefer to disavow.

A controversial but pertinent example of an approach that seeks to strengthen both cultural integrity and individual freedom is India's Muslim Women (Protection of Rights upon Divorce) Act. This piece of legislation was enacted following the famous Shah Banu case, in which the Supreme Court upheld the right of a divorced Muslim woman to alimony, prompting howls of outrage from Muslim traditionalists who claimed this violated their religious beliefs that divorced women were only entitled to the return of the bride price paid upon marriage. The Indian parliament then passed a law to override the court's judgment, under which Muslim women married under Muslim law would be obliged to accept the return of the bride price as the only payment of alimony, but that the official Muslim charity, the Waqf Board, would assist them.

Many Muslim women and feminists were outraged by this. But the interesting point is that if a Muslim woman does not want to be subject to the provisions of the act, she can marry under the civil code; if she marries under Muslim personal law, she will be subject to its provisions. That may be the kind of balance that can be struck between the rights of Muslims as a group to protect their traditional practices and the right of a particular Muslim woman, who may not choose to be subject to that particular law, to exempt herself from it.

It needs to be emphasized that the objections that are voiced to specific (allegedly Western) rights very frequently involve the rights of women, and are usually vociferously argued by men. Even conceding, for argument's sake, that child marriage, widow inheritance, female circumcision, and the like are not found reprehensible by many societies, how do the victims of these practices feel about them? How many teenage girls who have had their genitalia mutilated would have agreed to undergo circumcision if they had the human right to refuse to permit it? For me, the standard is simple: where coercion exists, rights are violated, and these violations must be condemned whatever the traditional justification. So it is not culture that is the test, it is coercion.

Not with Faith, But with the Faithful

Nor can religion be deployed to sanction the status quo. Every religion seeks to embody certain verities that are applicable to all mankind—justice, truth, mercy, compassion—though the details of their interpretation vary according to the historical and geographical context in which the religion originated. As U.N. secretary general Kofi Annan has often said, the problem is usually not with the faith, but with the faithful. In any case, freedom is not a value found only in Western faiths: it is highly prized in Buddhism and in different aspects of Hinduism and Islam.

If religion cannot be fairly used to sanction oppression, it should be equally obvious that authoritarianism promotes repression, not development. Development is about change, but repression prevents change. The Nobel Prize–winning economist Amartya Sen has pointed out in a number of interesting pieces that there is now a generally agreed-upon list of policies that are helpful to economic development— "openness to competition, the use of international markets, a high level of literacy and school education, successful land reforms, and public provision of incentives for investment, export and industrialization"—none of which requires authoritarianism; none is incompatible with human rights. Indeed, it is the availability of political and civil rights that gives people the opportunity to draw attention to their needs and to demand action from the government. Sen's work has established, for example, that no substantial famine has ever occurred in any independent and democratic country with a relatively free press. That is striking; though there may be cases where authoritarian societies have had success in achieving economic growth, a country like Botswana, an exemplar of democracy in Africa, has grown faster than most authoritarian states.

In any case, when one hears of the unsuitability or inapplicability or ethnocentrism of human rights, it is important to ask what the unstated assumptions of this view really are. What exactly are these human rights that it is so unreasonable to promote? If one picks up the more contentious covenant—the one on civil and political rights— and looks through the list, what can one find that someone in a developing country can easily do without? Not the right to life, one trusts. Freedom from torture? The right not to be enslaved, not to be physically assaulted, not to be arbitrarily arrested, imprisoned, executed? No one actually advocates in so many words the abridgement of any of these rights. As Kofi Annan asked at a speech in Tehran University in 1997: "When have you heard a free voice demand an end to freedom? Where have you heard a slave argue for slavery? When have you heard a victim of torture endorse the ways of the torturer? Where have you heard the tolerant cry out for intolerance?"

Tolerance and mercy have always, and in all cultures, been ideals of government rule and human behavior. If we do not unequivocally assert the universality of the rights that oppressive governments abuse, and if we admit that these rights can be diluted and changed, ultimately we risk giving oppressive governments an intellectual justification for the morally indefensible. Objections to the applicability of international human rights standards have all too frequently been voiced by authoritarian rulers and power elites to rationalize their violations of human rights—violations that serve primarily, if not solely, to sustain them in power. Just as the Devil can quote scripture for his purpose, Third World communitarianism can be the slogan of a deracinated tyrant trained, as in the case of Pol Pot, at the Sorbonne. The authentic voices of the Third World know how to cry out in pain. It is time to heed them.

The "Right to Development"

At the same time, particularly in a world in which market capitalism is triumphant, it is important to stress that the right to development is also a universal human right. The very concept of development evolved in tune with the concept of human rights; decolonization and self-determination advanced side by side with a consciousness of the need to improve the standards of living of subject peoples. The idea that human rights could be ensured merely by the state not interfering with individual freedom cannot survive confrontation with a billion hungry, deprived, illiterate, and jobless human beings around the globe. Human rights, in one memorable phrase, start with breakfast.

For the sake of the deprived, the notion of human rights has to be a positive, active one: not just protection from the state but also the protection of the state, to permit these human beings to fulfill the basic aspirations of growth and development that are frustrated by poverty and scarce resources. We have to accept that social deprivation and economic exploitation are just as evil as political oppression or racial persecution. This calls for a more profound approach to both human rights and to development. Without development, human rights could not be

truly universal, since universality must be predicated upon the most underprivileged in developing countries achieving empowerment. We can not exclude the poorest of the poor from the universality of the rich.

After all, do some societies have the right to deny human beings the opportunity to fulfill their aspirations for growth and fulfillment legally and in freedom, while other societies organize themselves in such a way as to permit and encourage human beings freely to fulfill the same needs? On what basis can we accept a double standard that says that an Australian's need to develop his own potential is a right, while an Angolan's or an Albanian's is a luxury?

Universality, Not Uniformity

But it is essential to recognize that universality does not presuppose uniformity. To assert the universality of human rights is not to suggest that our views of human rights transcend all possible philosophical, cultural, or religious differences or represent a magical aggregation of the world's ethical and philosophical systems. Rather, it is enough that they do not fundamentally contradict the ideals and aspirations of any society, and that they reflect our common universal humanity, from which no human being must be excluded.

Most basically, human rights derive from the mere fact of being human; they are not the gift of a particular government or legal code. But the standards being proclaimed internationally can become reality only when applied by countries within their own legal systems. The challenge is to work towards the "indigenization" of human rights, and their assertion within each country's traditions and history. If different approaches are welcomed within the established framework—if, in other words, eclecticism can be encouraged as part of the consensus and not be seen as a threat to it—this flexibility can guarantee universality, enrich the intellectual and philosophical debate, and so complement, rather than undermine, the concept of worldwide human rights. Paradoxical as it may seem, it is a universal idea of human rights that can in fact help make the world safe for diversity.

Note

This article was adapted from the first Mahbub-ul-Haq Memorial Lecture, South Asia Forum, October 1998.

Shashi Tharoor is Director of Communications and Special Projects in the Office of the Secretary General of the United Nations. The views expressed here are the author's own and do not necessarily reflect the positions of the United Nations.

The Grameen Bank

A small experiment begun in Bangladesh has turned into a major new concept in eradicating poverty

by Muhammad Yunus

Over many years, Amena Begum had become resigned to a life of grinding poverty and physical abuse. Her family was among the poorest in Bangladesh—one of thousands that own virtually nothing, surviving as squatters on desolate tracts of land and earning a living as day laborers.

In early 1993 Amena convinced her husband to move to the village of Kholshi, 112 kilometers (70 miles) west of Dhaka. She hoped the presence of a nearby relative would reduce the number and severity of the beatings that her husband inflicted on her. The abuse continued, however—until she joined the Grameen Bank. Oloka Ghosh, a neighbor, told Amena that Grameen was forming a new group in Kholshi and encouraged her to join. Amena doubted that anyone would want her in their group. But Oloka persisted with words of encouragement. "We're all poor—or at least we all were when we joined. I'll stick up for you because I know you'll succeed in business.

Amena's group joined a Grameen Bank Center in April 1993. When she received her first loan of $60, she used it to start her own business raising chickens and ducks. When she repaid her initial loan and began preparing a proposal for a second loan of $110, her friend Oloka gave her some sage advice: "Tell your husband that Grameen does not allow borrowers who are beaten by their spouses to remain members and take loans." From that day on, Amena suffered significantly less physical abuse at the hands of her husband. Today her business continues to grow and provide for the basic needs of her family.

Unlike Amena, the majority of people in Asia, Africa and Latin America have few opportunities to escape from poverty. According to the World Bank, more than 1.3 billion people live on less than a dollar a day. Poverty has not been eradicated in the 50 years since the Universal Declaration on Human Rights asserted that each individual has a right to:

> A standard of living adequate for the health and well-being of himself and of his family, including food, clothing, housing and medical care and

necessary social services, and the right to security in the event of unemployment, sickness, disability, widowhood, old age or other lack of livelihood in circumstances beyond his control.

Will poverty still be with us 50 years from now? My own experience suggests that it need not.

After completing my Ph.D. at Vanderbilt University, I returned to Bangladesh in 1972 to teach economics at Chittagong University. I was excited about the possibilities for my newly independent country. But in 1974 we were hit with a terrible famine. Faced with death and starvation outside my classroom, I began to question the very economic theories I was teaching. I started feeling there was a great distance between the actual life of poor and hungry people and the abstract world of economic theory.

I wanted to learn the real economics of the poor. Because Chittagong University is located in a rural area, it was easy for me to visit impoverished households in the neighboring village of Jobra. Over the course of many visits, I learned all about the lives of my struggling neighbors and much about economics that is never taught in the classroom. I was dismayed to see how the indigent in Jobra suffered because they could not come up with small amounts of working capital. Frequently they needed less than a dollar a person but could get that money only on extremely unfair terms. In most cases, people were required to sell their goods to moneylenders at prices fixed by the latter.

This daily tragedy moved me to action. With the help of my graduate students, I made a list of those who needed small amounts of money. We came up with 42 people. The total amount they needed was $27.

I was shocked. It was nothing for us to talk about millions of dollars in the classroom, but we were ignoring the minuscule capital needs of 42 hardworking, skilled people next door. From my own pocket, I lent $27 to those on my list.

Still, there were many others who could benefit from access to credit. I decided to approach the university's bank and try to persuade it to lend to the local poor. The branch manager said, however, that the bank could not give loans to the needy: the villagers, he argued, were not creditworthy.

I could not convince him otherwise. I met with higher officials in the banking hierarchy with similar results. Finally, I offered myself as a guarantor to get the loans.

In 1976 I took a loan from the local bank and distributed the money to poverty-stricken individuals in Jobra. Without exception, the villagers paid back their loans. Confronted with this evidence, the bank still refused to grant them loans directly. And so I tried my experiment in another village, and again it was successful. I kept expanding my work, from two to five, to 20, to 50, to 100 villages, all to convince the bankers that they should be lending to the poor. Although each time we expanded to a new village the loans were repaid, the bankers still would not change their view of those who had no collateral.

Because I could not change the banks, I decided to create a separate bank for the impoverished. After a great deal of work and negotiation with the government, the Grameen Bank ("village bank" in Bengali) was established in 1983.

From the outset, Grameen was built on principles that ran counter to the conventional wisdom of banking. We sought out the very poorest borrowers, and we required no collateral. The bank rests on the strength of its borrowers. They are required to join the bank in self-formed groups of five. The group members provide one another with peer support in the form of mutual assistance and advice. In addition, they allow for peer discipline by evaluating business viability and ensuring repayment. If one member fails to repay a loan, all members risk having their line of credit suspended or reduced.

The Power of Peers

Typically a new group submits loan proposals from two members, each requiring between $25 and $100. After these two borrowers successfully repay their first five weekly installments, the next two group members become eligible to apply for their own loans. Once they make five repayments, the final member of the group may apply. After 50 installments have been repaid, a borrower pays her interest, which is slightly above the commercial rate. The borrower is now eligible to apply for a larger loan.

The bank does not wait for borrowers to come to the bank; it brings the bank to the people. Loan payments are made in weekly meetings consisting of six to eight groups, held in the villages where the members live. Grameen staff attend these meetings and often visit individual borrowers' homes to see how the business—whether it be raising goats or growing vegetables or hawking utensils—is faring.

Today Grameen is established in nearly 39,000 villages in Bangladesh. It lends to approximately 2.4 million borrowers, 94 percent of whom are women. Grameen reached its first $1 billion in cumulative loans in March 1995, 18 years after it began in Jobra. It took only two more years to reach the $2-billion mark. After 20 years of work, Grameen's average loan size now stands at $180. The repayment rate hovers between 96 and 100 percent.

A year after joining the bank, a borrower becomes eligible to buy shares in Grameen. At present, 94 percent of the bank is owned by its borrowers. Of the 13 members of the board of directors, nine are elected from among the borrowers; the rest are government representatives, academics, myself and others.

A study carried out by Sydney R. Schuler of John Snow, Inc., a private research group, and her colleagues concluded that a Grameen loan empowers a woman by increasing her economic security and status within the family. In 1998 a study by Shahidur R. Khandker an economist with the World Bank, and others noted that participation in Grameen also has a significant positive effect on the schooling and nutrition of children—as long as women rather than men receive the loans. (Such a tendency was clear from the early days of the bank and is one reason Grameen lends primarily to women: all too often men spend the money on themselves.) In particular, a 10 percent increase in borrowing by women resulted in the arm circumference of girls—a common measure of nutritional status—expanding by 6 percent. And for every 10 percent increase in borrowing by a member the likelihood of her daughter being enrolled in school increased by almost 20 percent.

Not all the benefits derive directly from credit. When joining the bank, each member is required to memorize a list of 16 resolutions. These include commonsense items about hygiene and health—drinking clean water, growing and eating vegetables, digging and using a pit latrine, and so on—as well as social dictums such as refusing dowry and managing family size. The women usually recite the entire list at the weekly branch meetings, but the resolutions are not otherwise enforced.

Even so, Schuler's study revealed that women use contraception more consistently after joining the bank. Curiously, it appears that women who live in villages where Grameen operates, but who are not themselves members, are also more likely to adopt contraception. The population growth rate in Bangladesh has fallen dramatically in the past two decades, and it is possible that Grameen's influence has accelerated the trend.

In a typical year 5 percent of Grameen borrowers—representing 125,000 families—rise above the poverty level. Khandker concluded that among these borrowers extreme poverty (defined by consumption of less than 80 percent of the minimum requirement stipulated by the Food and Agriculture Organization of the United Nations) declined by more than 70 percent within five years of their joining the bank.

To be sure, making a microcredit program work well—so that it meets its social goals and also stays economi-

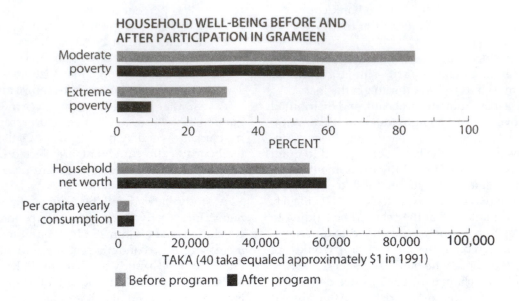

HOUSEHOLD WELL-BEING BEFORE AND
AFTER PARTICIPATION IN GRAMEEN

Before program   After program

cally sound—is not easy. We try to ensure that the bank serves the poorest: only those living at less than half the poverty line are eligible for loans. Mixing poor participants with those who are better off would lead to the latter dominating the groups. In practice, however, it can be hard to include the most abjectly poor, who might be excluded by their peers when the borrowing groups are being formed. And despite our best efforts, it does sometimes happen that the money lent to a woman is appropriated by her husband.

Given its size and spread, the Grameen Bank has had to evolve ways to monitor the performance of its branch managers and to guarantee honesty and transparency. A manager is not allowed to remain in the same village for long, for fear that he may develop local connections that impede his performance. Moreover, a manager is never posted near his home. Because of such constraints—and because managers are required to have university degrees—very few of them are women. As a result, Grameen has been accused of adhering to a paternalistic pattern. We are sensitive to this argument and are trying to change the situation by finding new ways to recruit women.

Grameen has also often been criticized for being not a charity but a profit-making institution. Yet that status, I am convinced, is essential to its viability. Last year a disastrous flood washed away the homes, cattle and most other belongings of hundreds of thousands of Grameen borrowers. We did not forgive the loans, although we did issue new ones, and give borrowers more time to repay. Writing off loans would banish accountability, a key factor in the bank's success.

Liberating Their Potential

The Grameen model has now been applied in 40 countries. The first replication, begun in Malaysia in 1986, currently serves 40,000 poor families; their repayment rate has consistently stayed near 100 percent. In Bolivia, microcredit has allowed women to make the transition from "food for work" programs to managing their own businesses. Within two years the majority of women in the program acquire enough credit history and financial skills to qualify for loans from mainstream banks. Similar success stories are coming in from programs in poor countries everywhere. These banks all target the most impoverished, lend to groups and usually lend primarily to women.

The Grameen Bank in Bangladesh has been economically self-sufficient since 1995. Similar institutions in other countries are slowly making their way toward self-reliance. A few small programs are also running in the U.S., such as in innercity Chicago. Unfortunately, because labor costs are much higher in the U.S. than in developing countries—which often have a large pool of educated unemployed who can serve as managers or accountants—the operations are more expensive there. As a result, the U.S. programs have had to be heavily subsidized.

In all, about 22 million poor people around the world now have access to small loans. Microcredit Summit, an institution based in Washington, D.C., serves as a resource center for the various regional microcredit institutions and organizes yearly conferences. Last year the attendees pledged to provide 100 million of the world's poorest families, especially their women, with credit by the year 2005. The campaign has grown to include more than 2,000 organizations, ranging from banks to religious institutions to nongovernmental organizations to United Nations agencies.

The standard scenario for economic development in a poor country calls for industrialization via investment. In this "topdown" view, creating opportunities for employment is the only way to end poverty. But for much of the developing world, increased employment exacerbates migration from the countryside to the cities and creates low-paying jobs in miserable conditions. I firmly believe that, instead, the eradication of poverty starts with people

IMPACT OF GRAMEEN ON NUTRITIONAL MEASURES OF CHILDREN

*Bars reflect changes accompanying a 10 percent increase in credit to women.

being able to control their own fates. It is not by creating jobs that we will save the poor but rather by providing them with the opportunity to realize their potential. Time and time again I have seen that the poor are poor not because they are lazy or untrained or illiterate but because they cannot keep the genuine returns on their labor.

Self-employment may be the only solution for such people, whom our economies refuse to hire and our taxpayers will not support. Microcredit views each person as a potential entrepreneur and turns on the tiny economic engines of a rejected portion of society. Once a large number of these engines start working, the stage can be set for enormous socioeconomic change.

Applying this philosophy, Grameen has established more than a dozen enterprises, often in partnership with other entrepreneurs. By assisting microborrowers and microsavers to take ownership of large enterprises and even infrastructure companies, we are trying to speed the process of overcoming poverty. Grameen Phone, for instance, is a cellular telephone company that aims to serve urban and rural Bangladesh. After a pilot study in 65 villages, Grameen Phone has taken a loan to extend its activities to all villages in which the bank is active. Some 50,000 women, many of whom have never seen a telephone or even an electric light, will become the providers of telephone service in their villages. Ultimately, they will become the owners of the company itself by buying its shares. Our latest innovation, Grameen Investments, allows U.S. individuals to support companies such as Grameen Phone while receiving interest on their investment. This is a significant step toward putting commercial funds to work to end poverty.

I believe it is the responsibility of any civilized society to ensure human dignity to all members and to offer each in-

dividual the best opportunity to reveal his or her creativity. Let us remember that poverty is not created by the poor but by the institutions and policies that we, the better off, have established. We can solve the problem not by means of the old concepts but by adopting radically new ones.

The Author

MUHAMMAD YUNUS, the founder and managing director of the Grameen Bank, was born in Bangladesh. He obtained a Ph.D. in economics from Vanderbilt University in 1970 and soon after returned to his home country to teach at Chittagong University. In 1976 he started the Grameen project, to which he has devoted all his time for the past decade. He has served on many advisory committees: for the government of Bangladesh, the United Nations, and other bodies concerned with poverty, women and health. He has received the World Food Prize, the Ramon Magsaysay Award, the Humanitarian Award, the Man for Peace Award and numerous other distinctions as well as six honorary degrees.

Further Reading

GRAMEEN BANK: PERFORMANCE AND SUSTAINABILITY. Shahidur R. Khandker, Baqui Khalily and Zahed Khan. World Bank Discussion Papers, No. 306. ISBN 0-8213-3463-8. World Bank, 1995.

GIVE US CREDIT. Alex Counts. Times Books (Random House), 1996.

FIGHTING POVERTY WITH MICROCREDIT: EXPERIENCE IN BANGLADESH. Shahidur R. Khandker. Oxford University Press, 1998.

Grameen Bank site is available at www.grameenfoundation.org on the World Wide Web.

Why Environmental Ethics Matters to International Relations

"Environmental ethics [should] not be seen as an add-on to be approached after the important issues of security and economics have been settled. Instead, we [should] recognize that all our important social choices are inherently about the 'natural' world we create."

JOHN BARKDULL

What challenge does environmental ethics pose for international relations? International relations is usually understood as the realm of power politics, a world in which military might and the quest to survive dominate. In this world, moral concern for other human beings, much less nature, is limited or entirely lacking. Environmental ethics—a set of principles to guide human interaction with the earth—calls on us to extend moral consideration beyond humans to other living things and to natural "wholes" such as bioregions and ecosystems. Is it possible to introduce environmental ethics' far-reaching moral claims into the competitive, militarized, economically unequal world political system?

Although explorations in environmental ethics now have a long resumé, the dialogue over the human debt to the natural environment has proceeded largely without reference to international politics, to international relations theory, or even to the literature on international ethics. Practical politics is thus often removed from consideration. And scholars of international relations have barely considered the relationship between their studies and environmental ethics.

Bringing the two fields into the same conversation is possible. International political theory has profound implications for understanding how humans ought to relate to the environment. Realism and liberal institutionalism (the mainstream of international relations theory), by suggesting what political, economic, and social goals are desirable, also imply what environmental values should prevail. They indicate what kind of world humans should or can create and thus tell us how we should relate to the environment. The question then is not whether environmental ethics should matter in world politics, but in which way: which environmental ethic does in fact matter, which should, and what obstacles prevent needed changes in political practices from being made?

WHICH ENVIRONMENTAL ETHICS?

Environmental ethics can be anthropocentric, biocentric, or ecocentric. Anthropocentric ethics is about what humans owe each other. It evaluates environmental policies with regard to how they affect human well-being. For example, exploitation of natural resources such as minerals can destroy forests on which indigenous peoples depend. Moral evaluation of the environmental destruction proceeds in terms of the rights, happiness, or just treatment of all human parties, including the displaced tribes and the consumers who benefit from the minerals. Anthropocentric environmental ethics generally calls for more environmental protection than we now undertake; current unsustainable resource-use patterns and conversion of land to agricultural or urban uses mean that existing practices do more harm to humans than good, especially when future generations are considered. Still, many observers find anthropocentric environmental ethics unsatisfactory because it appears not to recognize other creatures' inherent right to share the planet and considers only their value to human beings.

Biocentric environmental ethics seeks to correct this deficiency by according moral standing to non-human creatures. Humans have moral worth but only as one species among many living things that also have moral standing. The grizzly bear's right to sufficient domain for sustaining life and reproduction has as much moral weight (if not more) as a logging company's desire to make a profit in that domain. Even if maintaining the grizzly bear's habitat means some humans must live in somewhat less spacious homes, the loss of human utility by no means cancels the animal's moral claim to the forest. In short, animals have rights. Which animals have moral standing and whether plants do as well remain matters of dispute among biocentric theorists. Nonetheless, biocentric theory expands the moral realm beyond humans and hence implies greater moral obligations than anthropocentric ethics.[1]

Ecocentric theory tackles a problem at the heart of biocentric theory. In reality, ecosystems work on the principle of eat and be eaten. We may accord the grizzly "rights" but the bear survives by consuming salmon, rodents, and so forth, thus violating other living creatures' right to life. Humans are simply part of a complex food chain or web of life. Given this, ecocentric theory asserts that moral status should attach to ecological wholes, from bioregions to the planetary ecosystem (sometimes called Gaia). Ecocentric theorists are not concerned about particular animals or even species, but with the entire evolutionary process. Evolution involves the "land" broadly understood to include all its organic and nonorganic components. To disrupt or destroy the evolutionary process, reducing the diversity of life and the stability and beauty of the natural system, is unethical. As Aldo Leopold, the environmental philosopher who first developed the land ethic, put it in his 1949 book, *A Sand County Almanac and Sketches Here and There*, "A thing is right when it tends to preserve the integrity, stability, and beauty of the biotic community. It is wrong when it tends otherwise." The emphasis here is on the word "community."

Each of these approaches suggests the need for change in the practice of international politics. Anthropocentric environmental ethics implies the least extensive reform, although these still could be far-reaching, especially with regard to current economic arrangements. Developed industrial economies rely heavily on the global commons for "free" natural services, such as areas to dispose of pollutants. For example, reliance on fossil fuels leads to increases in CO_2 in the atmosphere, and in turn to global warming. Developing countries undergoing industrialization will draw on the atmosphere's capacity to absorb greenhouse gases. The added load, along with already high levels of emissions from developed countries, could push the environment beyond a critical threshold, setting off catastrophic climate changes because of global warming. These climatic upheavals could lead to crop failures and destructive storms battering coastal cities. What is fair under these circumstances? Should developed countries make radical changes—such as decentralizing and deindustrializing—in their economic arrangements? Should they refrain from adding the potentially disastrous increment of greenhouse gases that will push the climate over the threshold of climatic catastrophe? If yes, then anthropocentric environmental ethics calls for far-reaching social and economic reform.

Biocentric environmental ethics also implies considerable economic reform. If animals have moral standing, then killing them or destroying their habitat for human benefit is unacceptable. In particular, the massive species loss resulting from deforestation is a moral failure even if humans profit. Likewise, agricultural practices that rely on pesticides and fertilizers that harm nonhuman species should be curtailed. Warfare's effects on nonhuman living things would also need to be evaluated. Just-war theory generally evaluates collateral damage's significance in the context of civilians killed or injured due to military operations. Yet collateral damage also kills and injures animals that have even less stake and less say in the conflict than civilians. Should their right to life be considered? Biocentric ethics would say yes. If so, virtually the entire practice of modern war might be held as inherently immoral.

Ecocentric ethics implies the strongest critique of current practices. Disrupting the ecological cycle or the evolutionary process is morally unacceptable. Most current economic or military practices would not pass muster. Indeed, in its strong form, ecocentric ethics would require a major reduction in the human population, since the 6 billion people now on earth are already disrupting the evolutionary process and will continue to do so as world population grows to 10 billion or more. Political institutions must be replaced, either with one-world government capable of implementing ecocentric environmental policy, or with ecologically based bioregional political units (ecocentric theorists hold differing views on whether authoritarian government or more democracy is needed to make ecocentrism effective in practice). If bioregionalism were adopted, world trade would come to a halt since each bioregion would be self-sustaining. Wasteful resource use would be curtailed. Long-term sustainability in harmony with the needs of other living things would be the desired end. For some ecocentric thinkers, the model is a hunter-gatherer society or a peasant agriculture society.

The gap between what environmental ethics calls for and what international political theory postulates may find its bridge in the land ethic.

Environmental ethics in each form carries important implications for the practice of international politics. Yet the environmental ethics literature usually pays little attention to obvious features of the international system. This is not to say that environmental ethics bears no relationship to political realities. If realism (the theory of power politics) and liberal institutionalism (the theory emphasizing interdependence and the possibilities for cooperation) both contain implicit environmental ethics, then environmental ethics contains implicit political theory. Yet without explicit attention to international political theory, environmental ethics lacks the basis to determine which of its recommendations is feasible, and which utopian.

BRIDGING THE GAP

The gap between what environmental ethics calls for and what international political theory postulates may find its bridge in the land ethic. The land ethic, as formulated by J. Baird Callicott, recognizes that environmental obligations are only part of our moral world.[2] Although the land ethic implies significant change in existing practices, it does not necessarily call for abandoning the sovereign state, relinquishing national identity or authority to a world government, or even abolishing capitalism. Rather, it asks for balance between human needs and the requisites of preserving the diversity of life flowing from the evolutionary process. The land ethic simply states that which enhances the integrity, stability, and beauty of the land is good, and that which does not is bad. Human intervention can serve good purposes by this standard. (Indeed, Aldo Leopold was himself a hunter, and found no contradiction between that pursuit and his commitment to the land ethic.) Presumably, human-induced changes to the landscape must be evaluated in context, assessing positive and negative effects.

Yet if the land ethic is to provide guidance for international politics, it needs to identify the other values that must be balanced with its requirements. In international politics, these can be determined in terms of mainstream international political theory, realism and liberal institutionalism. Realism and liberal insitutionalism capture much about how the international system works and what values shape international political practices. Thus we can observe the world for clues as to what international theory entails for the kind of world we should create. Although the verdict is not positive for existing practices, this does not foreclose the possibility of change within either paradigm. But how specifically does realism and liberal institutionalism see the relationship of humans to nature?

REALISM AND THE ENVIRONMENT

Realism is generally understood to be amoral. States do as they must to survive. Survival can justify breaking agreements, lying, deception, violence, and theft. Those who fail to play the game disappear. Those who are best at the game dominate the others. Morality, when invoked, is usually a cover for state interests. Certainly, some prominent realists have said otherwise. Hans Morgenthau recognized the moral content of foreign policy, as did E. H. Carr and Reinhold Niebuhr. Nonetheless, realists usually observe the human capacity for "evil" when the stakes are high.

But this negative perspective on morality obscures realism's highly moral claims. Realism asserts that humans naturally form groups, which experience conflicts of interest because resources are scarce. Maintaining the group's autonomy and freedom is the highest good. On it depends the ability of a people to work out their destiny within the borders of the state. Implicitly, this moral project justifies the extreme measures states undertake. Environmental ethics must recognize this as an extremely powerful moral claim. At the same time, international relations theory must recognize that staking this claim, which superficially appears to be a social question, implies a view of how humans should relate to the natural world.

Realism assumes that the state system, or at least some form of power politics involving contending groups, will characterize human relations as long as humans inhabit the planet. The possibilities for environmental (or any other) ethics are limited by this evidently permanent institutional arrangement. Virtually every state action must be evaluated in terms of the relative gains it offers with other states. The struggle for survival and dominance is an endless game in which any minor advantage today could have profound consequences tomorrow. Moreover, as Machiavelli observed, chance plays such a large role in human affairs that immediate advantage is all the prudent policymaker can consider. Thus, to think about long-term environmental trends, for example, is impractical, because an actor that sacrifices present advantage for future gains may not be around to enjoy the fruits.

Perhaps the most significant implication of realism lies in its emphasis on military security. Military imperatives dictate that states develop and deploy the most effective military technology available. The effects on the land of the particular choices made are little considered. No military technology could be more environmentally damaging than nuclear weapons, but these weapons confer maximum national power. Thus the environmental effects of producing, storing, deploying, and dismantling them (not to mention the effects they would have on the environment if ever used) are considered secondary. Here we see how realism as an international political theory is at the same time an implicit environmental ethic: land has little or no moral worth. This is a choice about how humans are to relate to the natural world, not only a choice about how states (or humans) are to relate to one another. A similar argument could be made about the entire range of military technologies, from cluster bombs to napalm to defoliants to biological weapons.

According to realism, the economic institutions of a society must support the most effective military establishment. Societies that attempt to structure economic relations along other lines, such as long-term sustainability, will soon find themselves overwhelmed by other states that make choices geared toward military dominance. States that wish to survive will emulate the most successful economic systems of other states and constantly seek economic innovations that will give them the edge. Capitalism as practiced in the United States and other major Western nations seems to be most compatible with military preponderance.

Realism thus implicitly endorses capitalism, albeit only because it is the most successful economic system at present for enhancing national power (as the former Soviet Union discovered). Capitalism has put the United States at the top of the international order. Others fail to emulate the United States at their peril. Moreover, realism would suggest that because economic growth facilitates military preparedness, autonomous economic growth should override other goals, including environmental protection. Saving a wetland will not contribute as much to national security as producing goods for export. Consequently, realism's emphasis on security leads to embracing the market in its most environmentally heedless form.

Realism's attitude toward the land is that it is territory, an asset of the state, a form of property. The land's status is as a mere resource with no moral standing apart from human uses. It has no life of its own. Prudent management is the most that is morally required. Hence realism shares the modern notion of nature as a spiritless "other" that humans can rightfully manipulate to serve their own ends.

We see that realism's strong moral claim—that a people's right to determine their own destiny, to define and develop their own idea of freedom and the good society, without interference from others—contradicts the institutional arrangements that realism produces. In practice, few alternatives are available. A people who decide that their destiny is to live in harmony with the land, to follow the land ethic's central precept, would quickly lose the freedom to do so through conquest and domination by other states. Aside from the evident fact that many environmental problems require cooperation across political jurisdictions, the competitiveness of the state system ensures that environmental consciousness will not long guide state policy.[3] Realism thus implies an environmental ethic; unfortunately, it is a most pernicious one. Equally unfortunate for environmental ethics is that realism is an undoubtedly incomplete, but not inaccurate, description of how the international system works.

Yet realism's historical and social argument rests on the moral claim that the group is the highest value: that it is within the group that some conception of the good society can be pursued. But surely a good society is one that fosters environmental values, a goal that suffers when nations pursue national security at all costs. As environmental crises mount, the contradiction at the heart of realist ethics becomes more obvious. Perhaps this can lead to changed conceptions of morality.

Ethical standards change. Nationalism, which underpins today's state system, has not driven human behavior for all history (nationalism, for example, had little influence in feudal Europe). Humans can change their view of how best to pursue a vision of the good society. To the extent that the land ethic becomes part of the moral dialogue, institutional change to bring about a healthier human-environment relationship is possible. Nonetheless, realism reminds us that the road to ethical change toward a land ethic likely will be long and hard.

LIBERAL INSTITUTIONALISM AND THE ENVIRONMENT

Liberal institutionalism is far more ready to accept that universal values such as respect for the land exist. Unlike realism, the liberal perspective considers human rights standards to apply across boundaries and cultures. Individuals are the moral agents and moral objects of liberal thought. Individuals have rights that exist regardless of their cultural heritage.

Furthermore, liberalism asserts that these individuals have a particular character. Partly self-interested and partly altruistic, individuals are aware of their dependence on collective action to obtain the good life. The liberal individual also acts, or should, through enlightened self-interest, which is the best way to secure the means of life and protection against bodily harm. Liberal individuals are predisposed to make certain choices. But would they choose a different way of life, namely, one more in harmony with the land?

The question becomes pertinent because it is not at all clear that liberal society is sustainable. Liberalism as manifested in practice is strongly committed to the market system. Indeed, the entire point of liberal institutional international political economy is to find the means to open the world economy to free trade and investment. From this perspective, environmental problems become unintended side-effects of otherwise desirable industrialization and economic growth. The problem for liberalism is simply managing these unfortunate consequences in ways that maintain the open economy. But as the modern market system encompasses more of the globe and penetrates deeper into social life, profound social choices occur, the result of the incremental effects of countless discrete, uncoordinated individual actions. Liberals are comfortable with this way of making social choices due to their faith in progress; the mounting ecological catastrophe might speak against this optimistic view.

Liberal institutionalism cannot escape its entanglement with and commitment to the capitalist market system. In effect, this means that liberal institutionalism can only with difficulty critique that system as it has developed in history. Hence liberal institutionalism will continue to see normal diplomacy and statecraft, the operations of multinational corporations, the growth of free trade and investment, and rising interdependence as progress toward a better world. This in turn exhibits liberal institutionalism's environmental ethic: managerial, limited to mitigation of the market's worst effects, and committed to economic growth and development. The world we should build is on display. It is embodied in the more enlightened liberal states, those that combine commitment to individual liberty, representative democracy, and free enterprise with some degree of environmental awareness. It is industrialized, or postindustrial. It is technologically advanced. It provides a wide range of goods and services to consumers. Environmental concerns enter by way of interest groups devoted to the "issue" rather than as fundamental values that determine which practices to retain and which to abandon.

Like realism, liberal institutionalism captures a large part of the truth about how contemporary international politics operates. It suggests emphasizing certain trends in the hope of dampening others; strengthening the forces for globalization to reduce the impact of military competition. But its commitment to the predominant global economic institutions leaves little room for a land ethic. Liberal institutionalism's anthropocentrism and consequent emphasis on economic growth leads to a relative lack of concern for the stability, integrity, and beauty of the land. Nonetheless, liberal institutionalism is far more open to the possibility of value change and political transformation than is realism. Liberal theory's faith in progress can imply that liberalism itself eventually will be transcended in favor of more earth-centered ethics. Yet current liberal international theory does not recognize or embrace this possibility. To the extent that theory is practice, liberal international theory contributes to the worsening environmental crisis rather than offering a way out.

A NEW DIALOGUE

Both realism and liberal institutionalism are implicitly environmental ethics. They tell us the relationship humans should have with nature, even if they largely base their claims not on ethical choice but on what we must do under existing circumstances. But humans can make conscious choices about what kind of international order to create and maintain. The realist imperative to play the game of power politics or be eliminated from the system depends on a prior choice about ethics and practice. It precludes the possibility of collaboratively engaging the "other" in democratic dialogue aimed at discovering different social practices that do not, for example, lead to environmentally heedless arms races. The "other" must always remain other in realist thought, an assumption that is far from proven. Likewise, liberalism's imperative to rely on the market if we are to achieve individual liberty and social progress is open to question. If the individual is constituted in community—that is, by social practices—then the self-definition of the community can

change. Acquisitive individualism and consumerism need not define the individuals.

How is change to come about? More authentic democracy, based on unforced, open discourse, affords the opportunity to choose consciously the kind of world we are to build. The choice need not come about indirectly, as the result of more immediate decisions on how to achieve national security, nor need it occur unintentionally as individuals make the best of circumstances not of their choosing. Engaging in this dialogue will require abandoning the notion that nature and the social are distinct. The social and the natural are inextricable. We constitute nature through our practices at the same time that we constitute the social world. Thus an ecologically informed political discourse will be one that recognizes the environmental ethics embedded in all political worldviews.

Environmental ethics will not be seen as an add-on to be approached after the important issues of security and economics have been settled. Instead, we will recognize that all our important social choices are inherently about the "natural" world we create. We will consciously raise the question of what this particular action means for that world, and we will recognize that it is our responsibility, not something external to us.

NOTES

1. For more on anthropocentric and biocentric ethics, see J. Baird Callicott, *In Defense of the Land Ethic: Essays for Environmental Philosophy* (Albany: State University of New York Press, 1989).
2. See Callicott, *In Defense of the Land Ethic.*
3. This competition also influences the abilities of states to cooperate to deal with transnational environmental problems. States are expected to attempt to free ride or otherwise exploit the global "commons." If environmental cooperation occurs, it is likely due to a hegemonic power or small group of large powers imposing an international regime. Of course, the Hobbesian use of power to make and enforce law is the antithesis of democratic decision making (which could well be a major element of the society's vision of the good life). But because states are and must be short-sighted and self-interested, no alternative to coercive imposition of regimes exists. Whether such regimes would conform to the requisites of long-term environmental sustainability—much less to the integrity, beauty, and stability of the land—is doubtful.

JOHN BARKDULL *is an associate professor of political science at Texas Tech University. His research interests include international political theory, international ethics, and environmental policy.*

Exploring the "Singularity"

The point in time when current trends may go wildly off the charts—known as the "Singularity"—is now getting serious attention. What it suggests is that technological change will soon become so rapid that we cannot possibly envision its results.

by James John Bell

Technological change isn't just happening fast. It's happening at an exponential rate. Contrary to the commonsense, intuitive, linear view, we won't just experience 100 years of progress in the twenty-first century—it will be more like 20,000 years of progress.

The near-future results of exponential technological growth will be staggering: the merging of biological and nonbiological entities in biorobotics, plants and animals engineered to grow pharmaceutical drugs, software-based "life," smart robots, and atom-sized machines that self-replicate like living matter. Some individuals are even warning that we could lose control of this expanding techno-cornucopia and cause the total extinction of life as we know it. Others are researching how this permanent technological overdrive will affect us. They're trying to understand what this new world of ours will look like and how accelerating technology already impacts us.

A number of scientists believe machine intelligence will surpass human intelligence within a few decades, leading to what's come to be called the Singularity. Author and inventor Ray Kurzweil defines this phenomenon as "technological change so rapid and profound it could create a rupture in the very fabric of human history."

Singularity is technically a mathematical term, perhaps best described as akin to what happens on world maps in a standard atlas. Everything appears correct until we look at regions very close to the poles. In the standard Mercator projection, the poles appear not as points but as a straight line. Each line is a singularity: Everywhere along the top line contains the exact point

of the North Pole, and the bottom line is the entire South Pole.

The singularity on the edge of the map is nothing compared to the singularity at the center of a black hole. Here one finds the astrophysicist's singularity, a rift in the continuum of space and time where Einstein's rules no longer function. The approaching technological Singularity, like the singularities of black holes, marks a point of departure from reality. Explorers once wrote "Beyond here be dragons" on the edges of old maps of the known world, and the image of life as we approach these edges of change are proving to be just as mysterious, dangerous, and controversial.

There is no concise definition for the Singularity. Kurzweil and many transhumanists define it as "a future time when societal, scientific, and economic change is so fast we cannot even imagine what will happen from our present perspective." A range of dates is given for the advent of the Singularity. "I'd be surprised if it happened before 2004 or after 2030," writes author and computer science professor Vernor Vinge. A distinctive feature will be that machine intelligence will have exceeded and even merged with human intelligence. Another definition is used by extropians, who say it denotes "the singular time when technological development will be at its fastest." From an environmental perspective, the Singularity can be thought of as the point at which technology and nature become one. Whatever perspective one takes, at this juncture the world as we have known it will become extinct, and new definitions of *life, nature, and human* will take hold.

Many leading technology industries have been aware of the possibility of a Singularity for some time. There are concerns that, if the public understood its ramifications, they might panic over accepting new and untested technologies that bring us closer to Singularity. For now, the debate about the consequences of the Singularity has stayed within the halls of business and technology; the kinks are being worked out, avoiding "doomsday" hysteria. At this time, it appears to matter little if the Singularity ever truly comes to pass.

"The true believers call themselves extropians, posthumans, and transhumanists, and are actively organizing not just to bring the Singularity about, but to counter the technophobes and neo-Luddites."

What Will Singularity Look Like?

Kurzweil explains that central to the workings of the Singularity are a number of "laws," one of which is Moore's law. Intel cofounder Gordon E. Moore noted that the number of transistors that could fit on a single computer chip had doubled every year for six years from the beginnings of integrated circuits in 1959. Moore predicted that the trend would continue, and it has—although the doubling rate was later adjusted to an 18-month cycle.

Today, the smallest transistors in chips span only thousands of atoms (hundreds of nanometers). Chipmakers build such components using a process in which they apply semiconducting, metallic, and insulating layers to a semiconductor wafer to create microscopic circuitry. They accomplish the procedure using light for imprinting patterns onto the wafer. In order to keep Moore's law moving right along, researchers today have built circuits out of transistors, wires, and other components as tiny as a few atoms across that can carry out simple computations.

Kurzweil and Sun Microsystems' chief scientist Bill Joy agree that, circa 2030, the technology of the 1999 film *The Matrix* (which visualized a three-dimensional interface between humans and computers, calling conventional reality into question) will be within our grasp and that humanity will be teetering on the edge of the Singularity. (See their essays in *Taking the Red Pill: Science, Philosophy, and Religion in The Matrix*, edited by Glenn Yeffeth, 2003.) Kurzweil explains that this will become possible because Moore's law will be replaced by another computing paradigm over the next few decades. "Moore's law was not the first but the fifth paradigm to provide exponential growth of computing power," Kurzweil says. The first paradigm of computer technology was the data processing machinery used in the 1890 American census. This electromechanical computing technology was followed by the paradigms of relay-based technology, vacuum tubes, transistors, and eventually integrated circuits. "Every time a paradigm ran out of steam," states Kurzweil, "another paradigm came along and picked up where that paradigm left off." The sixth paradigm, the one that will enable technology á la *The Matrix*, will be here in 20 to 30 years. "It's obvious what the sixth paradigm will be—computing in three dimensions," says Kurzweil. "We will effectively merge with our technology."

Stewart Brand in his book *The Clock of the Long Now* discusses the Singularity and another related law, Monsanto's law, which states that the ability to identify and use genetic information doubles every 12 to 24 months. This exponential growth in biological knowledge is transforming agriculture, nutrition, and health care in the emerging life-sciences industry.

A field of research building on the exponential growth rate of biotechnology is nanotechnology—the science of building machines out of atoms. A nanometer is atomic in scale, a distance that's 0.001% of the width of human hair. One goal of this science is to change the atomic fabric of matter—to engineer machinelike atomic structures that reproduce like living matter. In this respect, it is similar to biotechnol-

ogy, except that nanotechnology needs to literally create something like an inorganic version of DNA to drive the building of its tiny machines. "We're working out the rules of biology in a realm where nature hasn't had the opportunity to work," states University of Texas biochemistry professor Angela Belcher. "What would take millions of years to evolve on its own takes about three weeks on the bench top."

Machine progress is knocking down the barriers between all the sciences. Chemists, biologists, engineers, and physicists are now finding themselves collaborating on more and more experimental research. This collaboration is best illustrated by the opening of Cornell University's Nanobiotechnology Center and other such facilities around the world. These scientists predict breakthroughs soon that will open the way to molecular-size computing and the quantum computer, creating new scientific paradigms where exponential technological progress will leap off the map. Those who have done the exponential math quickly realize the possibilities in numerous industries and scientific fields—and then they notice the anomaly of the Singularity happening within this century.

In 2005, IBM plans to introduce Blue Gene, a supercomputer that can perform at about 5% of the power of the human brain. This computer could transmit the entire contents of the Library of Congress in less than two seconds. Blue Gene/L, specifically developed to advance and serve the growing life-sciences industry, is expected to operate at about 200 teraflops (200 trillion floating-point operations per second), larger than the total computing power of the top 500 supercomputers in the world. It will be able to run extremely complex simulations, including breakthroughs in computers and information technology, creating new frontiers in biology, says IBM's Paul M. Horn. According to Moore's law, computer hardware will surpass human brainpower in the first decade of this century. Software that emulates the human mind—artificial intelligence—may take another decade to evolve.

XenoMouse, patented in 2002 by Abgenix in Fremont, California, was genetically engineered to have a human immune system. Biotechnology breaks the natural boundaries existing between species. Discussion on the consequences of such breakthroughs, at least in the United States, remains on the sidelines, says author Bell.

Nanotech Advances Promote Singularity

Physicists, mathematicians, and scientists like Vinge and Kurzweil have identified through their research the likely boundaries of the Singularity and have predicted with confidence various paths leading up to it over the next couple of decades. These scientists are currently debating what discovery could set off a chain reaction of Earth-altering technological events. They suggest that advancements in the fields of nanotechnology or the discovery of artificial intelligence could usher in the Singularity.

The majority of people closest to these theories and laws—the tech sector—can hardly wait for these technologies to arrive. The true believers call themselves extropians, posthumans, and transhumanists, and are actively organizing not just to bring the Singularity about, but to counter the technophobes and neo-Luddites who believe that unchecked technological progress will exceed our ability to reverse any destructive process that might unintentionally be set in motion.

The antithesis to neo-Luddite activists is the extropians. For example, the Progress Action Coalition, formed in 2001 by bio-artist, author, and extropian activist Natasha Vita-More, fantasizes about "the dream of true artificial intelligence…adding a new richness to the human landscape never before known." Pro-Act, AgBioworld, Biotechnology Progress, Foresight Institute, the Progress and Freedom Foundation, and other industry groups acknowledge, however, that the greatest threat to technological progress comes not just from environmental groups, but from a small faction of the scientific community.

Knowledge-Enabled Mass Destruction

In April 2000, a wrench was thrown into the arrival of the Singularity by an unlikely source: Sun Microsystems chief scientist Bill Joy. He is a neo-Luddite without being a Luddite, a technologist warning the world about technology. Joy co-founded Sun Microsystems, helped create the Unix computer operating system, and developed the Java and Jini software systems—systems that helped give the Internet "life."

In a now-infamous cover story in *Wired* magazine, "Why the Future Doesn't Need Us," Joy warned of the dangers posed by

Keys to Understanding the Singularity

Singularity is the postulated point in our future when human evolutionary development-powered by such developments as nanotechnology, neuroscience, and artificial intelligence—accelerates enormously so that nothing beyond that time can reliably be conceived. Typical developments include the merging of man and machine (cybernetic organisms-or cyborgs) and accelerated technology beyond our ability to control.

Nanotechnology is the development and use of devices that have a size of only a few nanometers, including building and manipulating complex structures on an atomic scale. As we approach the Singularity, nanodevices will be able to replicate themselves like living matter.

Biorobotics is the merging of living organisms with technologies. At a simple level, this includes implanting chips encoded with health or security information. Biorobotics also encompasses the development of cyborgs that seamlessly blend living tissue with mechanical devices.

Cloning is the growing of genetically identical cells, eliminating the natural role of human biology and bringing us closer to the Singularity.

Extropians await the Singularity, seeking to overcome human limits, live indefinitely long, and become more intelligent through technology. Related groups include transhumanists and posthumanists.

Neo-Luddites oppose the impending Singularity by raising questions about moral and ethical aspects of modern technology and the threat it may pose to humanity.

---

developments in genetics, nanotechnology, and robotics. Joy's warning of the impacts of exponential technological progress run amok gave new credence to the coming Singularity. Unless things change, Joy predicted, "We could be the last generation of humans." Joy warned that "knowledge alone will enable mass destruction" and termed this phenomenon "knowledge-enabled mass destruction."

The twentieth century gave rise to nuclear, biological, and chemical (NBC) technologies that, while powerful, require access to vast amounts of raw (and often rare) materials, technical information, and large-scale industries. The twenty-first-century technologies of genetics, nanotechnology, and robotics (GNR), however, will require neither large facilities nor rare raw materials.

The threat posed by GNR technologies becomes further amplified by the fact that some of these new technologies have been designed to be able to replicate—i.e., they can build new versions of themselves. Nuclear bombs did not sprout more bombs, and toxic spills did not grow more spills. If the new selfreplicating GNR technologies are released into the environment, they could be nearly impossible to recall or control.

Joy understands that the greatest dangers we face ultimately stem from a world where global corporations dominate—a future where much of the world has no voice in how the world is run. Twenty-first-century GNR technologies, he writes, "are being developed almost exclusively by corporate enterprises. We are aggressively pursuing the promises of these new technologies within the now-unchallenged system of global capitalism and its manifold financial incentives and competitive pressures."

Joy believes that the system of global capitalism, combined with our current rate of progress, gives the human race a 30% to 50% chance of going extinct around the time the Singularity is expected to happen, around 2030. "Not only are these estimates not encouraging," he adds, "but they do not include the probability of many horrid outcomes that lie short of extinction."

It is very likely that scientists and global corporations will miss key developments—or, worse, actively avoid discussion of them. A whole generation of biologists has left the field for the biotech and nanotech labs. Biologist Craig Holdredge, who has followed biotech since its beginnings in the 1970s, warns, "Biology is losing its connection with nature."

When Machines Make War

Cloning, biotechnology, nanotechnology, and robotics are blurring the lines between nature and machine. In his 1972 speech "The Android and the Human," science-fiction visionary Philip K. Dick told his audience, "Machines are becoming more human. Our environment, and I mean our man-made world of machines, is becoming alive in ways specifically and fundamentally analogous to ourselves." In the near future, Dick prophesied, a human might shoot a robot only to see it bleed from its wound. When the robot shoots back, it may be surprised to find the human gush smoke. "It would be rather a great moment of truth for both of them," Dick added.

In November 2001, Advanced Cell Technology of Massachusetts jarred the nation's focus away from recession and terrorism when it announced that it had succeeded in cloning early-stage human embryos. Debate on the topic stayed equally divided between those who support therapeutic cloning and those, like the American Medical Association, who want an outright ban.

Karel Capek coined the word robot (Czech for "forced labor") in the 1920 play *R.U.R.*, in which machines assume the drudgery of factory production, then develop feelings and proceed to wipe out humanity in a violent revolution. While the robots in *R.U.R.* could represent the "nightmare vision of the proletariat seen through middle-class eyes," as science-fiction author Thomas Disch has suggested, they also are testament to the persistent fears of man-made technology run amok.

Similar themes have manifested themselves in popular culture and folklore since at least medieval times. While some might dismiss these stories simply as popular paranoia, robots are already being deployed beyond Hollywood and are poised to take over the deadlier duties of the modern soldier. The Pentagon is replacing soldiers with sensors, vehicles, aircraft, and weapons that can be operated by remote control or are autonomous. Pilotless aircraft played an important role in the bombings of Afghanistan, and a model called the *Gnat* conducted surveillance flights in the Philippines in 2002.

Leading the Pentagon's remote-control warfare effort is the Defense Advanced Research Projects Agency (DARPA). Best known for creating the infrastructure that became the World Wide Web, DARPA is working with Boeing to develop the X-45 unmanned combat air vehicle. The 30-foot-long windowless planes will carry up to 12 bombs, each weighing 250 pounds. According to military analysts, the X-45 will be used to attack radar and antiaircraft installations as early as 2007. By 2010, it will be programmed to distinguish friends from foes without consulting humans and independently attack targets in designated areas. By 2020, robotic planes and vehicles will direct remote-controlled bombers toward targets, robotic helicopters will coordinate driverless convoys, and unmanned submarines will clear mines and launch cruise missiles.

Rising to the challenge of mixing man and machine, MIT's Institute for Soldier Nanotechnologies (backed by a five-year, $50-million U.S. Army grant) is busy innovating materials and designs to create military uniforms that rival the best science fiction. Human soldiers themselves are being transformed into modern cyborgs through robotic devices and nanotechnology.

The Biorobotic Arms Race

The 2002 International Conference on Robotics and Automation, hosted by the Institute of Electrical and Electronics Engineers, kicked off its technical session with a discussion on *biorobots*, the melding of living and artificial structures into a cybernetic organism or cyborg.

"In the past few years, the biosciences and robotics have been getting closer and closer," says Paolo Dario, founder of Italy's Advanced Robotics Technology and Systems Lab. "More and more, biological models are used for the design of biometric robots [and] robots are increasingly used by neuroscientists as clinical platforms for validating biological models." Artificial constructs are beginning to approach the scale and complexity of living systems.

Some of the scientific breakthroughs expected in the next few years promise to make cloning and robotics seem rather benign. The merging of technology and nature has already yielded some shocking progeny. Consider these examples:

- Researchers at the State University of New York Health Science Center at Brooklyn have turned a living rat into a radio-controlled automaton using three electrodes placed in the animal's brain. The animal can be remotely steered through an obstacle course, making it twist, turn, and jump on demand.
- In May 2002, eight elderly Florida residents were injected with microscopic silicon identification chips encoded with medical information. The *Los Angeles Times* reported that this made them "scannable just like a jar of peanut butter in the supermarket checkout line." Applied Digital Solutions Inc., the maker of the chip, will soon have a

prototype of an implantable device able to receive GPS satellite signals and transmit a person's location.
- Human embryos have been successfully implanted and grown in artificial wombs. The experiments were halted after a few days to avoid violating in vitro fertilization regulations.
- Researchers in Israel have fashioned a "bio-computer" out of DNA that can handle a billion operations per second with 99.8% accuracy. Reuters reports that these bio-computers are so minute that "a trillion of them could fit inside a test tube."
- In England, University of Reading Professor Kevin Warwick has implanted microchips in his body to remotely monitor and control his physical motions. During Warwick's Project Cyborg experiments, computers were able to remotely monitor his movements and open doors at his approach.
- Engineers at the U.S. Sandia National Labs have built a remote-controlled spy robot equipped with a scanner, microphone, and chemical microsensor. The robot weighs one ounce and is smaller than a dime. Lab scientists predict that the microbot could prove invaluable in protecting U.S. military and economic interests.

The next arms race is not based on replicating and perfecting a single deadly technology, like the nuclear bombs of the past or some space-based weapon of the future. This new arms race is about accelerating the development and integration of advanced autonomous, biotechnological, and human-robotic systems into the military apparatus. A mishap or a massive war using these new technologies could be more catastrophic than any nuclear war.

Where the Map Exceeds the Territory

The rate at which GNR technologies are being adopted by our society—without regard to long-term safety testing or researching the political, cultural, and economic ramifications—mirrors the development and proliferation of nuclear power and weapons. The human loss caused by experimentation, production,

and development is still being felt from the era of NBC technologies.

The discussion of the environmental impacts of GNR technologies, at least in the United States, has been relegated to the margins. Voices of concern and opposition have likewise been missing in discussions of the technological Singularity. The true cost of this technological progress and any coming Singularity will mean the unprecedented decline of the planet's inhabitants at an ever-increasing rate of global extinction.

The World Conservation Union, the International Botanical Congress, and a majority of the world's biologists believe that a global mass extinction already is under way. As a direct result of human activity (resource extraction, industrial agriculture, the introduction of non-native animals, and population growth), up to one-fifth of all living species are expected to disappear within 30 years. A 1998 Harris Poll of the 5,000 members of the American Institute of Biological Sciences found that 70% believed that what has been termed "The Sixth Extinction" is now under way. A simultaneous Harris Poll found that 60% of the public were totally unaware of the impending biological collapse.

At the same time that nature's ancient biological creation is on the decline, laboratory-created biotech life-forms—genetically modified soybeans, genetically engineered salmon, cloned sheep, drug-crops, biorobots—are on the rise.

Nature and technology are not just evolving; they are competing and combining with one another. Ultimately they will become one. We hear reports daily about these new technologies and new creations, while shreds of the ongoing biological collapse surface here and there. Past the edges of change, beyond the wall across the future, anything becomes possible. Beware the dragons.

James John Bell is writer/director and network administrator for the environmental communications firm Sustain (www.sustainusa.org). He recently wrote the foreword for the 2003 reprinting of John Brunner's science-fiction novel *The Sheep Look Up*, available in June. Portions of this essay were excerpted from his foreword. His address is 920 North Franklin Street, Suite 301, Chicago, Illinois 60610. E-mail jamesbell@sustain.org.

Originally published in the May/June 2003 issue of *The Futurist*. Used with permission from the World Future Society, 7910 Woodmont Avenue, Suite 450, Bethesda, MD 20814. Telephone: 301/656-8274; Fax: 301/951-0394; http://www.wfs.org. © 2003.

Women Waging Peace

You can't end wars simply by declaring peace. "Inclusive security" rests on the principle that fundamental social changes are necessary to prevent renewed hostilities. Women have proven time and again their unique ability to bridge seemingly insurmountable divides. So why aren't they at the negotiating table?

By Swanee Hunt and Cristina Posa

Allowing men who plan wars to plan peace is a bad habit. But international negotiators and policymakers can break that habit by including peace promoters, not just warriors, at the negotiating table. More often than not, those peace promoters are women. Certainly, some extraordinary men have changed the course of history with their peacemaking; likewise, a few belligerent women have made it to the top of the political ladder or, at the grass-roots level, have taken the roles of suicide bombers or soldiers. Exceptions aside, however, women are often the most powerful voices for moderation in times of conflict. While most men come to the negotiating table directly from the war room and battlefield, women usually arrive straight out of civil activism and—take a deep breath—family care.

Yet, traditional thinking about war and peace either ignores women or regards them as victims. This oversight costs the world dearly. The wars of the last decade have gripped the public conscience largely because civilians were not merely caught in the crossfire; they were targeted, deliberately and brutally, by military strategists. Just as warfare has become "inclusive"—with civilian deaths more common than soldiers'—so too must our approach toward ending conflict. Today, the goal is not simply the absence of war, but the creation of sustainable peace by fostering fundamental societal changes. In this respect, the United States and other countries could take a lesson from Canada, whose innovative "human security" initiative—by making human beings and their communities, rather than states, its point of reference—focuses on safety and protection, particularly of the most vulnerable segments of a population.

The concept of "inclusive security," a diverse, citizen-driven approach to global stability, emphasizes women's agency, not their vulnerability. Rather than motivated by gender fairness, this concept is driven by efficiency: Women are crucial to inclu-sive security since they are often at the center of nongovernmental organizations (NGOs), popular protests, electoral referendums, and other citizen-empowering movements whose influence has grown with the global spread of democracy. An inclusive security approach expands the array of tools available to police, military, and diplomatic structures by adding collaboration with local efforts to achieve peace. Every effort to bridge divides, even if unsuccessful, has value, both in lessons learned and links to be built on later. Local actors with crucial experience resolving conflicts, organizing political movements, managing relief efforts, or working with military forces bring that experience into ongoing peace processes.

International organizations are slowly recognizing the indispensable role that women play in preventing war and sustaining peace. On October 31, 2000, the United Nations Security Council issued Resolution 1325 urging the secretary-general to expand the role of women in U.N. field-based operations, especially among military observers, civilian police, human rights workers, and humanitarian personnel. The Organization for Security and Co-operation in Europe (OSCE) is working to move women off the gender sidelines and into the everyday activities of the organization—particularly in the Office for Democratic Institutions and Human Rights, which has been useful in monitoring elections and human rights throughout Europe and the former Soviet Union. Last November, the European Parliament passed a hard-hitting resolution calling on European Union members (and the European Commission and Council) to promote the equal participation of women in diplomatic conflict resolution; to ensure that women fill at least 40 percent of all reconciliation, peacekeeping, peace-enforcement, peace-building, and conflict-prevention posts; and to support the creation and strengthening of NGOs (including women's organizations) that focus on conflict prevention, peace building, and post-conflict reconstruction.

> Ironically, women's status as second-class citizens is a source of empowerment, since it has made women adept at finding innovative ways to cope with problems.

But such strides by international organizations have done little to correct the deplorable extent to which local women have been relegated to the margins of police, military, and diplomatic efforts. Consider that Bosnian women were not invited to participate in the Dayton talks, which ended the war in Bosnia, even though during the conflict 40 women's associations remained organized and active across ethnic lines. Not surprisingly, this exclusion has subsequently characterized—and undermined—the implementation of the Dayton accord. During a 1997 trip to Bosnia, U.S. President Bill Clinton, Secretary of State Madeleine Albright, and National Security Advisor Samuel Berger had a miserable meeting with intransigent politicians elected under the ethnic-based requirements of Dayton. During the same period, First Lady Hillary Rodham Clinton engaged a dozen women from across the country who shared story after story of their courageous and remarkably effective work to restore their communities. At the end of the day, a grim Berger faced the press, offering no encouraging word from the meetings with the political dinosaurs. The first lady's meeting with the energetic women activists was never mentioned.

We can ignore women's work as peacemakers, or we can harness its full force across a wide range of activities relevant to the security sphere: bridging the divide between groups in conflict, influencing local security forces, collaborating with international organizations, and seeking political office.

BRIDGING THE DIVIDE

The idea of women as peacemakers is not political correctness run amok. Social science research supports the stereotype of women as generally more collaborative than men and thus more inclined toward consensus and compromise. Ironically, women's status as second-class citizens is a source of empowerment, since it has made women adept at finding innovative ways to cope with problems. Because women are not ensconced within the mainstream, those in power consider them less threatening, allowing women to work unimpeded and "below the radar screen." Since they usually have not been behind a rifle, women, in contrast to men, have less psychological distance to reach across a conflict line. (They are also more accepted on the "other side," because it is assumed that they did not do any of the actual killing.) Women often choose an identity, notably that of mothers, that cuts across international borders and ethnic enclaves. Given their roles as family nurturers,

women have a huge investment in the stability of their communities. And since women know their communities, they can predict the acceptance of peace initiatives, as well as broker agreements in their own neighborhoods.

As U.N. Secretary-General Kofi Annan remarked in October 2000 to the Security Council, "For generations, women have served as peace educators, both in their families and in their societies. They have proved instrumental in building bridges rather than walls." Women have been able to bridge the divide even in situations where leaders have deemed conflict resolution futile in the face of so-called intractable ethnic hatreds. Striking examples of women making the impossible possible come from Sudan, a country splintered by decades of civil war. In the south, women working together in the New Sudan Council of Churches conducted their own version of shuttle diplomacy—perhaps without the panache of jetting between capitals—and organized the Wunlit tribal summit in February 1999 to bring an end to bloody hostilities between the Dinka and Nuer peoples. As a result, the Wunlit Covenant guaranteed peace between the Dinka and the Nuer, who agreed to share rights to water, fishing, and grazing land, which had been key points of disagreement. The covenant also returned prisoners and guaranteed freedom of movement for members of both tribes.

On another continent, women have bridged the seemingly insurmountable differences between India and Pakistan by organizing huge rallies to unite citizens from both countries. Since 1994, the Pakistan-India People's Forum for Peace and Democracy has worked to overcome the hysterics of the nationalist media and jingoistic governing elites by holding annual conventions where Indians and Pakistanis can affirm their shared histories, forge networks, and act together on specific initiatives. In 1995, for instance, activists joined forces on behalf of fishers and their children who were languishing in each side's jails because they had strayed across maritime boundaries. As a result, the adversarial governments released the prisoners and their boats.

In addition to laying the foundation for broader accords by tackling the smaller, everyday problems that keep people apart, women have also taken the initiative in drafting principles for comprehensive settlements. The platform of Jerusalem Link, a federation of Palestinian and Israeli women's groups, served as a blueprint for negotiations over the final status of Jerusalem during the Oslo process. Former President Clinton, the week of the failed Camp David talks in July 2000, remarked simply, "If we'd had women at Camp David, we'd have an agreement."

Sometimes conflict resolution requires unshackling the media. Journalists can nourish a fair and tolerant vision of society or feed the public poisonous, one-sided, and untruthful accounts of the "news" that stimulate violent conflict. Supreme Allied Commander of Europe Wesley Clark understood as much when he ordered NATO to bomb transmitters in Kosovo to prevent the Milosevic media machine from spewing ever more inflammatory rhetoric. One of the founders of the independent Kosovo radio station RTV-21 realized that there were "many instances of male colleagues reporting with anger, which served to raise the tensions rather than lower them." As a result, RTV-

21 now runs workshops in radio, print, and TV journalism to cultivate a core of female journalists with a noninflammatory style. The OSCE and the BBC, which train promising local journalists in Kosovo and Bosnia, would do well to seek out women, who generally bring with them a reputation for moderation in unstable situations.

Nelson Mandela suggested at last summer's Arusha peace talks that if Burundian men began fighting again, their women should withhold "conjugal rights" (like cooking, he added).

INFLUENCING SECURITY FORCES

The influence of women on warriors dates back to the ancient Greek play *Lysistrata*. Borrowing from that play's story, former South African President Nelson Mandela suggested at last summer's Arusha peace talks on the conflict in Burundi that if Burundian men began fighting again, their women should withhold "conjugal rights" (like cooking, he added).

Women can also act as a valuable interface between their countries' security forces (police and military) and the public, especially in cases when rapid response is necessary to head off violence. Women in Northern Ireland, for example, have helped calm the often deadly "marching season" by facilitating mediations between Protestant unionists and Catholic nationalists. The women bring together key members of each community, many of whom are released prisoners, as mediators to calm tensions. This circle of mediators works with local police throughout the marching season, meeting quietly and maintaining contacts on a 24-hour basis. This intervention provides a powerful extension of the limited tools of the local police and security forces.

Likewise, an early goal of the Sudanese Women's Voice for Peace was to meet and talk with the military leaders of the various rebel armies. These contacts secured women's access to areas controlled by the revolutionary movements, a critical variable in the success or failure of humanitarian efforts in war zones. Women have also worked with the military to search for missing people, a common element in the cycle of violence. In Colombia, for example, women were so persistent in their demands for information regarding 150 people abducted from a church in 1999 that the army eventually gave them space on a military base for an information and strategy center. The military worked alongside the women and their families trying to track down the missing people. In short, through moral suasion, local women often have influence where outsiders, such as international human rights agencies, do not.

That influence may have allowed a female investigative reporter like Maria Cristina Caballero to go where a man could not go, venturing on horseback alone, eight hours into the jungle

to tape a four-hour interview with the head of the paramilitary forces in Colombia. She also interviewed another guerilla leader and published an award-winning comparison of the transcripts, showing where the two mortal enemies shared the same vision. "This [was] bigger than a story," she later said, "this [was] hope for peace." Risking their lives to move back and forth across the divide, women like Caballero perform work that is just as important for regional stabilization as the grandest Plan Colombia.

INTERNATIONAL COLLABORATION

Given the nature of "inclusive" war, security forces are increasingly called upon to ensure the safe passage of humanitarian relief across conflict zones. Women serve as indispensable contacts between civilians, warring parties, and relief organizations. Without women's knowledge of the local scene, the mandate of the military to support NGOs would often be severely hindered, if not impossible.

In rebel-controlled areas of Sudan, women have worked closely with humanitarian organizations to prevent food from being diverted from those who need it most. According to Catherine Loria Duku Jeremano of Oxfam: "The normal pattern was to hand out relief to the men, who were then expected to take it home to be distributed to their family. However, many of the men did what they pleased with the food they received: either selling it directly, often in exchange for alcohol, or giving food to the wives they favored." Sudanese women worked closely with tribal chiefs and relief organizations to establish a system allowing women to pick up the food for their families, despite contrary cultural norms.

In Pristina, Kosovo, Vjosa Dobruna, a pediatric neurologist and human rights leader, is now the joint administrator for civil society for the U.N. Interim Administration Mission in Kosovo (UNMIK). In September 2000, at the request of NATO, she organized a multiethnic strategic planning session to integrate women throughout UNMIK. Before that gathering, women who had played very significant roles in their communities felt shunned by the international organizations that descended on Kosovo following the bombing campaign. Vjosa's conference pulled them back into the mainstream, bringing international players into the conference to hear from local women what stabilizing measures they were planning, rather than the other way around. There, as in Bosnia, the OSCE has created a quota system for elected office, mandating that women comprise one third of each party's candidate list; leaders like Vjosa helped turn that policy into reality.

In addition to helping aid organizations find better ways to distribute relief or helping the U.N. and OSCE implement their ambitious mandates, women also work closely with them to locate and exchange prisoners of war. As the peace processes in Northern Ireland, Bosnia, and the Middle East illustrate, a deadlock on the exchange and release of prisoners can be a major obstacle to achieving a final settlement. Women activists in Armenia and Azerbaijan have worked closely with the International Helsinki Citizens Assembly and the OSCE for the release

The Black and the Green

Grass-roots women's organizations in Israel come in two colors: black and green. The Women in Black, founded in 1988, and the Women in Green, founded in 1993, could not be further apart on the political spectrum, but both claim the mantle of "womanhood" and "motherhood" in the ongoing struggle to end the Israeli-Palestinian conflict.

One month after the Palestinian intifada broke out in December 1988, a small group of women decided to meet every Friday afternoon at a busy Jerusalem intersection wearing all black and holding hand-shaped signs that read: "Stop the Occupation." The weekly gatherings continued and soon spread across Israel to Europe, the United States, and then to Asia.

While the movement was originally dedicated to achieving peace in the Middle East, other groups soon protested against repression in the Balkans and India. For these activists, their status as women lends them a special authority when it comes to demanding peace. In the words of the Asian Women's Human Rights Council: "We are the Women in Black... women, unmasking the many horrific faces of more public 'legitimate' forms of violence—state repression, communalism, ethnic cleansing, nationalism, and wars...."

Today, the Women in Black in Israel continue their nonviolent opposition to the occupation in cooperation with the umbrella group Coalition of Women for a Just Peace. They have been demonstrating against the closures of various Palestinian cities, arguing that the blockades prevent pregnant women from accessing healthcare services and keep students from attending school. The group also calls for the full participation of women in peace negotiations.

While the Women in Black stood in silent protest worldwide, a group of "grandmothers, mothers, wives, and daughters; housewives and professionals; secular and religious" formed the far-right Women in Green in 1993 out of "a shared love, devotion and concern for Israel." Known for the signature green hats they wear at rallies, the Women in Green emerged as a protest to the Oslo accords on the grounds that Israel made too many concessions to Yasir Arafat's Palestinian Liberation Organization. The group opposes returning the Golan Heights to Syria, sharing sovereignty over Jerusalem with the Palestinians, and insists that "Israel remain a Jewish state."

The Women in Green boast some 15,000 members in Israel, and while they have not garnered the global support of the Women in Black, 15,000 Americans have joined their cause. An ardent supporter of Israeli Prime Minister Ariel Sharon, the group seeks to educate the Israeli electorate through weekly street theater and public demonstrations, as well as articles, posters, and newspaper advertisements.

White the groups' messages and methods diverge, their existence and influence demonstrate that women can mobilize support for political change—no matter what color they wear.

—*FP*

of hostages in the disputed region of Nagorno-Karabakh, where tens of thousands of people have been killed. In fact, these women's knowledge of the local players and the situation on the ground would make them indispensable in peace negotiations to end this 13-year-old conflict.

REACHING FOR POLITICAL OFFICE

In 1977, women organizers in Northern Ireland won the Nobel Peace Prize for their nonsectarian public demonstrations. Two decades later, Northern Irish women are showing how diligently women must still work not only to ensure a place at the negotiating table but also to sustain peace by reaching critical mass in political office. In 1996, peace activists Monica McWilliams (now a member of the Northern Ireland Assembly) and May Blood (now a member of the House of Lords) were told that only leaders of the top 10 political parties—all men—would be included in the peace talks. With only six weeks to organize, McWilliams and Blood gathered 10,000 signatures to create a new political party (the Northern Ireland Women's Coalition, or NIWC) and got themselves on the ballot. They were voted into the top 10 and earned a place at the table.

The grass-roots, get-out-the-vote work of Vox Femina convinced hesitant Yugoslav women to vote for change; those votes contributed to the margin that ousted President Slobodan Milosevic.

The NIWC's efforts paid off. The women drafted key clauses of the Good Friday Agreement regarding the importance of mixed housing, the particular difficulties of young people, and the need for resources to address these problems. The NIWC also lobbied for the early release and reintegration of political prisoners in order to combat social exclusion and pushed for a comprehensive review of the police service so that all members of society would accept it. Clearly, the women's prior work with individuals and families affected by "the Troubles" enabled them to formulate such salient contributions to the agreement. In the subsequent public referendum on the Good Friday Agreement, Mo Mowlam, then British secretary of state for Northern Ireland, attributed the overwhelming success of the YES Campaign to the NIWC's persistent canvassing and lobbying.

Women in the former Yugoslavia are also stepping forward to wrest the reins of political control from extremists (including women, such as ultranationalist Bosnian Serb President Biljana Plavsic) who destroyed their country. Last December, Zorica Trifunovic, founding member of the local Women in Black (an antiwar group formed in Belgrade in October 1991), led a meeting that united 90 women leaders of pro-democracy political campaigns across the former Yugoslavia. According to polling by the National Democratic Institute, the grass-roots, get-out-the-vote work of groups such as Vox Femina (a local NGO that participated in the December meeting) convinced hesitant women to vote for change; those votes contributed to the margin that ousted President Slobodan Milosevic.

International security forces and diplomats will find no better allies than these mobilized mothers, who are tackling the toughest, most hardened hostilities.

Argentina provides another example of women making the transition from protesters to politicians: Several leaders of the Madres de la Plaza de Mayo movement, formed in the 1970s to protest the "disappearances" of their children at the hands of the military regime, have now been elected to political office. And in Russia, the Committee of Soldiers' Mothers—a protest group founded in 1989 demanding their sons' rights amidst cruel conditions in the Russian military—has grown into a powerful organization with 300 chapters and official political status. In January, U.S. Ambassador to Moscow Jim Collins described the committee as a significant factor in countering the most aggressive voices promoting military force in Chechnya. Similar mothers' groups have sprung up across the former Soviet Union and beyond—including the Mothers of Tiananmen Square. International security forces and diplomats will find no better allies than these mobilized mothers, who are tackling the toughest, most hardened hostilities.

YOU'VE COME A LONG WAY, MAYBE

Common sense dictates that women should be central to peacemaking, where they can bring their experience in conflict resolution to bear. Yet, despite all of the instances where women have been able to play a role in peace negotiations, women remain relegated to the sidelines. Part of the problem is structural: Even though more and more women are legislators and soldiers, underrepresentation persists in the highest levels of political and military hierarchies. The presidents, prime ministers, party leaders, cabinet secretaries, and generals who typically negotiate peace settlements are overwhelmingly men. There is also a psychological barrier that precludes women from sitting in on negotiations: Waging war is still thought of as a "man's job," and as such, the task of stopping war often is delegated to men

(although if we could begin to think about the process not in terms of stopping war but promoting peace, women would emerge as the more logical choice). But the key reason behind women's marginalization may be that everyone recognizes just how good women are at forging peace. A U.N. official once stated that, in Africa, women are often excluded from negotiating teams because the war leaders "are afraid the women will compromise" and give away too much.

Some encouraging signs of change, however, are emerging. Rwandan President Paul Kagame, dismayed at his difficulty in attracting international aid to his genocide-ravaged country, recently distinguished Rwanda from the prevailing image of brutality in central Africa by appointing three women to his negotiating team for the conflict in the Democratic Republic of the Congo. In an unusually healthy tit for tat, the Ugandans responded by immediately appointing a woman to their team.

Will those women make a difference? Negotiators sometimes worry that having women participate in the discussion may change the tone of the meeting. They're right: A British participant in the Northern Ireland peace talks insightfully noted that when the parties became bogged down by abstract issues and past offenses, "the women would come and talk about their loved ones, their bereavement, their children and their hopes for the future." These deeply personal comments, rather than being a diversion, helped keep the talks focused. The women's experiences reminded the parties that security for all citizens was what really mattered.

The role of women as peacemakers can be expanded in many ways. Mediators can and should insist on gender balance among negotiators to ensure a peace plan that is workable at the community level. Cultural barriers can be overcome if high-level visitors require that a critical mass (usually one third) of the local interlocutors be women (and not simply present as wives). When drafting principles for negotiation, diplomats should determine whether women's groups have already agreed upon key conflict-bridging principles, and whether their approach can serve as a basis for general negotiations.

Moreover, to foster a larger pool of potential peacemakers, embassies in conflict areas should broaden their regular contact with local women leaders and sponsor women in training programs, both at home and abroad. Governments can also do their part by providing information technology and training to women activists through private and public partnerships. Internet communication allows women peace builders to network among themselves, as well as exchange tactics and strategies with their global counterparts.

"Women understood the cost of the war and were genuinely interested in peace," recalls retired Admiral Jonathan Howe, reflecting on his experience leading the U.N. mission in Somalia in the early 1990s. "They'd had it with their warrior husbands. They were a force willing to say enough is enough. The men were sitting around talking and chewing qat, while the women were working away. They were such a positive force.... You have to look at all elements in society and be ready to tap into those that will be constructive."

Want to Know More?

The Internet is invaluable in enabling the inclusive security approach advocated in this article. The Web offers not only a wealth of information but, just as important, relatively cheap and easy access for citizens worldwide. Most of the women's peace-building activities and strategies explored in this article can be found on the Web site of **Women Waging Peace**—a collaborative venture of Harvard University's John F. Kennedy School of Government and the nonprofit organization Hunt Alternatives, which recognize the essential role and contribution of women in preventing violent conflict, stopping war, reconstructing ravaged societies, and sustaining peace in fragile areas around the world. On the site, women active in conflict areas can communicate with each other without fear of retribution via a secure server. The women submit narratives detailing their strategies, which can then be read on the public Web site. The site also features a video archive of interviews with each of these women. You need a password to view these interviews, so contact Women Waging Peace online or call (617) 868-3910.

The Organization for Security and Co-operation in Europe (OSCE) is an outstanding resource for qualitative and quantitative studies of women's involvement in conflict prevention. Start with the final report of the *OSCE Supplementary Implementation Meeting: Gender Issues* (Vienna: UNIFEM, 1999), posted on the group's Web site. **The United Nations Development Fund for Women** (UNIFEM) also publishes reports on its colorful and easy-to-navigate site. The fund's informative book, *Women at the Peace Table: Making a Difference* (New York: UNIFEM, 2000), available online, features interviews with some of today's most prominent women peacemakers, including Hanan Ashrawi and Mo Mowlam.

For a look at how globalization is changing women's roles in governments, companies, and militaries, read Cynthia Enloe's *Bananas, Beaches and Bases: Making Feminist Sense of International Politics* (Berkeley: University of California Press, 2001). In *Maneuvers: The International Politics of Militarizing Women's Lives* (Berkeley: University of California Press, 2000), Enloe examines the military's effects on women, whether they are soldiers or soldiers' spouses. For a more general discussion of where feminism fits into academia and policymaking, see **"Searching for the Princess? Feminist Perspectives in International Relations"** (*The Harvard International Review*, Fall 1999) by J. Ann Tickner, associate professor of international relations at the University of Southern California.

The Fall 1997 issue of FOREIGN POLICY magazine features two articles that highlight how women worldwide are simultaneously gaining political clout but also bearing the brunt of poverty: **"Women in Power: From Tokenism to Critical Mass"** by Jane S. Jaquette and **"Women in Poverty: A New Global Underclass"** by Mayra Buvinic.

● For links to relevant Web sites, as well as a comprehensive index of related FOREIGN POLICY articles, access **www.foreignpolicy.com**.

Lasting peace must be homegrown. Inclusive security helps police forces, military leaders, and diplomats do their jobs more effectively by creating coalitions with the people most invested in stability and most adept at building peace. Women working on the ground are eager to join forces. Just let them in.

Swanee Hunt is director of the Women in Public Policy Program at Harvard University's John F. Kennedy School of Government. As the United States' ambassador to Austria (1993–97), she founded the "Vital Voices: Women in Democracy" initiative. Cristina Posa, a former judicial clerk at the United Nations International Criminal Tribunal for the former Yugoslavia, is an attorney at Cleary, Gottlieb, Steen & Hamilton in New York.

THE NEXT CHRISTIANITY

We stand at a historical turning point, the author argues—one that is as epochal for the Christian world as the original Reformation. Around the globe Christianity is growing and mutating in ways that observers in the West tend not to see. Tumultuous conflicts within Christianity will leave a mark deeper than Islam's on the century ahead

BY PHILIP JENKINS

Ever since the sexual-abuse crisis erupted in the U.S. Roman Catholic Church in the mid-1980s, with allegations of child molestation by priests, commentators have regularly compared the problems faced by the Church to those it faced in Europe at the start of the sixteenth century, on the eve of the Protestant Reformation—problems that included sexual laxity and financial malfeasance among the clergy, and clerical contempt for the interests of the laity. Calls for change have become increasingly urgent since January, when revelations of widespread sexual misconduct and grossly negligent responses to it emerged prominently in the Boston archdiocese. Similar, if less dramatic, problems have been brought to light in New Orleans, Providence, Palm Beach, Omaha, and many other dioceses. The reform agendas now under discussion within the U.S. hierarchy involve ideas about increased lay participation in governance—ideas of the sort heard when Martin Luther confronted the Roman Catholic orthodoxy of his day. They also include such ideas as admitting women to the priesthood and permitting priests to marry.

Explicit analogies to the Reformation have become commonplace not only among commentators but also among anticlerical activists, among victims' groups, and, significantly, among ordinary lay believers. One representative expert on sexual misconduct, much quoted, is Richard Sipe, a former monk who worked at the sexual-disorders clinic at Johns Hopkins University and is now a psychotherapist based in California. Over the years Sipe has spoken regularly of "a new Reformation." "We are at 1515," he has written, "between when Martin Luther went to Rome in 1510 and 1517 when he nailed his 95 theses on the door in Wittenberg." That act can reasonably be seen as the symbolic starting point of the Reformation, when a united Christendom was rent asunder.

Historians continue to debate the causes and consequences of the Reformation, and of the forces that it unleashed. Among other things, the Reformation broke the fetters that constrained certain aspects of intellectual life during the Middle Ages. Protestants, of course, honor the event as the source of their distinctive religious traditions; many Protestant denominations celebrate Reformation Day, at the end of October, commemorating the posting of the theses at Wittenberg. And liberal Catholics invoke the word these days to emphasize the urgency of reform—changes both broad and specific that they demand from the

Church. Their view is that the crisis, which exposes fault lines of both sexuality and power, is the most serious the Church has faced in 500 years—as serious as the one it faced in Luther's time.

The first Reformation was an epochal moment in the history of the Western world—and eventually, by extension, of the rest of the world. The status quo in religious affairs was brought to an end. Relations between religions and governments, not to mention among different denominations, took a variety of forms—sometimes symbiotic, often chaotic and violent. The transformations wrought in the human psyche by the Reformation, and by the Counter-Reformation it helped to provoke, continue to play themselves out. This complex historical episode, which is now often referred to simply as "the Reformation," touched everything. It altered not just the practice of religion but also the nature of society, economics, politics, education, and the law.

Commentators today, when speaking of the changes needed in the Catholic Church, generally do not have in mind the sweeping historical aftermath of the first Reformation—but they should. The Church has developed a fissure whose size most people do not fully appreciate. The steps that liberal Catholics would take to resolve some of the Church's urgent issues, steps that might quell unease or revolt in some places, would prove incendiary in others. The problem with reform, 500 years ago or today, is that people disagree—sometimes violently—on the direction it should take.

The fact is, we are at a moment as epochal as the Reformation itself—a Reformation moment not only for Catholics but for the entire Christian world. Christianity as a whole is both growing and mutating in ways that observers in the West tend not to see. For obvious reasons, news reports today are filled with material about the influence of a resurgent and sometimes angry Islam. But in its variety and vitality, in its global reach, in its association with the world's fastest-growing societies, in its shifting centers of gravity, in the way its values and practices vary from place to place—in these and other ways it is Christianity that will leave the deepest mark on the twenty-first century. The process will not necessarily be a peaceful one, and only the foolish would venture anything beyond the broadest predictions about the religious picture a century or two ahead. But the twenty-first century will almost certainly be regarded by future historians as a century in which religion replaced ideology as the prime animating and destructive force in human affairs, guiding attitudes to political liberty and obligation, concepts of nationhood, and, of course, conflicts and wars.

The original Reformation was far more than the rising up of irate lay people against corrupt and exploitative priests, and it was much more than a mere theological row. It was a far-reaching social movement that sought to return to the original sources of Christianity. It challenged the idea that divine authority should be mediated through institutions or hierarchies, and it denied the value of tradition. Instead it offered radical new notions of the supremacy of written texts (that is, the books of the Bible), interpreted by individual consciences. The Reformation made possible a religion that could be practiced privately, rather than mainly in a vast institutionalized community.

This move toward individualism, toward the privatization of religious belief, makes the spirit of the Reformation very attractive to educated people in the West. It stirs many liberal Catholic activists, who regard the aloof and arrogant hierarchy of the Church as not only an affront but something inherently corrupt. New concepts of governance sound exciting, even intoxicating, to reformers, and seem to mesh with likely social and technological trends. The invention of movable type and the printing press, in the fifteenth century, was a technological development that spurred mass literacy in the vernacular languages—and accelerated the forces of religious change. In the near future, many believe, the electronic media will have a comparably powerful impact on our ways of being religious. An ever greater reliance on individual choice, the argument goes, will help Catholicism to become much more inclusive and tolerant, less judgmental, and more willing to accept secular attitudes toward sexuality and gender roles. In the view of liberal Catholics, much of the current crisis derives directly from archaic if not primitive doctrines, including mandatory celibacy among the clergy, intolerance of homosexuality, and the prohibition of women from the priesthood, not to mention a more generalized fear of sexuality. In their view, anyone should be able to see that the idea that God, the creator and lord of the universe, is concerned about human sexuality is on its way out.

If we look beyond the liberal West, we see that another Christian revolution, quite different from the one being called for in affluent American suburbs and upscale urban parishes, is already in progress. Worldwide, Christianity is actually moving toward supernaturalism and neo-orthodoxy, and in many ways toward the ancient world view expressed in the New Testament: a vision of Jesus as the embodiment of divine power, who overcomes the evil forces that inflict calamity and sickness upon the human race. In the global South (the areas that we often think of primarily as the Third World) huge and growing Christian populations—currently 480 million in Latin America, 360 million in Africa, and 313 million in Asia, compared with 260 million in North America—now make up what the Catholic scholar Walbert Buhlmann has called the Third Church, a form of Christianity as distinct as Protestantism or Orthodoxy, and one that is likely to become dominant in the faith. The revolution taking place in Africa, Asia, and Latin America is far more sweeping in its implications than any current shifts in North American religion, whether Catholic or Protestant. There is increasing tension between what one might call a liberal Northern Reformation and the surging Southern religious revolution, which one might equate with the Counter-Reformation, the internal Catholic reforms that took place at the same time as the Reformation—although in references to the past and the present the term "Counter-Reformation" misleadingly implies a simple reaction instead of a social and spiritual explosion. No matter what the terminology, however, an enormous rift seems inevitable.

Although Northern governments are still struggling to come to terms with the notion that Islam might provide a powerful and threatening supranational ideology, few seem to realize the potential political role of ascendant Southern Christianity. The religious rift between Northern and Southern Europe in the sixteenth century suggests just how dramatic the political consequences of a North-South divide in the contemporary Christian world might be. The Reformation led to nothing less than the creation of the modern European states and the international order we recognize today. For more than a century Europe was rent by sectarian wars between Protestants and Catholics, which by the 1680s had ended in stalemate. Out of this impasse, this failure to impose a monolithic religious order across the Continent, there arose such fundamental ideas of modern society as the state's obligation to tolerate minorities and the need to justify political authority without constantly invoking God and religion. The Enlightenment—and, indeed, Western modernity—could have occurred only as a consequence of the clash, military and ideological, between Protestants and Catholics.

Today across the global South a rising religious fervor is coinciding with declining autonomy for nation-states, making useful an analogy with the medieval concept of Christendom—the Res Publica Christiana—as an overarching source of unity and a focus of loyalty transcending mere kingdoms or empires. Kingdoms might last for only a century or two before being supplanted by new states or dynasties, but rational people knew that Christendom simply endured. The laws of individual nations lasted only as long as the nations themselves; Christendom offered a higher set of standards and mores that could claim to be universal. Christendom was a primary cultural reference, and it may well re-emerge as such in the Christian South—as a new transnational order in which political, social, and personal identities are defined chiefly by religious loyalties.

The first Reformation was a lot less straightforward than some histories suggest. The sixteenth-century Catholic Church, after all, did not collapse after Luther kicked in the door. The Counter-Reformation was moving in a diametrically opposite direction, reasserting older forms of devotion and tradition, and reformulating the Church's controversial claims for hierarchy and spiritual authority. The Counter-Reformation was not just survivalist and defensive, as is commonly assumed; it was also innovative and dynamic. For at least a century after Luther's Reformation, in fact, the true political, cultural, and social centers of Europe were as much in the Catholic South as in the Protestant North. The Catholic states—Spain, Portugal, and France—were launching missionary ventures into Africa, Asia, North and South America. By the 1570s Catholic missionaries were creating a transoceanic Church structure: the see of Manila was an offshoot of the archdiocese of Mexico City.

By about 1600 the Catholic Church had become the first religious body—indeed, the first institution of any sort—to operate on a global scale. Even in the Protestant heartlands of Northern and Western Europe—England, Sweden, and the German lands—the heirs of the Reformation had to spend many years discouraging their people from succumbing to the attractions of Catholicism. Conversions to Catholicism were steady throughout the century or so after 1580. It looked as if the Reformation had effectively cut Protestant Europe off from the mainstream of the Christian world. Only in the eighteenth century would Protestantism find a secure and then strategically preponderant place on the global stage, through the success of booming commercial states such as England and the Netherlands, whose political triumphs ultimately contained and in some cases pushed back the earlier empires.

The changes that Catholic and other reformers today are trying to inspire in North America and Europe (and that seem essential if Christianity is to be preserved as a modern, relevant force on those continents) run utterly contrary to the dominant cultural movements in the rest of the Christian world, which look very much like the Counter-Reformation. But this century is unlike the sixteenth in that we are not facing a roughly equal division of Christendom between two competing groups. Rather, Christians are facing a shrinking population in the liberal West and a growing majority of the traditional Rest. During the past half century the critical centers of the Christian world have moved decisively to Africa, to Latin America, and to Asia. The balance will never shift back.

The growth in Africa has been relentless. In 1900 Africa had just 10 million Christians out of a continental population of 107 million—about nine percent. Today the Christian total stands at 360 million out of 784 million, or 46 percent. And that percentage is likely to continue rising, because Christian African countries have some of the world's most dramatic rates of population growth. Meanwhile, the advanced industrial countries are experiencing a dramatic birth dearth. Within the next twenty-five years the population of the world's Christians is expected to grow to 2.6 billion (making Christianity by far the world's largest faith). By 2025, 50 percent of the Christian population will be in Africa and Latin America, and another 17 percent will be in Asia. Those proportions will grow steadily. By about 2050 the United States will still have the largest single contingent of Christians, but all the other leading nations will be Southern: Mexico, Brazil, Nigeria, the Democratic Republic of the Congo, Ethiopia, and the Philippines. By then the proportion of non-Latino whites among the world's Christians will have fallen to perhaps one in five.

The population shift is even more marked in the specifically Catholic world, where Euro-Americans are already in the minority. Africa had about 16 million Catholics in the early 1950s; it has 120 million today, and is expected to have 228 million by 2025. The *World Christian Encyclopedia* suggests that by 2025 almost three quarters of all Catholics will be found in Africa, Asia, and Latin America. The likely map of twenty-first-century Catholicism represents an unmistakable legacy of the Counter-Reformation and its global missionary ventures.

These figures actually understate the Southern predominance within Catholicism, and within world Christianity more generally, because they fail to take account of Southern emigrants to Europe and North America. Even as this migration continues, established white communities in Europe are declining demographically, and their religious beliefs and practices are moving further away from traditional Christian roots. The result is that skins of other hues are increasingly evident in European churches; half of all London churchgoers are now black. African and West Indian churches in Britain are reaching out to whites, though members complain that their religion is often seen as "a black thing" rather than "a God thing."

In the United States a growing proportion of Roman Catholics are Latinos, who should represent a quarter of the nation by 2050 or so. Asian communities in the United States have sizable Catholic populations. Current trends suggest that the religious values of Catholics with a Southern ethnic and cultural heritage will long remain quite distinct from those of other U.S. populations. In terms of liturgy and worship Latino Catholics are strikingly different from Anglo believers, not least in maintaining a fervent devotion to the Virgin Mary and the saints.

European and Euro-American Catholics will within a few decades be a smaller and smaller fragment of a worldwide Church. Of the 18 million Catholic baptisms recorded in 1998, eight million took place in Central and South America, three million in Africa, and just under three million in Asia. (In other words, these three regions already account for more than three quarters of all Catholic baptisms.) The annual baptism total for the Philippines is higher than the totals for Italy, France, Spain, and Poland combined. The number of Filipino Catholics could grow to 90 million by 2025, and perhaps to 130 million by 2050.

The demographic changes within Christianity have many implications for theology and religious practice, and for global society and politics. The most significant point is that in terms of both theology and moral teaching, Southern Christianity is more conservative than the Northern—especially the American—version. Northern reformers, even if otherwise sympathetic to the indigenous cultures of non-Northern peoples, obviously do not like this fact. The liberal Catholic writer James Carroll has complained that "world Christianity [is falling] increasingly under the sway of anti-intellectual fundamentalism." But the cultural pressures may be hard to resist.

The denominations that are triumphing across the global South—radical Protestant sects, either evangelical or Pentecostal, and Roman Catholicism of an orthodox kind—are stalwartly traditional or even reactionary by the standards of the economically advanced nations. The Catholic faith that is rising rapidly in Africa and Asia looks very much like a pre-Vatican II faith, being more traditional in its respect for the power of bishops and priests and in its preference for older devotions. African Catholicism in particular is far more comfortable with notions of authority and spiritual charisma than with newer ideas of consultation and democracy.

This kind of faith is personified by Nigeria's Francis Cardinal Arinze, who is sometimes touted as a future Pope. He is sharp and articulate, with an attractively self-deprecating style, and he has served as the president of the Pontifical Council for Inter-Religious Dialogue, which has given him invaluable experience in talking with Muslims, Hindus, Jews, and members of other faiths. By liberal Northern standards, however, Arinze is rigidly conservative, and even repressive on matters such as academic freedom and the need for strict orthodoxy. In his theology as much as his social views he is a loyal follower of Pope John Paul II. Anyone less promising for Northern notions of reform is difficult to imagine.

Meanwhile, a full-scale Reformation is taking place among Pentecostal Christians—whose ideas are shared by many Catholics. Pentecostal believers reject tradition and hierarchy, but they also rely on direct spiritual revelation to supplement or replace biblical authority. And it is Pentecostals who stand in the vanguard of the Southern Counter-Reformation. Though Pentecostalism emerged as a movement only at the start of the twentieth century, chiefly in North America, Pentecostals today are at least 400 million strong, and heavily concentrated in the global South. By 2040 or so there could be as many as a billion, at which point Pentecostal Christians alone will far outnumber the world's Buddhists and will enjoy rough numerical parity with the world's Hindus.

The booming Pentecostal churches of Africa, Asia, and Latin America are thoroughly committed to re-creating their version of an idealized early Christianity (often described as the restoration of "primitive" Christianity). The most successful Southern churches preach a deep personal faith, communal orthodoxy, mysticism, and puritanism, all founded on obedience to spiritual authority, from whatever source it is believed to stem. Pentecostals—and their Catholic counterparts—preach messages that may appear simplistically charismatic, visionary, and apocalyptic to a Northern liberal. For them prophecy is an everyday reality, and many independent denominations trace their foundation to direct prophetic authority. Scholars of religion customarily speak of these proliferating congregations simply as the "prophetic churches."

Of course, American reformers also dream of a restored early Church; but whereas Americans imagine a Church freed from hierarchy, superstition, and dogma, Southerners look back to one filled with spiritual power and able to exorcise the demonic forces that cause sickness and poverty. And yes, "demonic" is the word. The most successful Southern churches today speak openly of spiritual healing and exorcism. One controversial sect in the process of developing an international following is the Brazilian-based Universal Church of the Kingdom of God, which claims to offer "strong prayer to destroy witchcraft, demon possession, bad luck, bad dreams, all spiritual problems," and promises that members will gain "prosperity and financial breakthrough." The Cherubim and Seraphim movement of West Africa claims to have "conscious knowledge of the evil spirits which sow the seeds of discomfort, set afloat ill-luck, diseases, induce barrenness, sterility and the like."

Americans and Europeans usually associate such religious ideas with primitive and rural conditions, and assume that the older world view will disappear with the coming of modernization and urbanization. In the contemporary South, however, the success of highly supernatural churches should rather be seen as a direct by-product of urbanization. (This should come as no surprise to Americans; look at the Pentecostal storefronts in America's inner cities.) As predominantly rural societies have become more urban over the past thirty or forty years, millions of migrants have been attracted to ever larger urban areas, which lack the resources and the infrastructure to meet the needs of these wanderers. Sometimes people travel to cities within the same nation, but often they find themselves in different countries and cultures, suffering a still greater sense of estrangement. In such settings religious communities emerge to provide health, welfare, and education.

This sort of alternative social system, which played an enormous role in the earliest days of Christianity, has been a potent means of winning mass support for the most committed religious groups and is likely to grow in importance as the gap between people's needs and government's capacities to fill them becomes wider. Looking at the success of Christianity in the Roman Empire, the historian Peter Brown has written, "The Christian community suddenly came to appeal to men who felt deserted... Plainly, to be a Christian in 250 brought more protection from one's fellows than to be a *civis Romanus*." Being a member of an active Christian church today may well bring more tangible benefits than being a mere citizen of Nigeria or Peru.

Often the new churches gain support because of the way they deal with the demons of oppression and want: they interpret the horrors of everyday urban life in supernatural terms. In many cases these churches seek to prove their spiritual powers in struggles against witchcraft. The intensity of belief in witchcraft across much of Africa can be startling. As recently as last year at least 1,000 alleged witches were hacked to death in a single "purge" in the Democratic Republic of the Congo. Far from declining with urbanization, fear of witches has intensified. Since the collapse of South Africa's apartheid regime, in 1994, witchcraft has emerged as a primary social fear in Soweto, with its three million impoverished residents.

The desperate public-health situation in the booming mega-cities of the South goes far toward explaining the emphasis of the new churches on healing mind and body. In Africa in the early twentieth century an explosion of Christian healing movements and new prophets coincided with a dreadful series of epidemics, and the religious upsurge of those years was in part a quest for bodily health. Today African churches stand or fall by their success in healing, and elaborate rituals have formed around healing practices (though church members disagree on whether believers should rely entirely on spiritual assistance). The same interest in spiritual healing is found in what were once the mission churches—bodies such as the Anglicans and the Lutherans. Nowhere in the global South do the various spiritual healers find serious competition from modern scientific medicine: it is simply beyond the reach of most of the poor.

Disease, exploitation, pollution, drink, drugs, and violence, taken together, can account for why people might easily accept that they are under siege from demonic forces, and that only divine intervention can save them. Even radical liberation theologians use apocalyptic language on occasion. When a Northerner asks, in effect, where the Southern churches are getting such ideas, the answer is not hard to find: they're getting them from the Bible. Southern Christians are reading the New Testament and taking it very seriously; in it they see the power of Jesus fundamentally expressed through his confrontations with demonic powers, particularly those causing sickness and insanity. "Go back and report to John what you hear and see," Jesus says in the Gospel according to Matthew (11: 4–5). "The blind receive sight, the lame walk, those who have leprosy are cured, the deaf hear, the dead are raised, and the good news is preached to the poor." For the past two hundred years Northern liberals have employed various nonliteral interpretations of these healing passages—perhaps Jesus had a good sense of the causes and treatment of psychosomatic ailments? But that is not, of course, how such scenes are understood within the Third Church.

Today, as in the early sixteenth century, a literal interpretation of the Bible can be tremendously appealing. To quote a modern-day follower of the African prophet Johane Masowe, cited in Elizabeth Isichei's *A History of Christianity in Africa*, "When we were in these synagogues [the European churches], we used to read about the works of Jesus Christ... cripples were made to walk and the dead were brought to life... evil spirits driven out... That was what was being done in Jerusalem. We Africans, however, who were being instructed by white people, never did anything like that... We were taught to read the Bible, but we ourselves never did what the people of the Bible used to do."

Alongside the fast-growing churches have emerged apocalyptic and messianic movements that try to bring in the kingdom of God through armed violence. Some try to establish the thousand-year reign of Jesus Christ on earth, as prophesied in the Book of Revelation. This phenomenon would have been instantly familiar to Europeans 500 years ago, when the Anabaptists and other millenarian groups flourished. Perhaps the most traumatic event of the Reformation occurred in the German city of Münster in 1534–1535, when Anabaptist rebels established a radical social order that abolished property and monogamy; a homicidal king-messiah held dictatorial power until the forces of state authority conquered and annihilated the fanatics. Then as now, it was difficult to set bounds to religious enthusiasm.

Extremist Christian movements have appeared regularly across parts of Africa where the mechanisms of the state are weak. They include groups such as the Lumpa Church, in Zambia, and the terrifying Lord's Resistance Army (LRA), in Uganda. In 2000 more than a thousand people in another Ugandan sect, the Movement for the Restoration of the Ten Commandments of God, perished in an apparent mass suicide. In each case a group emerged from orthodox roots and then gravitated toward apocalyptic fanaticism. The Ten Commandments sect grew out of orthodox Catholicism. The Lumpa Church began, in the 1950s, with Alice Lenshina, a Presbyterian convert who claimed to receive divine visions urging her to fight witchcraft. She became the *lenshina*, or queen, of her new church, whose name, Lumpa, means "better than all others." The group attracted a hundred thousand followers, who formed a utopian community in order to await the Second Coming of Jesus Christ. Since it rejected worldly regimes to the point of refusing to pay taxes, the Lumpa became increasingly engaged in confrontations with the Zambian government, leading to open rebellion in the 1960s.

Another prophetic Alice appeared in Uganda during the chaotic civil wars that swept that country in the 1980s. Alice Lakwena was a former Catholic whose visions led her to establish the Holy Spirit Mobile Force, also pledged to fight witches. She refused to accept the national peace settlement established under President Yoweri Museveni, and engaged in a holy war against his regime. Holy Spirit soldiers, many of them children and young teenagers, were ritually anointed with butter on the understanding that it would make them bulletproof. When Lakwena's army was crushed, in 1991, most of her followers merged with the LRA, which is notorious for filling its ranks by abducting children. Atrocities committed by the group include mass murder, rape, and forced cannibalism. Today as in the sixteenth century, an absolute conviction that one is fighting for God's cause makes moot the laws of war.

The changing demographic balance between North and South helps to explain the current shape of world Catholicism, including the fact that the Church has been headed by Pope John Paul II. In the papal election of 1978 the Polish candidate won the support of Latin American cardinals, who were not prepared to accept yet another Western European. In turn, John Paul has recognized the growing Southern presence in the Church. Last year he elevated forty-four new cardinals, of whom eleven were Latin American, two Indian, and three African. The next time a papal election takes place, fifty-seven of the 135 cardinals eligible to vote, or more than 40 percent, will be from Southern nations. Early this century they will constitute a majority.

It may be true that from the liberal Northern perspective, pressure for a Reformation-style solution to critical problems in the Church—the crisis in clerical celibacy, the shortage of priests, the sense that the laity's concerns are ignored—seems overwhelming. Poll after poll in the United States and Europe indicates significant distrust of clerical authority and support for greater lay participation and women's equality. The obvious question in the parishes of the developed world seems to be how long the aloof hierarchy can stave off the forces of history.

From Rome, however, the picture looks different, as do the "natural" directions that history is going to take. The Roman church operates on a global scale and has done so for centuries. Long before the French and British governments had become aware of global politics—and well before their empires came into being—papal diplomats were thinking through their approaches to China, their policies in Peru, their views on African affairs, their stances on the issues facing Japan and Mexico. To adapt a popular activist slogan, the Catholic Church not only thinks globally, it acts globally. That approach is going to have weighty consequences. On present evidence, a Southern-dominated Catholic Church is likely to react traditionally to the issues that most concern American and European reformers: matters of theology and devotion, sexual ethics and gender roles, and, most fundamentally, issues of authority within the Church.

Neatly illustrating the cultural gulf that separates Northern and Southern churches is an incident involving Moses Tay, the Anglican archbishop of Southeast Asia, whose see is based in Singapore. In the early 1990s Tay traveled to Vancouver, where he encountered the totem poles that are a local tourist attraction. To him, they were idols possessed by evil spirits, and he concluded that they required handling by prayer and exorcism. This horrified the local Anglican Church, which was committed to building good relationships with local Native American communities, and which regarded exorcism as absurd superstition. The Canadians, like other good liberal Christians throughout the North, were long past dismissing alien religions as diabolically inspired. It's difficult not to feel some sympathy with the archbishop, however. He was quite correct to see the totems as authentic religious symbols, and considering the long history of Christian writing on exorcism and possession, he could also summon many precedents to support his position. On that occasion Tay personified the global Christian confrontation.

The cultural gap between Christians of the North and the South will increase rather than diminish in the coming decades, for reasons that recall Luther's time. During the early modern period Northern and Southern Europe were divided between the Protestantism of the word and the Catholicism of the senses—between a religious culture of preaching, hymns, and Bible reading, and one of statues, rituals, and processions. Today we might see as a parallel the impact of electronic technologies, which is being felt at very different rates in the Northern and Southern worlds. The new-media revolution is occurring in Europe, North America, and the Pacific Rim while other parts of the globe are focusing on—indeed, still catching up with—the traditional world of book learning. Northern communities will move to ever more decentralized and privatized forms of faith as Southerners maintain older ideals of community and traditional authority.

On moral issues, too, Southern churches are far out of step with liberal Northern churches. African and Latin American churches tend to be very conservative on issues such as homosexuality and abortion. Such disagreement can pose real political difficulties for churches that aspire to a global identity and that try to balance diverse opinions. At present this is scarcely an issue for the Roman Catholic Church, which at least officially preaches the same conservatism for all regions. If, however, Church officials in North America or Europe proclaimed a moral stance more in keeping with progressive secular values, they would be divided from the growing Catholic churches of the South by a de facto schism, if not a formal breach.

For thirty years Northern liberals have dreamed of a Third Vatican Council to complete the revolution launched by Pope John XXIII—one that would usher in a new age of ecclesiastical democracy and lay empowerment. It would be a bitter irony for the liberals if the council were convened but turned out to be a conservative, Southern-dominated affair that imposed moral and theological litmus tests intolerable to North Americans and Europeans—if, in other words, it tried to implement not a new Reformation but a new Counter-Reformation. (In that sense we would be witnessing not a new Wittenberg but, rather, a new Council of Trent—that is, a strongly traditional gathering that would restate the Church's older ideology and attempt to set it in stone for all future ages.) If a future Southern Pope struggled to impose a new vision of orthodoxy on America's Catholic bishops, universities, and seminaries, the result could well be an actual rather than a de facto schism.

The experience of the world's Anglicans and Episcopalians may foretell the direction of conflicts within the Roman Catholic Church. In the Anglican Communion, which is also torn by a global cultural conflict over issues of gender and sexuality, orthodox Southerners seek to re-evangelize a Euro-American world that they view as coming close to open heresy. This uncannily recalls the situation in sixteenth-century Europe, in which Counter-Reformation Catholics sent Jesuits and missionary priests to reconvert those regions that had fallen into Protestantism.

Anglicans in the North tend to be very liberal on homosexuality and the ordination of women. In recent years, however, liberal clerics have been appalled to find themselves outnumbered and regularly outvoted. In these votes the bishops of Africa and Asia have emerged as a rock-solid conservative bloc. The most ferocious battle to date occurred at the Lambeth World Conference in 1998, which adopted, over the objections of the liberal bishops, a forthright traditional statement proclaiming the impossibility of reconciling homosexual conduct with Christian ministry. As in the Roman Catholic Church, the predominance of Southerners at future events of this kind will only increase. Nigeria already has more practicing Anglicans than any other country, far more than Britain itself, and Uganda is not far behind. By mid-century the global total of Anglicans could approach 150 million, of whom only a small minority will be white Europeans or North Americans. The shifting balance with-in the church could become a critical issue very shortly, since the new Archbishop of Canterbury, Rowan Williams, is notably gay-friendly and has already ordained a practicing homosexual as a priest.

The Lambeth debate also initiated a series of events that Catholic reformers should study carefully. Briefly, American conservatives who were disenchanted with the liberal establishment in the U.S. Episcopal Church realized that they had powerful friends overseas, and transferred their religious allegiance to more-conservative authorities in the global South. Since 2000 some conservative American Episcopalians have traveled to Moses Tay's cathedral in Singapore, where they were consecrated as bishops by Asian and African Anglican prelates, including the Rwandan archbishop Emmanuel Kolini. By tradition an Anglican archbishop is free to ordain whomever he pleases within his province, so although the Americans live and work in South Carolina, Pennsylvania, and other states, they are now technically bishops within the province of Rwanda. They have become missionary bishops, charged with ministering to conservative congregations in the United States, where they support a dissident "virtual province" within the church. They and their conservative colleagues are now part of the Anglican Mission in America, which is intended officially to "lead the Episcopal Church back to its biblical foundations." The mission aims to restore traditional teachings and combat what it sees as the "manifest heresy" and even open apostasy of the U.S. Church leadership. Just this past summer Archbishop Kolini offered his protection to dissident Anglicans in the Vancouver area, who were rebelling against liberal proposals to allow same-sex couples to receive a formal Church blessing.

Ultimately, the first Christendom—the politico–religious order that dominated Europe from the sixth century through the sixteenth—collapsed in the face of secular nationalism, under the overwhelming force of what Thomas Carlyle described as "the three great elements of modern civilization, gunpowder, printing, and the Protestant religion." Nation-states have dominated the world ever since. Today, however, the whole concept of national autonomy is under challenge, partly as a result of new technologies. In the coming decades, according to a recent CIA report, "Governments will have less and less control over flows of information, technology, diseases, migrants, arms, and financial transactions, whether licit or illicit, across their borders. The very concept of 'belonging' to a particular state will probably erode." If a once unquestionable construct like Great Britain is under threat, it is not surprising that people are questioning the existence of newer and more artificial entities in Africa and Asia.

For a quarter of a century social scientists analyzing the decline of the nation-state have drawn parallels between the world today and the politically fragmented yet cosmopolitan world of the Middle Ages. Some scholars have even predicted the emergence of some secular movement or ideology that would command loyalty across nations like the Christendom of old. Yet the more we look at the Southern Hemisphere, the more we see that although supranational ideas are flourishing, they are not in the least secular. The parallels to the Middle Ages may be closer than anyone has guessed.

Across the global South cardinals and bishops have become national moral leaders in a way essentially unseen in the West since the seventeenth century. The struggles of South African churches under apartheid spring to mind, but just as impressive were the pro-democracy campaigns of many churches and denominations elsewhere in Africa during the 1980s and 1990s. Prelates know that they are expected to speak for their people, even though if they speak boldly, they may well pay with their lives. Important and widely revered modern martyrs include Archbishop Luwum, of Uganda; Archbishop Munzihirwa, of Zaire; and Cardinal Biayenda, of Congo-Brazzaville.

As this sense of moral leadership grows, we might reasonably ask whether Christianity will also provide a guiding political ideology for much of the world. We might even imagine a new wave of Christian states, in which political life is inextricably bound up with religious belief. Zambia declared itself a Christian nation in 1991, and similar ideas have been bruited in Zimbabwe, Kenya, and Liberia. If this ideal does gain popularity, the Christian South will soon be dealing with some debates, of long standing in the North, over the proper relationship between Church and State and between rival churches under the law. Other inevitable questions involve tolerance and diversity, the relationship between majority and minority communities, and the extent to which religiously inspired laws can (or should) regulate private morality and behavior. These issues were all at the core of the Reformation.

Across the regions of the world that will be the most populous in the twenty-first century, vast religious contests are already in progress, though so far they have impinged little on Western opinion. The most significant conflict is in Nigeria, a nation that by rights should be a major regional power in this century and perhaps even a global power; but recent violence between Muslims and Christians raises the danger that Nigerian society might be brought to ruin by the clash of *jihad* and crusade. Muslims and Christians are at each other's throats in Indonesia, the Philippines, Sudan, and a growing number of other African nations; Hindu extremists persecute Christians in India. Demographic projections suggest that these feuds will simply worsen. Present-day battles in Africa and Asia may anticipate the political outlines to come, and the roots of future great-power alliances. These battles are analogous to the ideological conflicts of the twentieth century, the alternating hot and cold wars between advocates of fascism and of democracy, of socialism and of capitalism. This time, however, the competing ideologies are explicitly religious, promising their followers a literal rather than merely a metaphorical kingdom of God on earth.

Let us imagine Africa in the throes of fiery religious revivals, as Muslim and Christian states jostle for political influence. Demographic change alone could provoke more-aggressive international policies, as countries with swollen populations tried to appropriate living space or natural resources. But religious tensions could make the situation far worse. If mega-cities are not to implode through social unrest and riot, governments have to find some way to mobilize the teeming masses of unemployed teenagers and young adults. Persuading them to fight for God is a proven way of siphoning off internal tension, especially if the religion in question already has a powerful ideal of martyrdom. Liberia, Uganda, and Sierra Leone have given rise to ruthless militias ready to kill or die for whatever warlord directs them, often following some notionally religious imperative. In the 1980s the hard-line Shiite mullahs of Iran secured their authority by sending hundreds of thousands of young men to martyr themselves in human-wave assaults against the Iraqi front lines. In contemporary Indonesia, Islamist militias can readily find thousands of poor recruits to fight against the nation's Christian minorities.

Some of the likely winners in the religious economy of the new century are precisely those groups with a strongly apocalyptic mindset, in which the triumph of righteousness is associated with the vision of a world devastated by fire and plague. This could be a perilously convenient ideology for certain countries with weapons of mass destruction. (The candidates that come to mind include not only Iraq and Iran but also future regional powers such as Indonesia, Nigeria, the Congo, Uganda, and South Africa.) All this means that our political leaders and diplomats should pay at least as much attention to religions and sectarian frontiers as they ever have to the location of oil fields.

Perhaps the most remarkable point about these potential conflicts is that the trends pointing toward them have registered so little on the consciousness of even well-informed Northern observers. What, after all, do most Americans know about the distribution of Christians worldwide? I suspect that most see Christianity very much as it was a century ago—a predominantly European and North American faith. In discussions of the recent sexual-abuse crisis "the Catholic Church" and "the American Church" have been used more or less synonymously.

As the media have striven in recent years to present Islam in a more sympathetic light, they have tended to suggest that Islam, not Christianity, is the rising faith of Africa and Asia, the authentic or default religion of the world's huddled masses. But Christianity is not only surviving in the global South, it is enjoying a radical revival, a return to scriptural roots. We are living in revolutionary times.

But we aren't participating in them. By any reasonable assessment of numbers, the most significant transformation of Christianity in the world today is not the liberal Reformation that is so much desired in the North. It is the Counter-Reformation coming from the global South. And it's very likely that in a decade or two neither component of global Christianity will recognize its counterpart as fully or authentically Christian.

Philip Jenkins is a Distinguished Professor of History and Religious Studies at Pennsylvania State University. His most recent book is The Next Christendom *(2002).*

Modernization's Challenge to Traditional Values: Who's Afraid of Ronald McDonald?

"Modernization" means "Americanization" to many who fear a coming McWorld. But a study by two social researchers indicates that traditional values will keep most countries from becoming clones of the United States.

By Ronald Inglehart and Wayne E. Baker

The World Values Survey—a two-decade-long examination of the values of 65 societies coordinated by the University of Michigan's Institute for Social Research—is the largest investigation ever conducted of attitudes, values, and beliefs around the world. This study has carried out three waves of representative national surveys: the first in 1981–1982, the second in 1990–1991; and the third in 1995–1998. The fourth wave is being completed in 1999–2001. The study now represents some 80% of the world's population. These societies have per capita GNPs ranging from $300 to more than $30,000. Their political systems range from long-established stable democracies to authoritarian states.

The World Values Survey data have been used by researchers around the world for hundreds of publications in more than a dozen languages. Studies that have been based on the data cover a wide range of topics, including volunteerism in Europe, political partisanship and social class in Ireland, democratization in Korea, liberalization in Mexico, future values in Japan, and the religious vote in Western Europe.

This article examines the relationship between cultural values and economic globalization and modernization: What impact does economic development have on the values of a culture, and vice versa? Is a future "McWorld" inevitable?

Rich Values, Poor Values

The World Values Survey data show us that the world views of the people of rich societies differ systematically from those of low-income societies across a wide range of political, social, and religious norms and beliefs. The two most significant dimensions that emerged reflected, first, a polarization between *traditional* and *secular-rational* orientations toward authority and, second, a polarization between *survival* and *self-expres-*

sion values. By *traditional* we mean those societies that are relatively authoritarian, place strong emphasis on religion, and exhibit a mainstream version of preindustrial values such as an emphasis on male dominance in economic and political life, respect for authority, and relatively low levels of tolerance for abortion and divorce. Advanced societies, or *secular-rational*, tend to have the opposite characteristics.

A central component of the survival vs. self-expression dimension involves the polarization between materialist and post-materialist values. Massive evidence indicates that a cultural shift throughout advanced industrial society is emerging among generations who have grown up taking survival for granted. Values among this group emphasize environmental protection, the women's movement, and rising demand for participation in decision making in economic and political life. During the past 25 years, these values have become increasingly widespread in almost all advanced industrial societies for which extensive time-series evidence is available.

Economic development brings with it sweeping cultural change, some modernization theorists tell us. Others argue that cultural values are enduring and exert more influence on society than does economic change. Who's right?

One goal of the World Values Survey is to study links between economic development and changes in values. A key question that we ask is whether the globalization of the economy will necessarily produce a homogenization (or, more specifically, an Americanization) of culture—a so-called "McWorld."

In the nineteenth century, modernization theorists such as Karl Marx and Friedrich Nietzsche made bold predictions about the future of industrial society, such as the rise of labor and the decline of religion. In the twentieth century, non-Western societies were expected to abandon their traditional cultures and assimilate the technologically and morally "superior" ways of the West.

Clearly now, at the start of the twenty-first century, we need to rethink "modernization." Few people today anticipate a proletarian revolution, and non-western societies such as East Asia have surpassed their Western role models in key aspects of modernization, such as rates of economic growth. And few observers today attribute moral superiority to the West.

Two Dimensions of Cross-Cultural Variation

1. Traditional vs. Secular-Rational Values
   Traditional values emphasize the following:
   - God is very important in respondent's life.
   - Respondent believes it is more important for a child to learn obedience and religious faith than independence and determination.
   - Respondent believes abortion is never justifiable.
   - Respondent has strong sense of national pride.
   - Respondent favors more respect for authority.
   Secular-Rational values emphasize the opposite.
2. Survival vs. Self-Expression Values
   Survival values emphasize the following:
   - Respondent gives priority to economic and physical security over self-expression and quality of life.
   - Respondent describes self as not very happy.
   - Respondent has not signed and would not sign a petition.
   - Respondent believes homosexuality is never justifiable.
   - Respondent believes you have to be very careful about trusting people.
   Self-Expression values emphasize the opposite.

Source: World Values Survey (http://wvs.isr.umich.edu)

On the other hand, one core concept of modernization theory still seems valid: Industrialization produces pervasive social and cultural consequences, such as rising educational levels, shifting attitudes toward authority, broader political participation, declining fertility rates, and changing gender roles. On the basis of the World Values Surveys, we believe that economic development has systematic and, to some extent, predictable cultural and political consequences. Once a society has embarked on industrialization—the central element of the modernization process—certain changes are highly likely to occur. But economic development is not the *only* force at work.

In the past few decades, modernization has become associated with *post*-industrialization: the rise of the knowledge and service-oriented economy. These changes in the nature of work had major political and cultural consequences, too. Rather than growing more materialistic with increased prosperity, postindustrial societies are experiencing an increasing emphasis on quality-of-life issues, environmental protection, and self-expression.

While industrialization increased human dominance over the environment—and consequently created a dwindling role for religious belief—the emergence of postindustrial society is stimulating further evolution of prevailing world views in a different direction. Life in postindustrial societies centers on services rather than material objects, and more effort is focused on communicating and processing information. Most people spend their productive hours dealing with other people and symbols.

Thus, the rise of postindustrial society leads to a growing emphasis on self-expression. Today's unprecedented wealth in advanced societies means an increasing share of the population grows up taking survival for granted. Their value priorities shift from an overwhelming emphasis on economic and physical security toward an increasing emphasis on subjective well-being and quality of life. "Modernization," thus, is not linear—it moves in new directions.

How Values Shape Culture

Different societies follow different trajectories even when they are subjected to the same forces of economic development, in part because situation-specific factors, such as a society's cultural heritage, also shape how a particular society develops. Recently, Samuel Huntington, author of *The Clash of Civilizations* (Simon & Schuster, 1996), has focused on the role of religion in shaping the world's eight major civilizations or "cultural zones": Western Christianity, Orthodox, Islam, Confucian, Japanese, Hindu, African, and Latin American. These zones were shaped by religious traditions that are still powerful today, despite the forces of modernization.

Distinctive cultural zones persist two centuries after the industrial revolution began.

Other scholars observe other distinctive cultural traits that endure over long periods of time and continue to shape a society's political and economic performance. For example, the regions of Italy in which democratic institutions function most successfully today are those in which civil society was relatively well developed in the nineteenth century and even earlier, as Robert Putnam notes in *Making Democracy Work* (Princeton University Press, 1993). And a cultural heritage of "low trust" puts a society at a competitive disadvantage in global markets because it is less able to develop large and complex social institutions, Francis Fukuyama argues in *Trust: The Social Virtues and the Creation of Prosperity* (Free Press, 1995).

The impression that we are moving toward a uniform "McWorld" is partly an illusion. The seemingly identical McDonald's restaurants that have spread throughout the world actually have different social meanings and fulfill dif-

Less than $2,000 GNP per capita A   C $5,000 to $15,000 GNP per capita

$2,000 to $5,000 GNP per capita B   D   More than $15,000 GNP per capita

ferent social functions in different cultural zones. Eating in a McDonald's restaurant in Japan is a different social experience from eating in one in the United States, Europe, or China.

Likewise, the globalization of communication is unmistakable, but its effects may be overestimated. It is certainly apparent that young people around the world are wearing jeans and listening to U.S. pop music; what is less apparent is the persistence of underlying value differences.

Mapping and Predicting Values

Using the 1995–1998 wave of the World Values Survey, we produced a map of the world's values, showing the locations of 65 societies on the two cross-cultural dimensions—traditional vs. secular-rational values and survival vs. self-expression values.

What the map shows us is that cross-cultural variation is highly constrained. That is, if the people of a given society place a strong emphasis on religion, that society's relative position on many other variables can be predicted—such as attitudes toward abortion, national pride, respect for authority, and child-drearing. Similarly, survival vs. self-expression values reflect wide-ranging but tightly correlated clusters of values: Materialistic (survival-oriented) societies can be predicted to value maintaining order and fighting inflation, while postmaterialistic (self-expression-oriented) societies can be predicted to value freedom, interpersonal trust, and tolerance of outgroups.

Economic development seems to have a powerful impact on cultural values: The value systems of rich countries differ systematically from those of poor countries. If we superimpose an income "map" over the values map, we see that all 19 societies with an annual per capita GNP of over $15,000 rank relatively high on both dimensions, placing them in the upper right-hand corner. This economic zone cuts across the boundaries of the Protestant, ex-Communist, Confucian, Catholic, and English-speaking cultural zones.

On the other hand, all societies with per capita GNPs below $2,000 fall into a cluster at the lower left of the map, in an economic zone that cuts across the African, South Asian, ex-Communist, and Orthodox cultural zones. The remaining societies fall into two intermediate cultural-economic zones. Economic development seems to move societies in a common direction, regardless of their cultural heritage. Nevertheless, distinctive cultural zones persist two centuries after the industrial revolution began.

Of course, per capita GNP is only one indicator of a society's level of economic development. Another might be the percentage of the labor force engaged in the agricultural sector, the industrial sector, or the service sector. The shift from an agrarian mode of production to industrial production seems to bring with it a shift from traditional values toward increasing rationalization and secularization.

But a society's cultural heritage also plays a role: All four of the Confucian-influenced societies (China, Taiwan, South Korea, and Japan) have relatively secular values, regardless of the proportion of their labor forces in the industrial sector. Conversely, the historically Roman Catholic societies (e.g., Italy, Portugal, and Spain) display relatively traditional values when compared with Confucian or ex-Communist societies with the same proportion of industrial workers. And virtually all of the historically Protestant societies (e.g., West Germany, Denmark, Norway, and Sweden) rank higher on the survival/self-expression dimension than do all of the historically Roman Catholic societies, regardless of the extent to which their labor forces are engaged in the service sector.

We can conclude from this that changes in GNP and occupational structure have important influences on prevailing world views, but traditional cultural influences persist.

Religious traditions appear to have had an enduring impact on the contemporary value systems of the 65 societies. But a society's culture reflects its entire historical heritage. A central historical event of the twentieth century was the rise and fall of a Communist empire that once ruled one-third of the world's population. Communism left a clear imprint on the value systems of those who lived under it. East Germany remains culturally close to West Germany despite four decades of Communist rule, but its value system has been drawn toward the Communist zone. And although China is a member of the Confucian zone, it also falls within a broad Communist-influenced zone. Similarly, Azerbaijan, though part of the Islamic cluster, also falls within the Communist superzone that dominated it for decades.

The Deviant U.S.

The World Value Map clearly shows that the United States is a deviant case. We do not believe it is a prototype of cultural modernization for other societies to follow, as some postwar modernization theorists have naively assumed. The United States has a much more traditional value system than any other advanced industrial society.

On the traditional/secular-rational dimension, the United States ranks far below other rich societies, with levels of religiosity and national pride comparable to those found in developing societies. The United States does rank among the most advanced societies along the survival/self-expression dimension, but even here it does not lead the world. The Swedes and the Dutch seem closer to the cutting edge of cultural change than do the Americans.

Modernization theory implies that as societies develop economically their cultures tend to shift in a predictable direction. Our data supports this prediction. Economic differences are linked with large and pervasive cultural differences. But we find clear evidence of the influence of long-established cultural zones.

Do these cultural clusters simply reflect economic differences? For example, do the societies of Protestant Europe have similar values simply because they are rich? No. The impact of a society's historical-cultural heritage persists when we control for GDP per capita and the structure of the labor force. On a value such as *interpersonal trust* (a variable on the surival / self-expression dimension), even rich Catholic societies rank lower than rich Protestant ones.

Within a given society, however, Catholics rank about as high on *interpersonal trust* as do Protestants. The shared historical experience of given nations, not individual personality, is crucial. Once established, the cross-cultural differences linked with religion have become part of a national culture that is transmitted by the educational institutions and mass media of given societies to the people of that nation. Despite globalization, the nation remains a key unit of shared experience, and its educational and cultural institutions shape the values of almost everyone in that society.

The Persistence of Religious and Spiritual Beliefs

As a society shifts from an agrarian to an industrial economy and survival comes to be taken for granted, traditional religious beliefs tend to decline. Nevertheless, as the twenty-first century opens, cleavages along religious lines remain strong. Why has religion been so slow to disappear?

History has taken an ironic turn: Communist-style industrialization was especially favorable to secularization, but the collapse of Communism has given rise to pervasive insecurity—

and a return to religious beliefs. Five of the seven ex-Communist societies for which we have time-series data show rising church attendance.

Throughout advanced industrial societies we see two contrasting trends: the decline of attendance at religious services on the one hand, and on the other the persistence of religious beliefs and the rise of spirituality. The need for answers to spiritual questions such as why we are here and where we are going does not die out in postindustrial society. Spiritual concerns will probably always be part of the human outlook. In fact, in the three successive waves of the World Values Survey, concern for the meaning and purpose of life became *stronger* in most advanced industrial societies.

Conclusion: Whither Modernization?

Economic development is associated with pervasive, and to an extent predictable, cultural changes. Industrialization promotes a shift from traditional to secular-rational values; postindustrialization promotes a shift toward more trust, tolerance, and emphasis on well-being. Economic collapse propels societies in the opposite direction.

Economic development tends to push societies in a common direction, but rather than converging they seem to move along paths shaped by their cultural heritages. Therefore, we doubt that the forces of modernization will produce a homogenized world culture in the foreseeable future.

Certainly it is misleading to view cultural change as "Americanization." Industrializing societies in general are not becoming like the United States. In fact, the United States seems to be a deviant case: Its people hold much more traditional values and beliefs than do those in any other equally prosperous society. If any societies exemplify the cutting edge of cultural change, it would be the Nordic countries.

Finally, modernization is probabilistic, not deterministic. Economic development tends to transform a given society in a predictable direction, but the process and path are not inevitable. Many factors are involved, so any prediction must be contingent on the historical and cultural context of the society in question.

Modernization and McDonald's

McDonald's restaurants have become a dominant symbol of the globalization of the economy and target of the wrath of globalization's many opponents. But local values still wield great influence on culture, so don't look for McWorld to emerge anytime soon, say social researchers Ronald Inglehart and Wayne E. Baker.

Nevertheless, the central prediction of modernization theory finds broad support: Economic development is associated with major changes in prevailing values and beliefs. The world views of rich societies differ markedly from those of poor societies. This does not necessarily imply cultural convergence, but it does predict the general direction of cultural change and (insofar as the process is based on intergenerational population replacement) even gives some idea of the rate at which such change is likely to occur.

In short, economic development will cause shifts in the values of people in developing nations, but it will not produce a uniform global culture. The future may *look* like McWorld, but it won't feel like one.

About the Authors

Ronald Inglehart is professor of political science and program director at the Institute for Social Research, University of Michigan, Ann Arbor, Michigan 48106. E-mail RFI@umich.edu. The World Values Survey Web site is http://wvs.isr.umich.edu/.

Wayne E. Baker is professor of organizational behavior and director of the Center for Society and Economy, University of Michigan Business School, and faculty associate at the Institute for Social Research. He may be reached by e-mail at wayneb@umich.edu; his Web site is www.bus.umich.edu/cse.

This article draws on their paper "Modernization, Cultural Change, and the Persistence of Traditional Values" in the American Sociological Review (February 2000).

Originally published in the March/April 2001 issue of *The Futurist*, pp. 16-21. Used with permission from the World Future Society, 7910 Woodmont Avenue, Suite 450, Bethesda, Maryland 20814. Telephone: 310/656-8274; Fax: 301/951-0394; (http://www.wfs.org).

Index

Index

Test Your Knowledge Form

We encourage you to photocopy and use this page as a tool to assess how the articles in *Annual Editions* expand on the information in your textbook. By reflecting on the articles you will gain enhanced text information. You can also access this useful form on a product's book support Web site at *http://www.dushkin.com/online/*.

NAME: _____

DATE: _____

TITLE AND NUMBER OF ARTICLE:
_____

BRIEFLY STATE THE MAIN IDEA OF THIS ARTICLE:
_____

LIST THREE IMPORTANT FACTS THAT THE AUTHOR USES TO SUPPORT THE MAIN IDEA:

_____

WHAT INFORMATION OR IDEAS DISCUSSED IN THIS ARTICLE ARE ALSO DISCUSSED IN YOUR TEXTBOOK OR OTHER READINGS THAT YOU HAVE DONE? LIST THE TEXTBOOK CHAPTERS AND PAGE NUMBERS:

_____

LIST ANY EXAMPLES OF BIAS OR FAULTY REASONING THAT YOU FOUND IN THE ARTICLE:

_____

LIST ANY NEW TERMS/CONCEPTS THAT WERE DISCUSSED IN THE ARTICLE, AND WRITE A SHORT DEFINITION:

We Want Your Advice

ANNUAL EDITIONS revisions depend on two major opinion sources: one is our Advisory Board, listed in the front of this volume, which works with us in scanning the thousands of articles published in the public press each year; the other is you—the person actually using the book. Please help us and the users of the next edition by completing the prepaid article rating form on this page and returning it to us. Thank you for your help!

ANNUAL EDITIONS: Global Issues 04/05

ARTICLE RATING FORM

Here is an opportunity for you to have direct input into the next revision of this volume.
We would like you to rate each of the articles listed below, using the following scale:

1. **Excellent: should definitely be retained**
2. **Above average: should probably be retained**
3. **Below average: should probably be deleted**
4. **Poor: should definitely be deleted**

Your ratings will play a vital part in the next revision.
Please mail this prepaid form to us as soon as possible.
Thanks for your help!

RATING	ARTICLE	RATING	ARTICLE
	1. A Special Moment in History		33. Aerial War Against Disease: Satellite Tracking of Epidemics Is Soaring
	2. Clash of Globalizations		34. Are Human Rights Universal?
	3. A New Grand Strategy		35. The Grameen Bank
	4. Mr. Order Meets Mr. Chaos		36. Why Environmental Ethics Matters to International Relations
	5. The Big Crunch		37. Exploring the "Singularity"
	6. SARS Has Lots of Company as Emerging Disease		38. Women Waging Peace
	7. Bittersweet Harvest: The Debate Over Genetically Modified Crops		39. The Next Christianity
	8. The Challenges We Face		40. Modernization's Challenge to Traditional Values: Who's Afraid of Ronald McDonald?
	9. Shifting the Pain: World's Resources Feed California's Growing Appetite		
	10. Water Scarcity Could Overwhelm the Next Generation		
	11. We Can Build a Sustainable Economy		
	12. The Complexities and Contradictions of Globalization		
	13. The Five Wars of Globalization		
	14. Will Globalization Go Bankrupt?		
	15. Croesus and Caesar		
	16. Render Unto Caesar: Putin and the Oligarchs		
	17. The Fall of the House of Saud		
	18. Thirty Years of Petro-Politics		
	19. India and China: Two Systems, One Grand Rivalry		
	20. Overcoming Japan's China Syndrome		
	21. What's Wrong With This Picture?		
	22. Leasing the Rain		
	23. The Transformation of National Security		
	24. Nuclear Nightmares		
	25. "Why Do They Hate Us?"		
	26. Changing Course on China		
	27. The Korea Crisis		
	28. India, Pakistan, and the Bomb		
	29. Strategies for World Peace: The View of the UN Secretary-General		
	30. Europe Drafting Its Constitution		
	31. The New Containment: An Alliance Against Nuclear Terrorism		
	32. Countdown to Eradication		

(Continued on next page)

BUSINESS REPLY MAIL
FIRST CLASS MAIL PERMIT NO. 551 DUBUQUE IA

POSTAGE WILL BE PAID BY ADDRESEE

McGraw-Hill/Dushkin
2460 KERPER BLVD
DUBUQUE, IA 52001-9902

ABOUT YOU

Name

Date

Are you a teacher? ☐ A student? ☐
Your school's name

Department

Address   City   State   Zip

School telephone #

YOUR COMMENTS ARE IMPORTANT TO US!

Please fill in the following information:
For which course did you use this book?

Did you use a text with this ANNUAL EDITION? ☐ yes ☐ no
What was the title of the text?

What are your general reactions to the *Annual Editions* concept?

Have you read any pertinent articles recently that you think should be included in the next edition? Explain.

Are there any articles that you feel should be replaced in the next edition? Why?

Are there any World Wide Web sites that you feel should be included in the next edition? Please annotate.

May we contact you for editorial input? ☐ yes ☐ no
May we quote your comments? ☐ yes ☐ no